SIGNALS: SWIFT AND SURE

A HISTORY OF THE ROYAL AUSTRALIAN CORPS OF SIGNALS 1947-1972

JOHN BLAXLAND

First Published in 1998 by The Royal Australian
Corps of Signals Corps Committee

Published in 2026 by Echo Books
Echo Books is an imprint of Superscript Publishing Pty Ltd.
ABN 76 644 812 395
35 Keeley Lane, Princes Hill, Victoria, 3054
www.echobooks.com.au
Copyright © 2026 John Blaxland
Book Design: Ronchon Villaester

ISBN: 978-1-923689-90-9 (paperback)

CONTENTS

ABSTRACT TO THE FIRST EDITION

THIS BOOK ANALYSES the impact of communications technology on the conduct of Australian military operations in the post-war period. This work is both a corps history and a story about military communications technology. It focuses particularly on the operations from 1947 to 1972 in Japan, Korea, Malaya, Borneo, Singapore and Vietnam.

Technological improvements have increased the range, capacity and reliability of communications systems. These, in turn, have improved the ability of commanders to exercise control over even fleeting opportunities during conflict while increasing commanders' vulnerability to information overload. Today, a reliable, effective and efficient communications system is acknowledged as central to a commander's ability to gain and retain the initiative in battle.

Perhaps the most significant area in which communications technology has had an impact on the conduct of military operations is in signals intelligence. In Borneo and Vietnam, it acted as a combat power multiplier. In Vietnam, the enemy also benefited from poor allied communications security and thus undermined the benefits of superior technology.

For field force Signal units, improved equipment enabled commanders to reduce the combat ratio of forces required to ensure survivability. Forces could rapidly be deployed to remote locations with the assurance that, through reliable communications, they could be provided with an unprecedented amount of combat support in a very short time frame.

The strategic communications network also played a key role. It helped to maximise the political advantage from the smallest of deployed elements whilst retaining national control without having to rely on another nation's communications facilities to do so.

The result of these capabilities combined was to make military communications increasingly swift and sure.

FOREWORD TO THE SECOND EDITION

THE FOCUS OF this book is on the work of the Royal Australian Corps of Signals and the implications of communications technology for the conduct of military operations. With the advent of the Fourth Industrial Revolution and an explosion in advanced information and communications technology (ICT) systems, let alone Artificial Intelligence, the significance of reliable, advanced, secure and robust communications in support of military operations is greater than ever. History doesn't repeat but understanding the past sheds light on the way forward. Yet the original version of this book, published in 1998, is now out of print and hard to find. With this in mind, we felt it important to ensure this work is available to a wider contemporary readership by having it republished.

The book covers the period of Forward Defence, following the end of World War II through to the election of Gough Whitlam as Prime Minister in November 1972. Thirty years after this book was first being prepared for publication, there appears to be much continuity amongst the change, adding momentum for this second edition being published.

The context is of relevance for a wider audience than purely communications practitioners. While a number of the circumstances may have changed, nowadays, the interest is as much on engagement in Southeast Asia and the Pacific as ever. For this reason, this will be of interest not just to practitioners and members of the Royal Australian Corps of Signals, but to those concerned to have a better understanding of Australia's military experience during the Cold War - spanning deployments as part of the British Commonwealth Occupation Force (BCOF) in Japan, during the Korean War, the Malayan Emergency, Confrontation, Indonesia, Singapore, Vietnam and more.

The book also points to the need for the publication of another study on the Royal Australian Corps of Signals experience on post-Cold War operations in places such as Namibia, Rwanda, Somalia, Cambodia, East Timor, Bougainville, Irian Jaya (West Papua), Solomon Islands, Sumatra as well as further afield in Afghanistan, and Iraq. Hopefully this work will provide inspiration and add momentum for a subsequent volume to review the Corps' activities from 1972 to today.

The book addresses a range of enduring challenges. One of the problems identified concerned lax communications security. A certain self-assured arrogance towards the enemy and cockiness about technological advantages can lead to catastrophic errors and undermine the prospects of victory, as this work attests. Another challenge concerned the nature of being a junior coalition partner alongside a major power conducting operations in which one's own interests do not necessarily feature prominently. With Australia as dependent as ever on its connections with the United States, reviewing Australia's close working experience alongside US forces in the past warrants attention.

This work does not set out to rework the argument set out in the original volume. Some may contend more recent works should be cited. I am aware that much has been written since this was first published but having sought to remain abreast of much of this genre, I do not consider that including references to more recent secondary works would make a material difference to the conclusions reached in this volume.

Since 1972 Australia has developed closer and more trusting relationships with a range of partner nations in the Pacific and Southeast Asia. In doing so, there is merit in having a clear understanding of how things have transpired. I hope you agree that reflecting on and understanding Australia's experience in providing swift and sure communications support to Australia's forces conducting military operations far and wide remains as relevant as ever.

I would like to add my heartfelt thanks to my colleague, friend, and Echo Books owner, Marcus Fielding, for seeing this through.

ABOUT THE AUTHOR

JOHN WROTE THIS on a part-time basis while serving in the Australian Army.

After the first edition was published in 1998, he spent a further twelve years in the Army before leaving the Army to join the Strategic and Defence Studies Centre at the Australian National University.

Writing this book spurred him to continue work as a historian.

He is now the author, co-author or editor of a dozen or so works on Australian military history (spanning Australia's wars since federation), Security Intelligence, Signals Intelligence, strategy and international relations.

His works can be found here
https://www.johnblaxland.au/

ACKNOWLEDGMENTS

THIS VOLUME OWES its origins to the Royal Australian Signals Corps Committee's desire to see a book written covering the Corps' history in the post-World War II conflicts. Theo Barker made a valuable contribution to the Corps with his Volume I of the Corps' history up to 1947, and this takes up the story where he left off. Others have also written brief histories covering the period up to 1972, but it was evident that a more detailed account covering the breadth of the Corps should be written.

In 1986, I graduated from Duntroon into the Australian Intelligence Corps and subsequently completed an Honours Degree in History through the University of New South Wales at the Australian Defence Force Academy. Afterwards, I was allocated to Signals for my regimental training. During my time with the Corps, I heard about the Corps Committee's plans and volunteered my services. Several years later, after having fitted my research and writing in between and during my subsequent Army postings, a manuscript emerged ready for the reader. I completed a preliminary draft some time before completing the final manuscript, and this was read by several retired and serving members of the Corps whose positive contributions and comments have been incorporated where appropriate. In writing this book, I have realised that it could not have been possible without the ready and generous assistance of many people along the way.

Particular thanks are due to the late Major General D. Vincent, Major General B.H. Hockney, Major General R.P. Woollard, Brigadier

K.R. Colwill, Colonel I.C. Gordon, Colonel M.A. Swan, Colonel B.P. O'Day, Lieutenant Colonel N.R. Churches, Lieutenant Colonel K.D. Munro and Lieutenant Colonel R.E. Cowley. I am indebted to many other members of the Corps as well, and their contributions are noted throughout in the footnotes and bibliography.

I would also like to thank the staff of the Australian Archives and, in particular, Dr Paul Mansfield and Mrs Lorraine MacKnight of the Written Records Section, who were at the Australian War Memorial at the time, for their generous assistance.

Special thanks are also due to Dr Noel MacLachlan from the Department of History at the University of Melbourne, Ian Hancock from the Department of History at the Australian National University and Dr David Horner from the Australian National University's Strategic and Defence Studies Centre for their wise counsel and support.

Finally, I admit that in such a technology-oriented Corps as the Royal Australian Corps of Signals, there is much to keep abreast with. As a non-technical officer, I hope I will be excused for any errors that may remain for which I am alone responsible.

LIST OF ABBREVIATIONS

AAAGV	Australian Army Assistance Group Vietnam
AAFV	Australian Army Force Vietnam
AATTV	Australian Army Training Team Vietnam
ABCA	America, Britain, Canada, Australia
ACAN	Army Command and Administration Network (US)
ADE	Army Design Establishment
AFV	Australian Force Vietnam
AGRA	Army Group Royal Artillery
AHQ	Army Headquarters
AIDFAC	Australian Inter-Departmental Frequencies Advisory Committee
AIDTAC	Australian Inter-Departmental Telecommunications Advisory Committee
ALSG	Australian Logistic Support Group
AN/	Army Navy (Prefix)
ANARE	Australian National Antarctic Research Expedition
ANZAC	Australian New Zealand Army Corps
ANZAM	Australia New Zealand in the Anglo Malayan Region
ANZUS	Australia New Zealand United States
ARA	Australian Regular Army

ARDF	Airborne Radar Direction Finder
ARVN	Army of the Republic of Vietnam
ASSU	Air Support Signal Unit
1 ATF	1 Australian Task Force
AUSTCAN	Australian Communications Army Network
AUSTFIKMA	Australian Forces in Korea Maintenance Area
AWA	Amalgamated Wireless Australasia
AWC	Army Wireless Chain
AWM	Australian War Memorial
BAOR	British Army on the Rhine
Baudot	5 point code
BCFESR	British Commonwealth Far Eastern Strategic Reserve
BCFK	British Commonwealth Forces Korea
BCOF	British Commonwealth Occupation Force (Japan)
BJCEB	British Joint Communications Electronics Board
Britcom	British Commonwealth
CANUKUS	Canada, United Kingdon, United States
CARA	Communications Assets Recovery Agency
CCK	Commonwealth Contingent Korea
CFS	Carrier Frequency Shift
CMF	Citizen Miltary Force
CO	Commanding Officer
COMAFV	Commander Australian Force Vietnam
COMCAN	Commonwealth Communications Army Network
Comm Z	Communication Zone
COMSEC	Communications Security
CRS	Commonwealth Record Series

CSO	Chief Signal Officer
CT	Communist Terrorist (Malaya)
CW	Continuous Wave (morse)
DC	Direct Current
DR	Despatch Rider
DSO	Duty Signals Officer
EARC	Extraordinary Administrative Radio Conference
FARC	Frequency Allocation Review Committee
FARELF	Far Eastern Land Forces
FPDA	Five Power Defence Arrangements
FSK	Frequency Shift Keying
FWMAO	Free World Military Assistance Organisation
GHQ	General Headquarters
GOC	General Officer Commanding
GRC	Ground Radio Communications
HMAS	Her Majesty's Australian Ship
HF	High Frequency
HQ	Headquarters
Hz	Hertz
KHz	Kilohertz
L of C	Line of Communication
MF	Medium Frequency
MGO	Master General of the Ordnance
MRC	Mobile Radio Communications
MRS	Major Relay Station
NATO	North Atlantic Treaty Organisation
NCO	Non-Commissioned Officer
OCS	Officer Cadet School
OKC	Operator Keyboard and Cipher
OTC	Overseas Telecommunications Commission
OWL	Operator, Wireless and Line

PMG	Post Master Generals' Department
PRC	Portable Radio Communications
RAAF	Royal Australian Air Force
RAEME	Royal Australian Electrical and Mechanical Engineers
RAE	Royal Australian Engineers
RAF	Royal Air Force
RAN	Royal Australian Navy
RAR	Royal Australian Regiment
RASCM	Royal Australian Signals Corps Museum, Macleod, Victoria
R & C	Rest and Convalescence
R & R	Rest and Recreation
RMC	Royal Military College, Duntroon
RSL	Returned and Services League
RSO	Regimental Signals Officer
RTT	Radio Teletype
SAS	Special Air Service
SCO	Systems Control Officer
SDS	Special Despatch Service
SEATO	South East Asia Treaty Organisation
SHF	Super High Frequency
SOI	Signal Operating Instructions
SSB	Single Side Band
STRAD	Simultaneous Transmit Receive and Distribute
STRATCOM	Strategic Communications Network (US Army)
TAOR	Tactical Area of Responsibility
TRC	Transportable Radio Communications
UHF	Ultra High Frequency
UMNO	United Malay Nationalist Organisation

CSO	Chief Signals Officer
URC	Utility Radio Communications
USMAAG	United States Military Assistance Advisory Group
USMACV	United States Military Assistance Command Vietnam
VHF	Very High Frequency
VRC	Vehicular Radio Communications
UK	United Kingdom
US	United States
VF	Voice Frequency
WRAAC	Women's Royal Australian Army Corps
WS	Wireless Set
WSO	Washington Standardisation Officers

GLOSSARY OF TERMS

Area of Operations (AO)	That portion of an area of conflict necessary for the conduct and administration of military operations.
Battalion	A unit normally part of a brigade or task force consisting of a headquarters and two or more companies, batteries or comparable units.
Battery	A tactical unit of an artillery regiment (or battalion) equivalent to an infantry company: Consisting of two or more pieces of artillery.
Baud	The unit of modulation rate. It corresponds to a rate of one unit interval per second.
Brigade	A tactical organisation larger than a battalion and smaller than a division, usually consisting of two or more battalions with supporting arms and services attached. The term was replaced by 'Task Force' from 1960 to the end of the Vietnam War.

Call Sign	Any combination of characters or pronounceable word(s) to identify a unit, facility or command which is used primarily for establishing and maintaining communications.
Circuit	an electronic path between two or more points capable of providing a number of channels.
COMCEN	Communications/Signal Centre. An agency charged with the responsibility for the receipt, transmission and delivery of messages.
Company	A body of men commanded by a Major or Captain, larger than a platoon and smaller than a battalion: the basic military unit equivalent to a battery or squadron.
COMSEC	Communications Security: the protection resulting from the application of crypto security, transmission security, and emission security measures to telecommunications and from the application of physical security measures to COMSEC information.
Communications Zone	Rear part of the theatre of operations (behind but contiguous to the combat zone), which contains the lines of communications, establishments for supply and evacuation, and other agencies required for the immediate support and maintenance of the field forces.

Corps (Army)	A tactical organisation larger than a division and smaller than a field army; usually consisting of two or more divisions together with supporting arms and services.
Corps (Regimental)	A specialised sub-division or branch of the Army eg, Signals, Armour, Aviation, Infantry etc.
Continuous Wave (CW)	A continuous signal, not pulsed on and off. A CW signal may be amplitude, phase or frequency modulated.
Cryptography	The encryption (encoding)and/or decryption (decoding) of messages.
Dipole	An antenna consisting of two elements each approximately one quarter wave length in length and fed with radio frequency energy of opposing polarity at adjacent ends of the elements.
Direction Finding (DF)	The process of determining the bearing of an electromagnetic emission.
Duplex Operation	Duplex or full duplex operation refers to communication between two points in both directions simultaneously.
Electronic Warfare (EW)	The military action involving the use of electromagnetic energy to determine, exploit, reduce or prevent hostile use of the electromagnetic spectrum and action that retains friendly use of the magnetic spectrum.
Encrypt	(Encipher/Encode) To convert a plain text message into a disguised form by means of a cryptosystem.

Formation	A grouping of units in a brigade, a task force, a division, or a corps of multiple divisions.
Frequency	The number of recurrences of a periodic phenomenon in a unit of time. In specifying the electrical frequency, the unit of time is the second (a hertz). Radio frequencies are usually expressed in kilohertz (KHz) at and below 30,000 kilohertz, and in megahertz (MHz) above this frequency). The radio frequency bands are as follows: MF 300 KHz - 3 MHz HF 3-30 MHz VHF 30-300 MHz UHF 300-3000 MHz SHF 3 - 30 GHz EHF 30-300 GHz
Frequency Shift Telegraphy (Frequency Shift Keying - FSK)	Telegraphy by frequency modulation in which the telegraph signal shifts the frequency of the carrier between predetermined values. There is phase continuity during the shift from one frequency to the other.
Fullerphone	An instrument that employs a very low direct current in the line, but converts this direct current into an intermittent current of audible frequency at the receiver and thus enabling hand speed Morse telegraphy over good or bad lines with the least chance of remote reception.

Heliograph	A mirror device for signalling by means of the sun's rays.
Half Duplex Operation	Communication between two points in a single direction only. A half-duplex facility is exactly half of a full-duplex facility and is not the same as a simplex facility.
Ionosphere	The region of the atmosphere, extending from roughly 40 to 250 miles altitude, in which there is appreciable ionisation. The presence of charged particles in this region profoundly affects the propagation of electromagnetic radiations of long wavelengths (radio and radar waves).
Jammer	A transmitter designed specifically to prevent or reduce the enemy's effective use of the electromagnetic spectrum.
Line of Sight	In communications, a direct propagation path that does not go beyond the radio horizon.
Modulation	The process in which the amplitude, frequency or phase of a carrier wave is varied with time in accordance with the waveform of superimposed intelligence. The principal types of modulation are: a. Amplitude Modulation - modulation in whch the amplitude of a carrier is modulated. b. Angle Modulation - modulation in which the phase angle of a sine wave carrrier is the characteristic varied.

	c. Phase Modulation - angle modulation of a sine-wave carrier in which the phase of the modulated wave differs from that of the carrier by an amount proportional to the instantaneous value of the modulating wave. d. Pulse Code Modulation - a form of pulse modulation in which the amplitude of the modulating signal is sampled, and the value of the magnitude of the sample (rounded off value) is transmitted as a coded group of constant amplitude pulses. e. Pulse Modulation - Modulation of one or more characteristics of a pulse carrier. In this sense the term is used to describe methods of transmitting information.
Multichannel Radio	A radio equipment designed to provide several channels of communications simultaneously.
Multiplex	Denotes the simultaneous use of a number of channels on a single circuit.
Net (Communications)	An organisation of stations capable of direct communications on a common channel or frequency.
Network	An organisation of stations capable of intercommunication but not necessarily on the same channel.
Off-Line Cipher	A method of operation in which the process of encryption and transmission (or reception and decryption) is performed in separate steps rather than automatically and simultaneously.

On-Line	A method of transmission by which signals from telecommunication equipment are passed direct to a channel/circuit to operate automatically, compatible equipment at one or more distant stations.
Oscillator	A device that produces an electrical signal of relatively constant frequency and amplitude.
Phased Array	An array of dipoles in which the phase of the signal feeding each dipole is varied in such a way that antenna beams can be formed and scanned very rapidly in azimuth and elevation without requiring physical movement of the antenna.
Platoon	A subdivision of a company commanded by a lieutenant.
Precedence	A designation assigned to a message by the originator to indicate to communication personnel the relative order of handling and to the addressee the order in which the message is to be noted. The communication precedence designations are in descending order: Flash, Immediate, Priority and Routine.
Radio	A descriptive term (used generally as a noun) applied to the use of electromagnetic waves between 10 kilohertz and 3,000,000 megahertz.
Radio Relay	A radio system for point-to-point transmission of multi-channel duplex signals employing carrier waves in the VHF or higher bands.
Radio Relay System	A communication system used to perform a radio relay function.

Radio Teletype	The system of communication by teletypewriter or teleprinter over radio circuits.
Receiver, Radio	A device connected to an antenna or other source of radio frequency signals in order to make available in some desired form the required information content of the signals.
Regiment	(1) A battalion-sized unit in the armoured, engineer, artillery, survey, signal, aviation and transport corps. (2) A brigade-sized formation (UK). (3) A Corps title equivalent for artillery. (4) An administrative or historical grouping of infantry battalions - eg Royal Australian Regiment (RAR) and Royal Victorian Regiment (RVR).
Relay Station, Major	A tape relay station is designated as a major tape relay station when two or more trunk circuits connected thereto provide an alternate route; and to meet command requirements.
Relay Station, Minor	A tape relay station is designated as a minor relay station when it has tape relay responsibility but does not provide an alternate tape relay route.
Retransmission	The repetition of a message that was previously sent by tape, card, or similar stored mode of communications.

Routing	The process of determining and prescribing the path or method to be used in forwarding messages.
Squadron	An organisation generally of company equivalent size, in the armoured, engineer, survey, signal, aviation and transport corps.
Secure	A generic term referring to a method of communicating which denies information to unauthorised recipients. The channel/circuit/net is secured by physical means or by the provision of online crypto equipment.
Sideband	A sideband is the frequency band, above or below the carrier, produced by the process of modulation.
Signal Security	A generic term that includes both communications security and electronic security.
Signal Intelligence (SIGINT)	A generic term that includes both Communication Intelligence (COMINT) and Electronic Intelligence (ELINT).
Simplex Operation	Simplex operation refers to communication between two points in both directions, but not simultaneously.
Single Sideband Transmission	A system of carrier transmission in which one sideband is transmitted and the other sideband is suppressed. The carrier wave may be either transmitted or suppressed.
Sky Wave	That portion of a radiated wave which travels in space and is returned to earth by refraction in the ionosphere.

System	An overall term used to describe communication facilities from an engineering aspect, including all the associated equipment.
Switching, Automatic	A method of operation that affects the automatic interconnection of channels, circuits and trunks and/or handling of traffic through a switching facility.
Switchboard	A manually operated apparatus at an exchange on which the various circuits from subscribers and other exchanges are terminated to enable operators to establish communication either between two subscribers on the same exchange or between subscribers on different exchanges.
Switchboard, Magneto	A manual exchange at which the subscribers and operators call and clear by means of magneto generators.
Tape Relay	A method of receiving and retransmitting messages in tape form.
Tape Relay, Automatic	A system of tape relay which embodies automatic switching.
Tape Relay, Semi Auto	(semi Auto Continuous Tape - Switching) A method of teletypewriter operation whereby incoming messages are received in continuous printed/perforated tape and given onward electrical transmission according to routing requirements through the pushbutton panel connection of a transmitter distributor into the appropriate outgoing channel(s).

Tape Relay, Torn Tape	A method of teletypewriter operation whereby incoming messages are received in printed/perforated tape and are separated by tearing the tape so that individual messages can be processed and hand-carried to the appropriate outgoing channel(s) in accordance with routing requirements.
Task Force	A tactical organisation larger than a battalion and smaller than a division, usually consisting of two or more battalions with supporting arms and services attached. The term was used in place of 'Brigade' from 1960 to the end of the Vietnam War.
Telecommunication	Any transmission, emission, or reception of signals, signs, writing, images, and sounds, or intelligence of any nature by wire, radio, visual, or other electromagnetic systems.
Telegraphy	A system of telecommunication for the transmission of intelligence by the use of a signal code - the transmission by wire or radio of alphanumeric codes.
Transmitter (Radio)	Apparatus producing radio frequency energy for radio communication.
Troop	A sub-division of a squadron, normally commanded by a lieutenant.
Tropospheric Scatter	The propagation of radio waves by scattering as a result of irregularities or discontinuities in the physical properties of the troposphere.
Tuning	The process of adjusting a circuit so that it resonates at the desired frequency.

Voice Frequency	Any frequency within the part of the audio frequency range essential to the transmission of speech of commercial quality, ie 300-3,000 hertz.
Wave Length	The distance travelled in one period or cycle by a periodic disturbance. It is the distance between corresponding phases of two consecutive waves of a wave train. A wavelength is the quotient of velocity divided by frequency.

MAPS AND DIAGRAMS

Maps:

Diagrams:

MAPS

Japan: BCOF Zone 1947

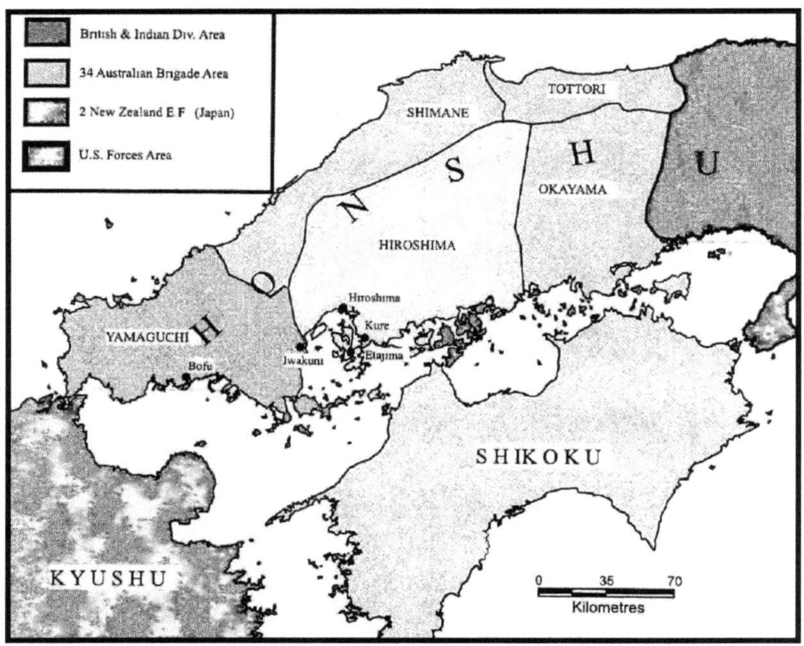

Map 1

This map shows the allocation of occupation zones assigned to BCOF formations is 1947. RA Signals elements operated predominantly in the Hiroshima Prefecture.

Korea

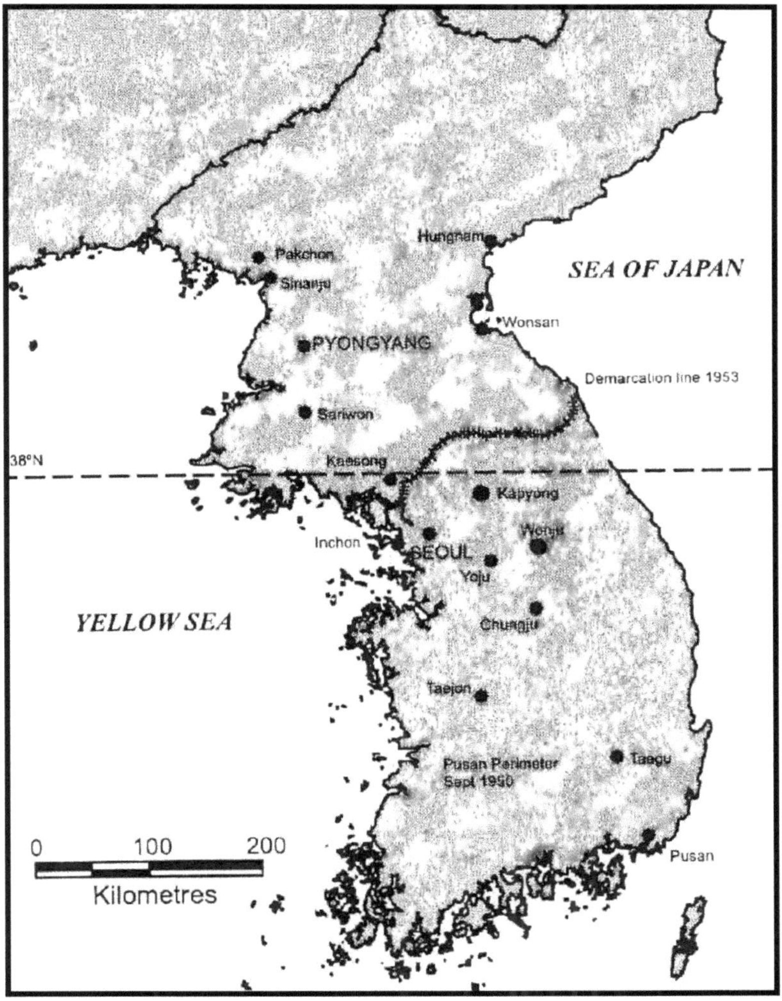
Hungnam
SEA OF JAPAN
Pakchon
Sinanju
Wonsan
PYONGYANG
Demarcation line 1953
Sariwon
38°N
Kaesong
Kapyong
Wonju
Inchon
SEOUL
Yoju
YELLOW SEA
Chungju
Taejon
Pusan Perimeter
Sept 1950
Taegu
0
100
200
Kilometres
Pusan

Map 2

Singapore

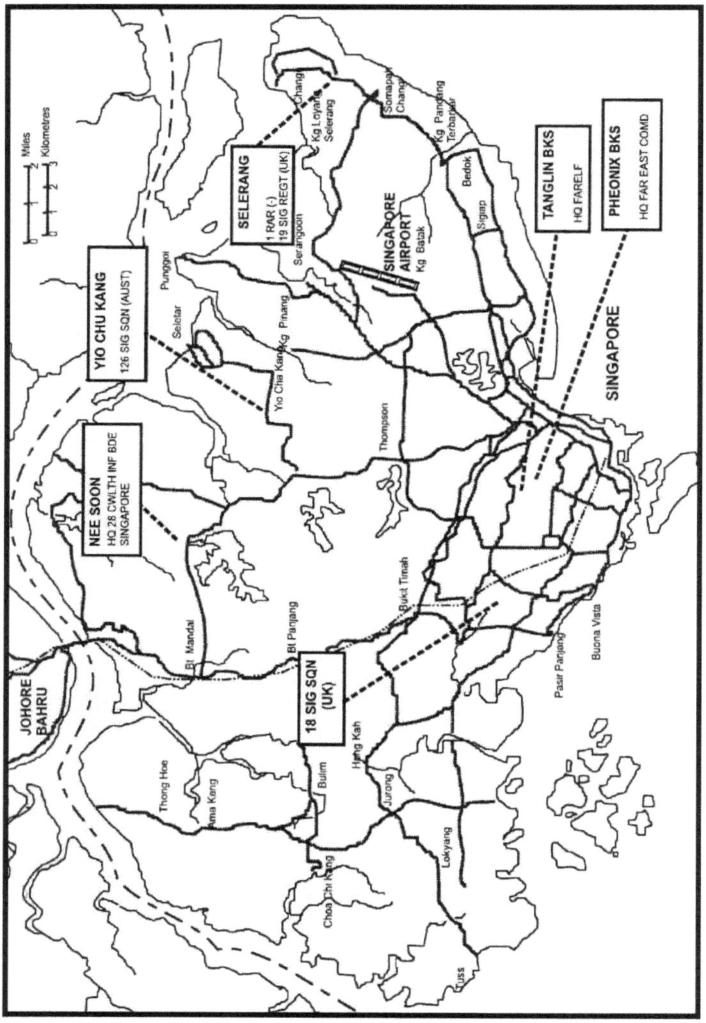

Map 3

This map shows the Commonwealth Forces in Singapore as at 1970.

Malaya 1957

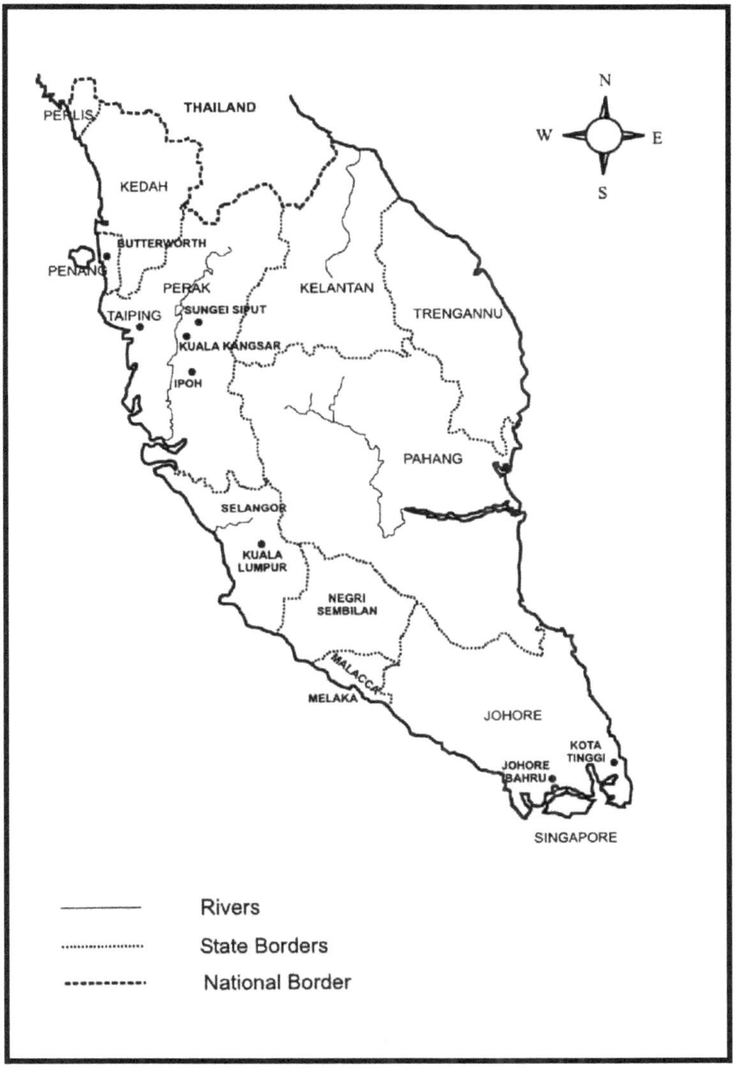

THAILAND
PERLIS
KEDAH
BUTTERWORTH
PENANG
PERAK
KELANTAN
TRENGANNU
TAIPING
SUNGEI SIPUT
KUALA KANGSAR
IPOH
PAHANG
SELANGOR
KUALA LUMPUR
NEGRI SEMBILAN
MALACCA
MELAKA
JOHORE
KOTA TINGGI
JOHORE BAHRU
SINGAPORE
N
W
E
S

Rivers
State Borders
National Border

Map 4

Sarawak, Brunei, Sabah, North Kalimantan (Borneo)

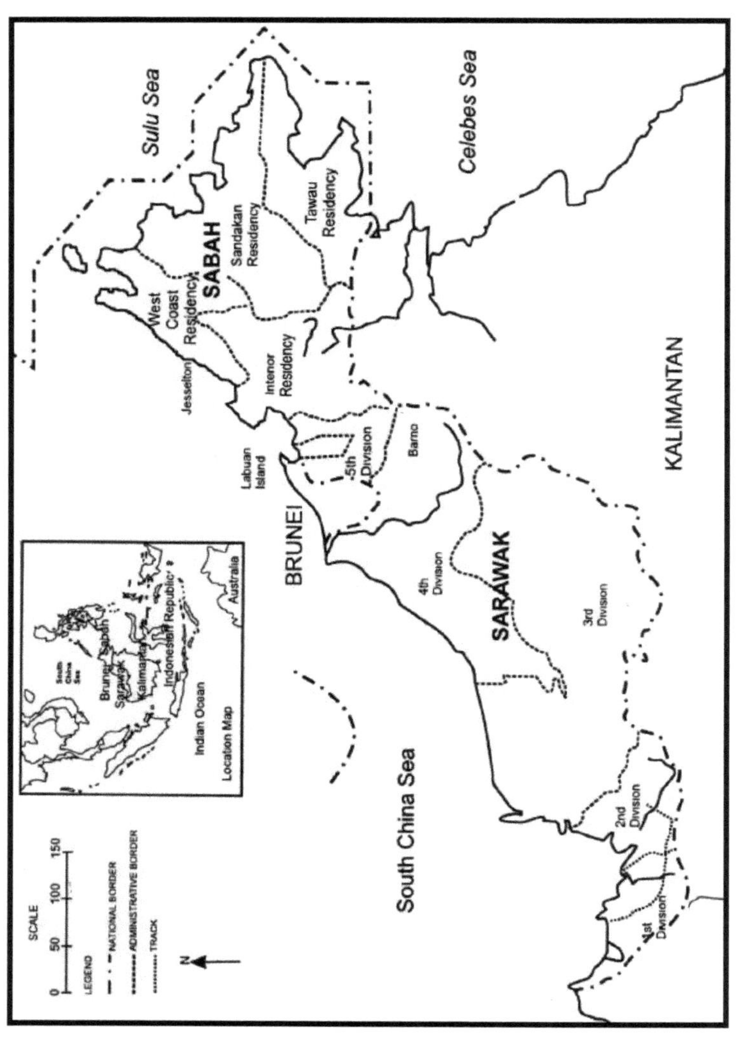

Map 5

South Vietnam

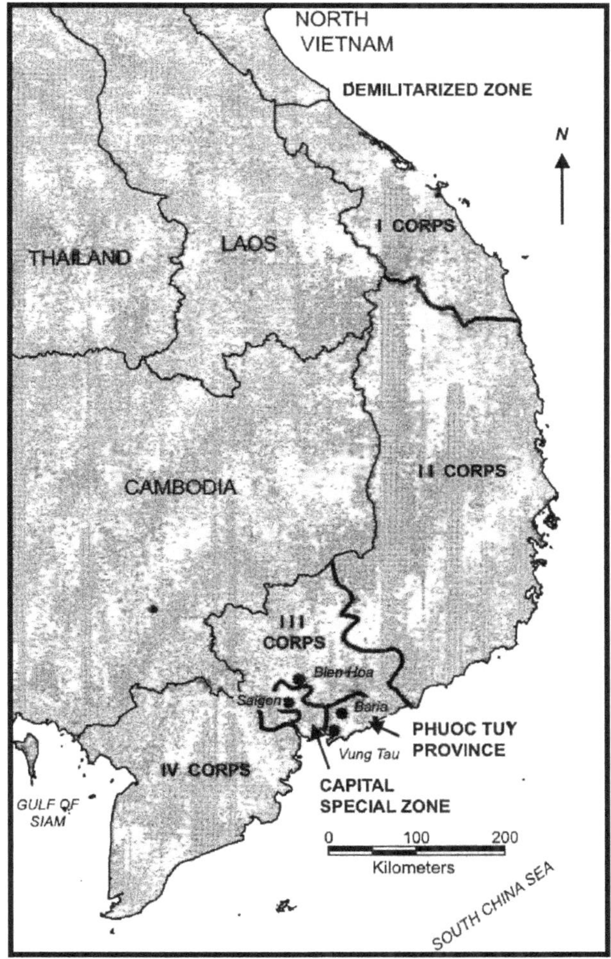

Map 6

This map shows the four Corps military geographic boundaries that were in use in South Vietnam at thee height of the US involvement. RA Sigs elements operated predominantly in the Phuoc Tuy Province and the Capital Special Zone.

Integrated Wideband Communications System 1969, South Vietnam

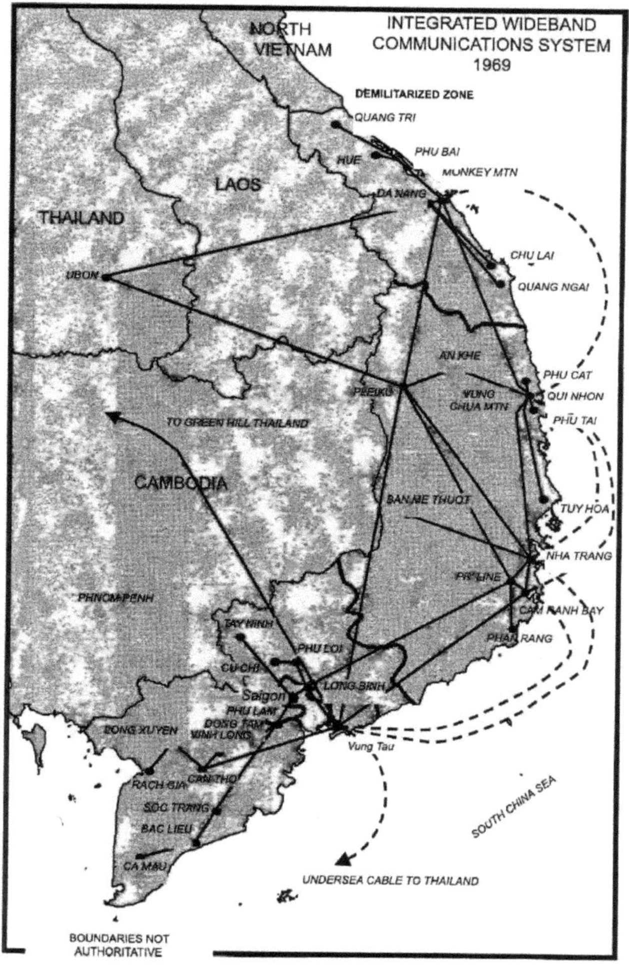

Map 7

This map shows the wideband communications links operated by the US armed forces in Vietnam in 1969. The RA Sigs elements in Vietnam were linked in this network.

Phuoc Tuy – South Vietnam

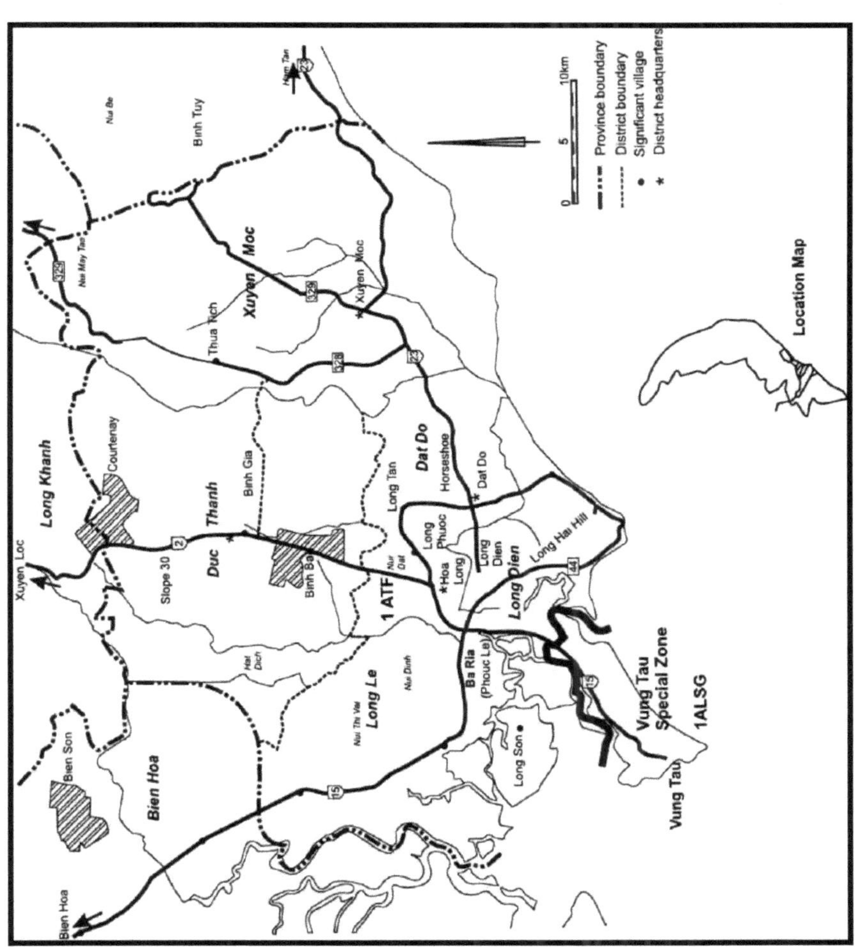

Map 8

Diagrams

Divitional Signal Regiment Structure as at 1954

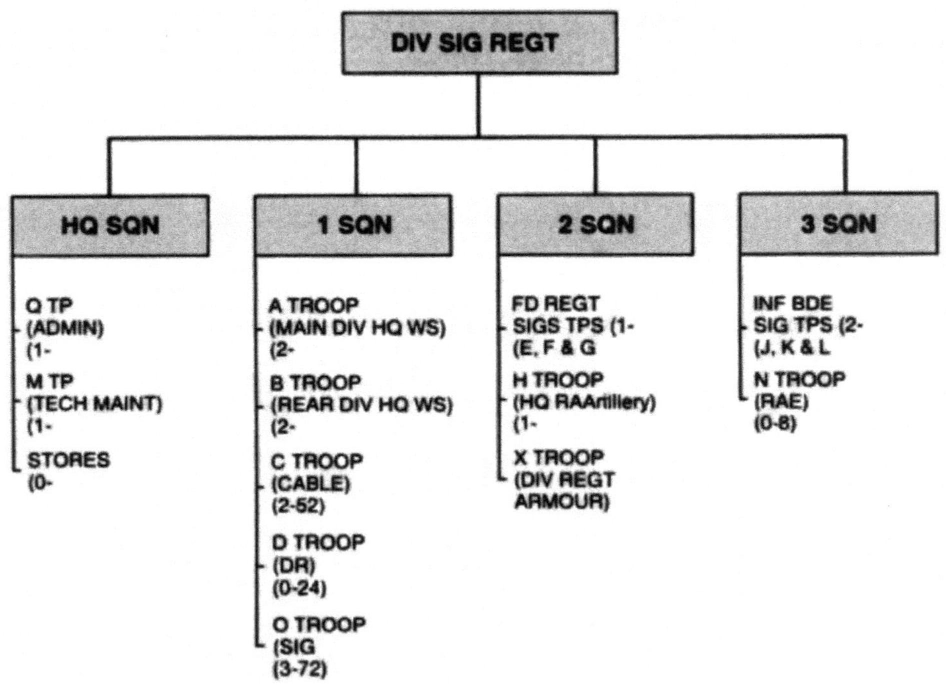

Diagram 1

Note:

1. RASCM, Australian Military Forces Study Precis, January 1954, issued by Director of Military Training, AHQ

Pentropic Division Signal Regiment Structure (1960 - 1964)

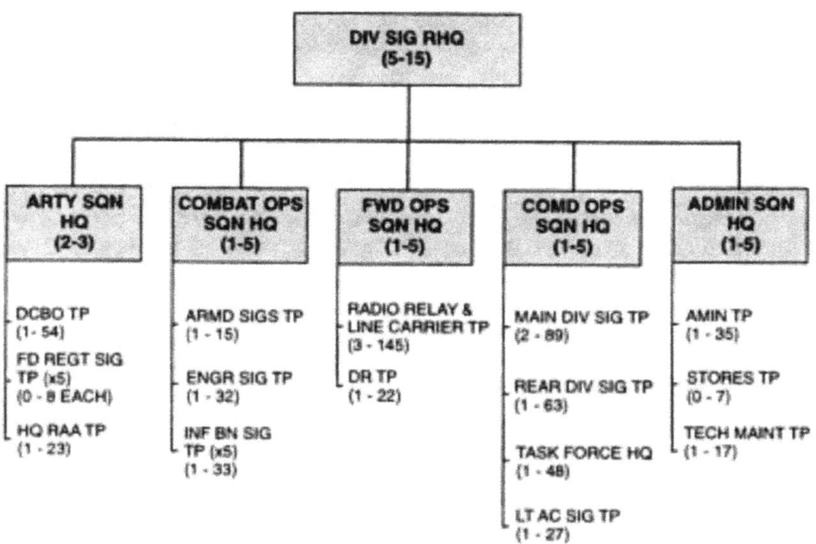

Diagram 2

Note:
RASCM, Notes by Director of Signals for visit by Brigadier Mohammed Suleman - Director of Signals Pakistan, February 1962, p. 17

Divisional Signal Regiment Organisation from 1965

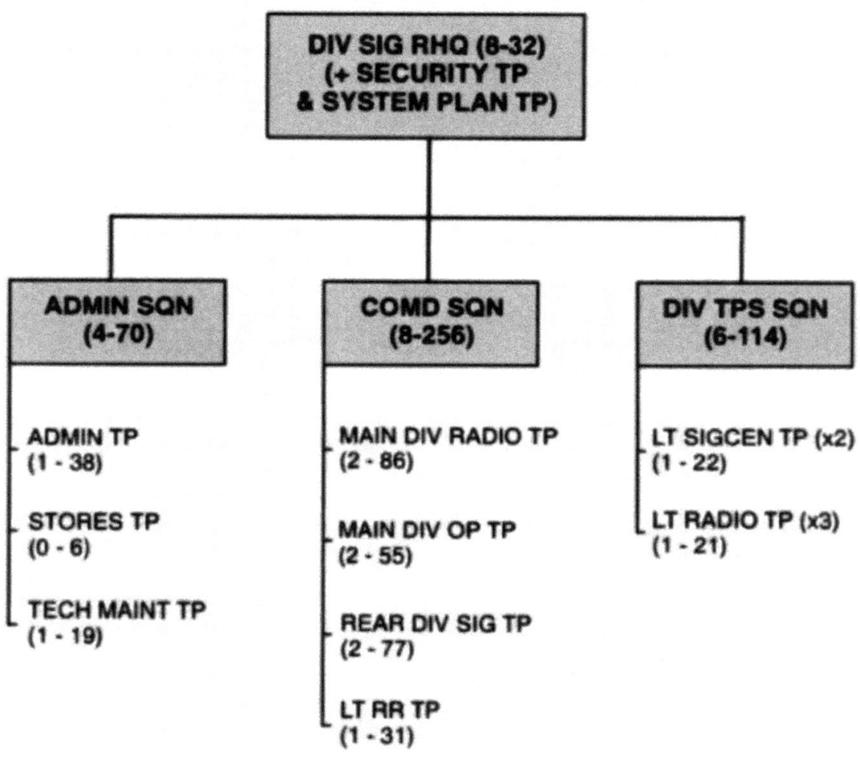

Diagram 3

Note:
Department of the Army, *The Royal Australian Corps of Signals Reference Manual*, 1968, Section 2.

Signals Symbols

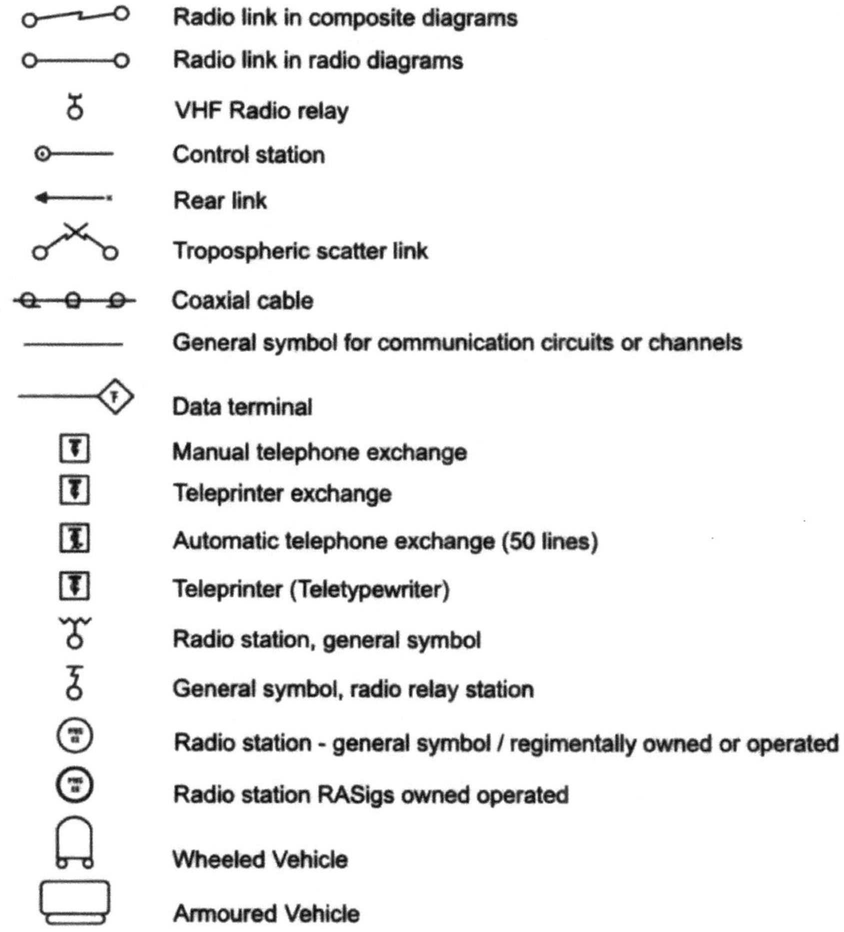

Radio link in composite diagrams

Radio link in radio diagrams

VHF Radio relay

Control station

Rear link

Tropospheric scatter link

Coaxial cable

General symbol for communication circuits or channels

Data terminal

Manual telephone exchange

Teleprinter exchange

Automatic telephone exchange (50 lines)

Teleprinter (Teletypewriter)

Radio station, general symbol

General symbol, radio relay station

Radio station - general symbol / regimentally owned or operated

Radio station RASigs owned operated

Wheeled Vehicle

Armoured Vehicle

Diagram 4

Diagram of Victorian Communications Layout

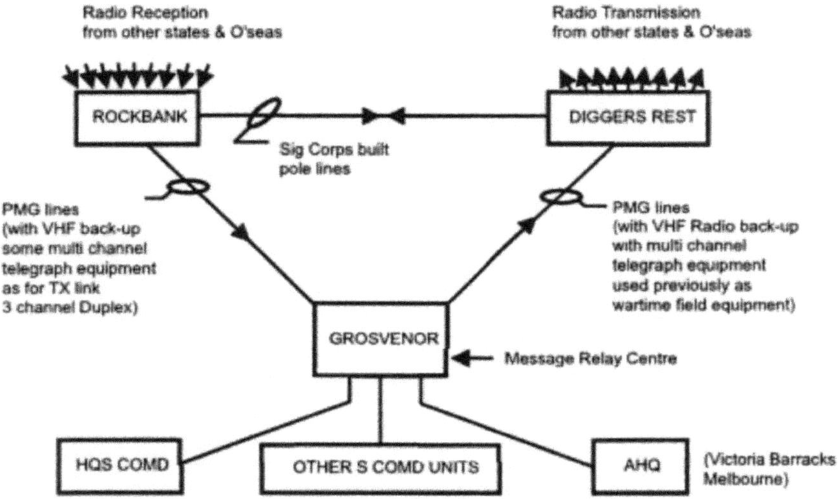

Diagram 5

This diagram illustrates the communications links in use in Southern Command in 1947 between AHQ and supported headquarters and units.

Basic Communication Framework for the Army's Strategic Commuunications Network – 1947

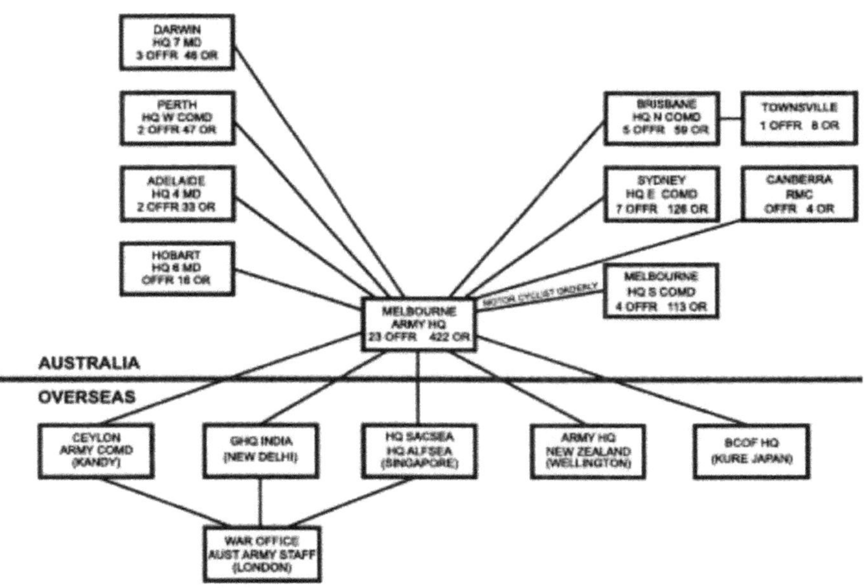

Diagram 6

This diagram illustrates the main internal Australian wireless links of the Army Wireless Chain (later known as AUSTCAN) and international wireless links to other British Commonwealth strategic communications hubs which formed the COMCAN.

Proposed Army WRLS Chain Network (Aug 48) (With main relay station at NAIROBI)

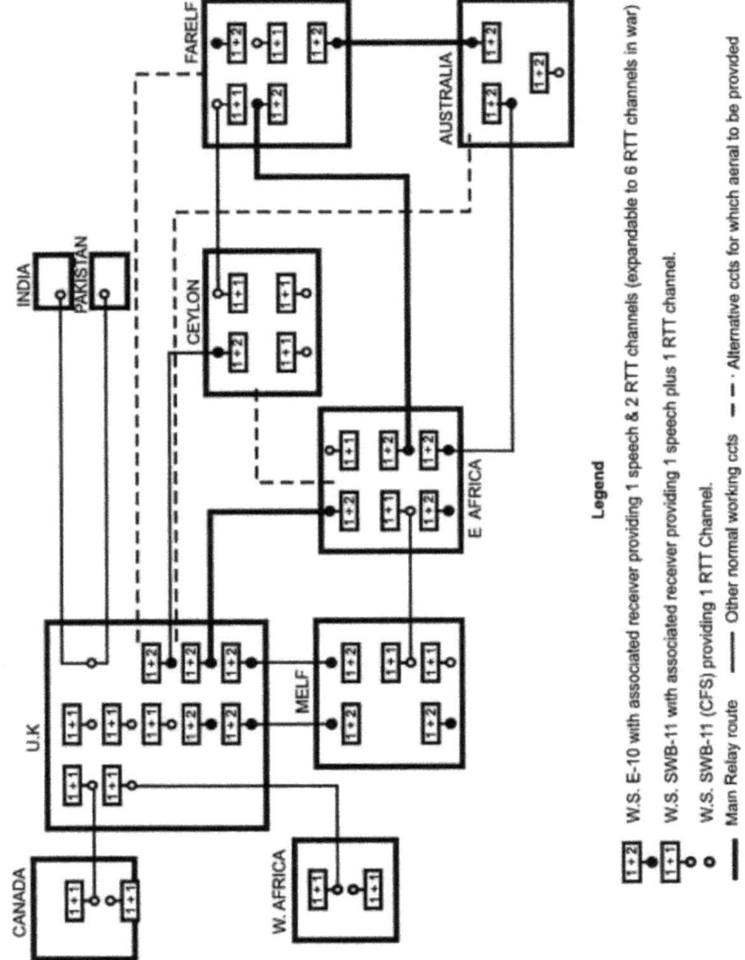

Diagram 7

Plan for a BCF to Participate in Occupation of Japan Joint Sig Comms – BCOF (1946)

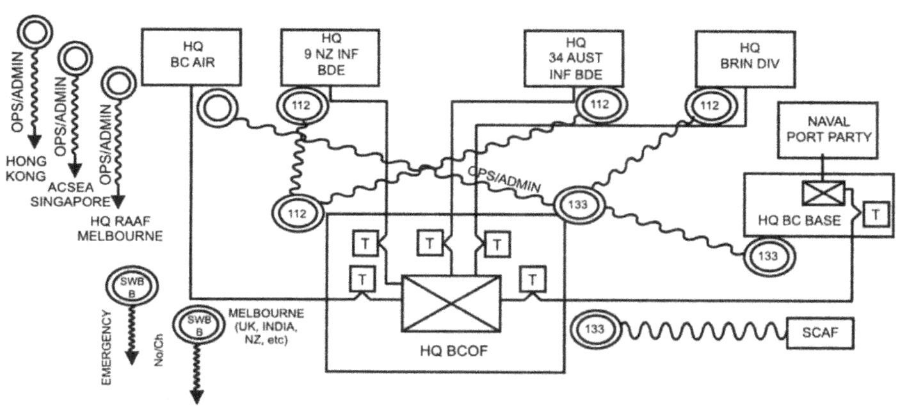

Diagram 8

This diagram illustrates the main proposed internal line and external wireless links for BCOF in 1946.

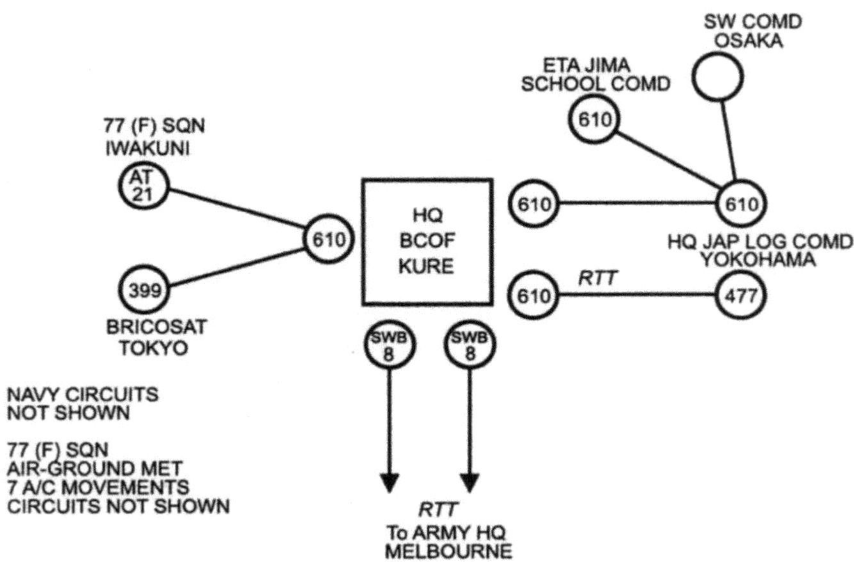

Diagram 9

This diagram shows the BCOF radio circuits prior to the deployment of Australian forces in Korea.

BCOF Signal Diagram (Wireless)
31 Dec 50

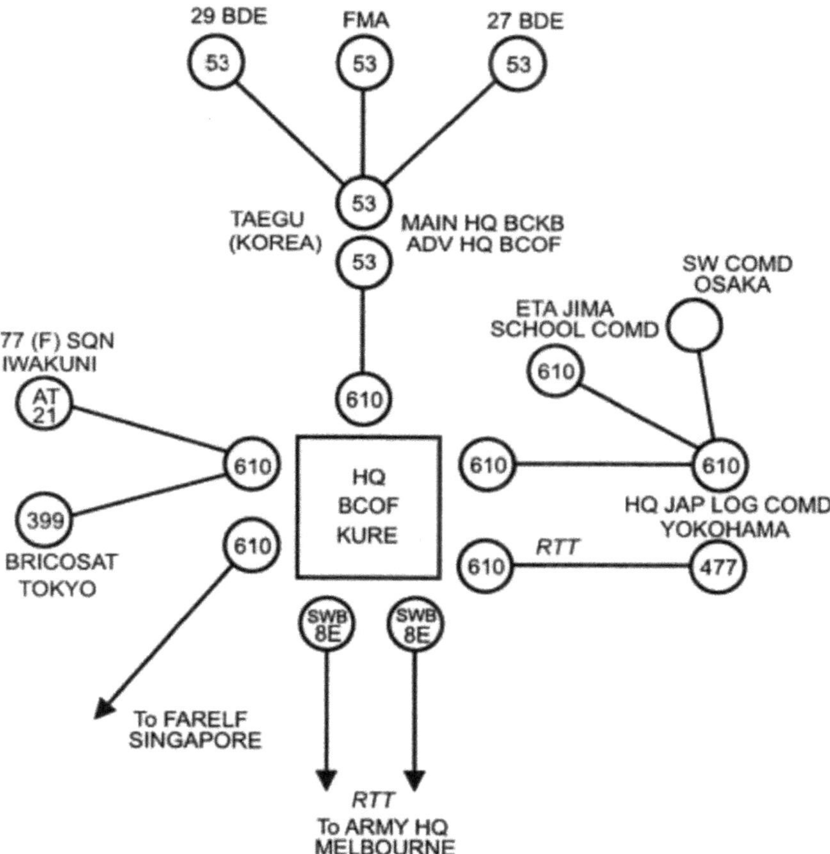

Diagram 10

This diagram shows the BCOF radio circuits to the deployed Australian and Commonwealth forces in Korea by the end of 1950.

BCOF Signal Diagram (Wireless) 30 Jun 51

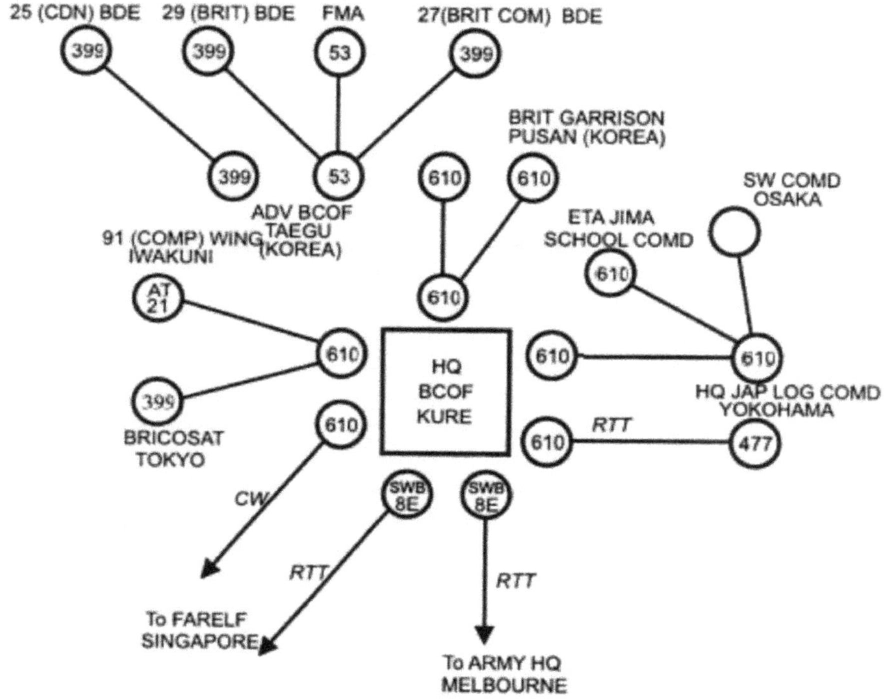

Diagram 11

This diagram shows the BCOF radio circuits to the deployed Australian and Commonwealth forces in Korea by the middle of 1951.

Tropospheric Scater and Line-of-sight Communiications

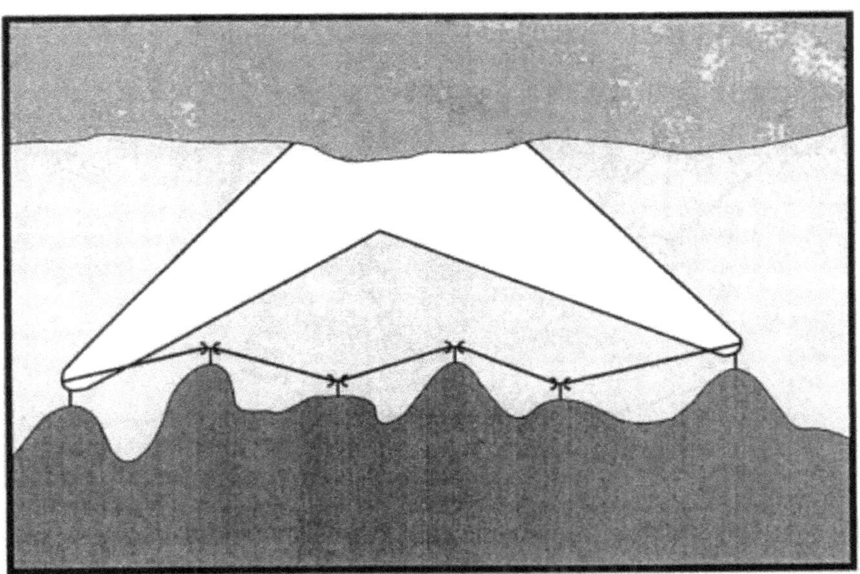

Diagram 12

This diagram illustrates the way in which tropospheric scatter radios rely on signals bouncing off the troposphere back down to a receiving station byond thee horizon. In contrast, line-of-sight communications signals require direct links; thus, repeater sites may be required to cover the longer distances that are more easily reached by tropospheric scatter radios.

1 RAR VHF Battalion Command Net Aug 65

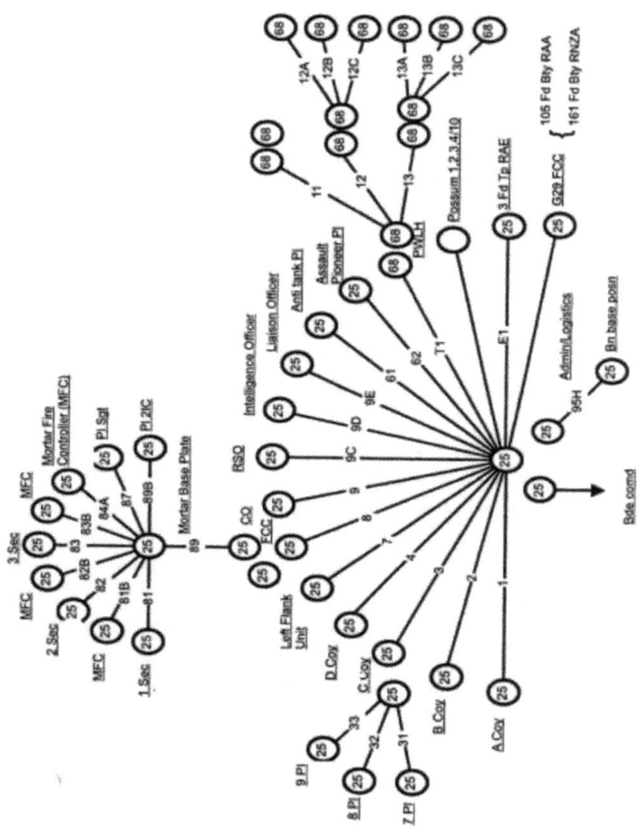

Diagram 13

This diagram illustrates the number of VHF radio links in use by 1 RAR in 1965. The introduction of the AN/PRC 25 radio set throughout the battalion increased the traffic capcity and reliability of the communications links. The central radio symbol represents the net control station through whom the battalion commander could exercise command over the disparate subordinate and attached elements.

Telephone Trunk Circuits – AFV (Aug 67)

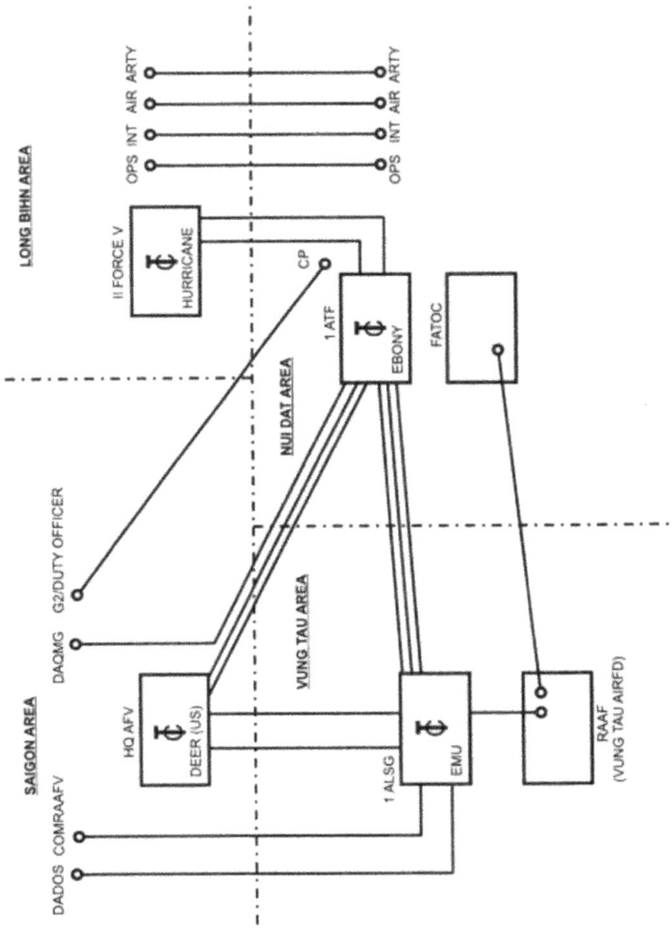

Diagram 14

This diagram illstrates the telephone links between the main sites where Australian personnel were deployed in Vietnam. Note the main telephone switchboards at the Australian bases at Nui Dat (Ebony) and Vung Tau (Emu)

Teletypewriter Circuits
Vietnam – 1967

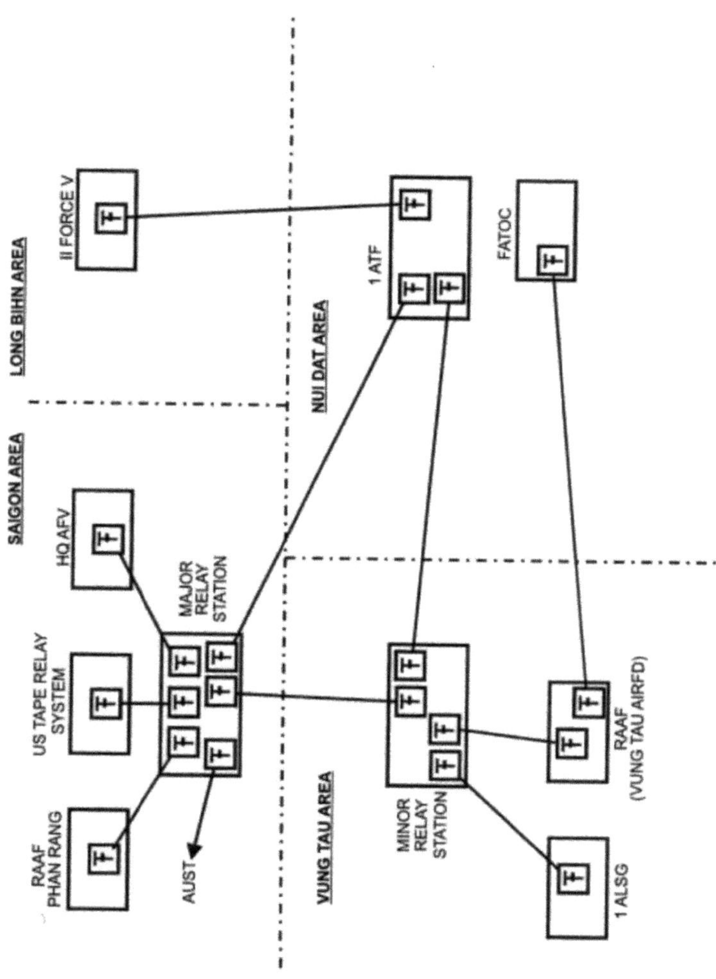

Diagram 15

This diagram illustrates the teletypewriter links between the main sites where Australian personnel were deployed in Vietnam.

Communications Diagram
Australian Forces Vietnam, 1970

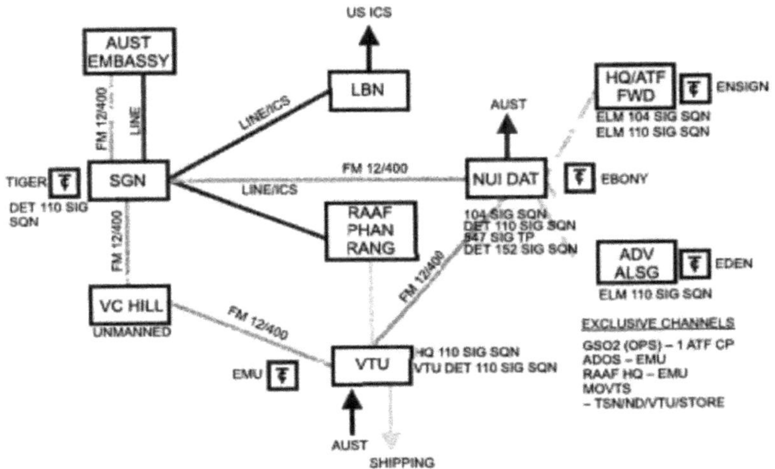

(IN COUNTRY CIRCUITS BACKED UP BY HF RADIO)

LEGEND (Not shown on diagram)

SGN	Saigon – Location of HQ AFV (including CSO)
NUI DAT	Location of 1 ATF
VTU	Vung Tau – Location of 1 ALSG
LBN	Long Binh (US Base)
	Siemens FM 12/400 Radio relay system
	US line/in-country system (ICS)
	HF tactical radio – typically AN/GRC-106
	AUSTCAN HF SSB/ISB Transmitter and receiver locations (TX station at Nui Dat and RX station at Vung Tau)
	Forward deployment as required – typically AN/TRC-24 radio relay system
	Switchboard

Diagram 16

Task Force Headquarters, Nui Dat
Vietnam - 1966
(Signal Sector)

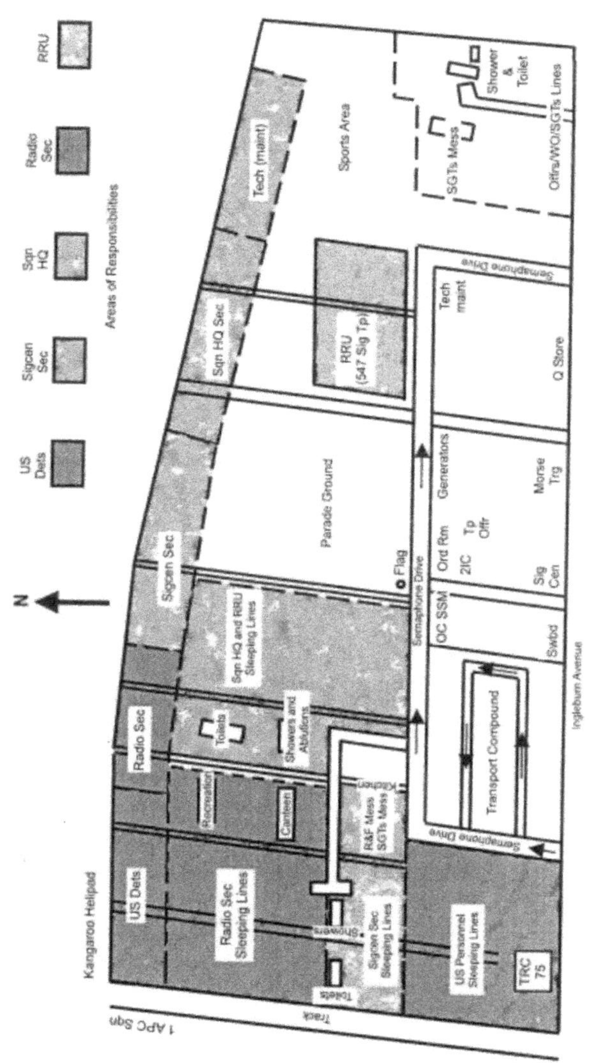

Diagram 17

110 Signals Squadron Functional Organisation, 1968

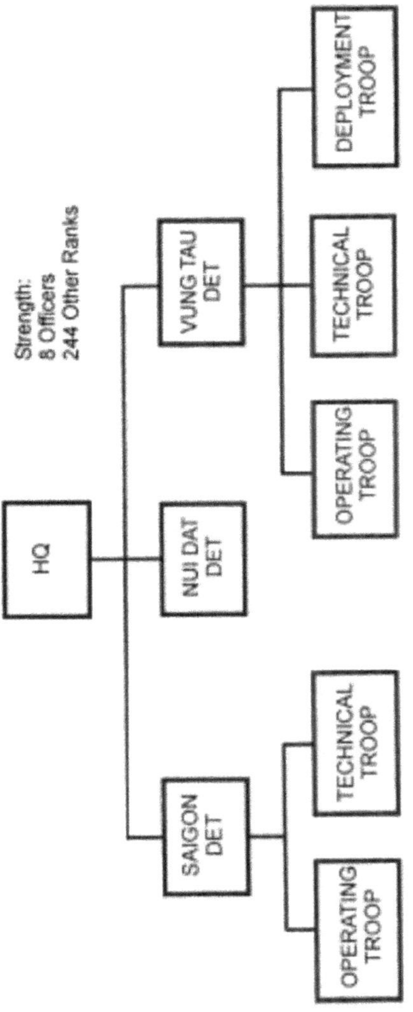

Diagram 18

This diagram shows the outline organisation of 110 Signals Squadron prior to the relocation from in 1968.

104 Signals Squadron Functional Organisation, 1968

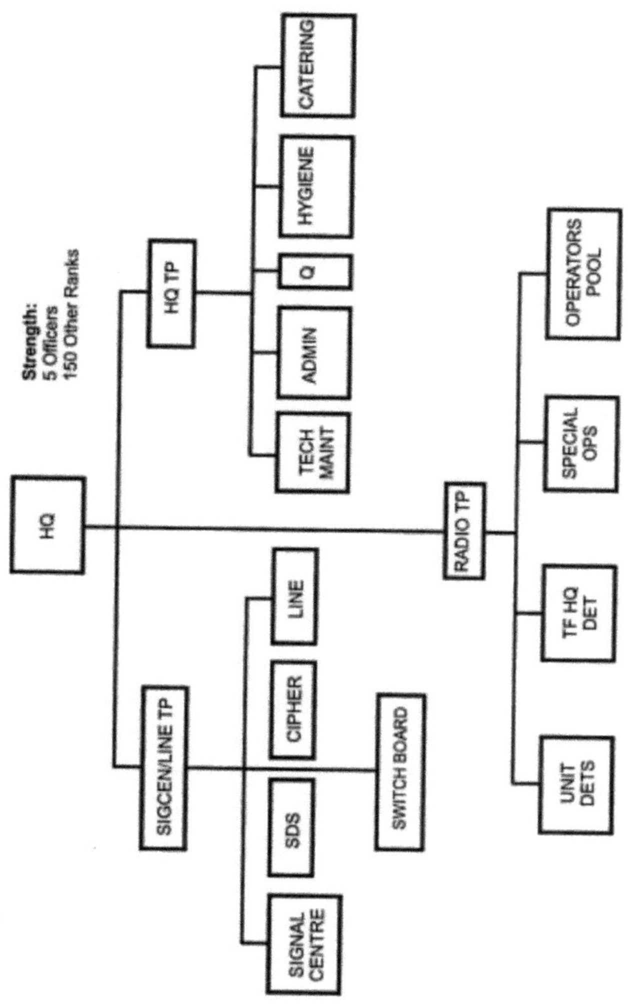

Diagram 19

This diagram shows the outline organisation of 104 Signal Squadron during operations in 1968.

INTRODUCTION

AT THE CONCLUSION of World War II, the Australian Corps of Signals was a complex organisation which adequately catered for the requirements of the vast and diverse communications requirements of the widely dispersed elements of the Australian Army then serving in Australia, Papua and New Guinea and beyond. Yet by the time the forces had been demobilised, when the rundown of military forces had reached completion in 1947, the only remaining active Australian Signals Corps units were those stationed with the occupation force in Japan. What emerged in the following twenty five years was the product of, among other things, Australia's strategic posture, its geography, and the adaptation and implementation of newly emerging communications technology then becoming available - technology which was to have a significant impact on the Army and the conduct of Australia's military operations in the South East Asian conflicts in which it was to become involved.

Literature covering the Royal Australian Corps of Signals in the period in question is limited to some unpublished manuscripts dealing with key events of individual units within the Corps, as well as various brief overviews of overseas deployments. There was no comprehensive record of the contribution made by the Signals Corps (covering the period from the demobilisation at the end of World War II through to the end of the Vietnam War) or a comprehensive analysis of the impact that the Corps has had on the wider Army and its methods of conducting military operations. A volume, written by Theo Barker, covering the history of the Signals Corps, entitled *Signals: A History of the Royal Australian Corps of Signals 1788- 1947* provides a sound long-term perspective on developments in the field of military telecommunications in Australia from 1788 to 1947.[1] His work sets the starting point for this study.

This Volume looks at the Royal Australian Corps of Signals in the period from 1947 to 1972, focusing in particular on the impact of communications technology on the Corps and the Australian Army and its conduct of military operations, while also presenting an anecdotal expression of life in the Corps. This period was chosen as it commences where Barker's *Signals* concluded - with the deployment of the British Commonwealth Occupation Force (BCOF) in Japan, and it concludes with the de facto abandonment of the strategy of Forward Defence with the withdrawal of Australia's military forces from Vietnam in 1972. The period in between saw the end of post-war demobilisation and the establishment of the Post War Army, including the Australian Regular Army (ARA) and the Citizen Military Force (CMF). It also saw the introduction of the National Service Training Scheme and the accompanying expansion of the CMF, the increase of the ARA, the establishment of a strategic military communications network in peacetime, and the gradual shift in equipment procurement and training from the United Kingdom towards the United States. The culmination came with the sending of troops to fight alongside the Americans, for the first time without British support, in operations in Vietnam.

The period chosen has further significance in that developments from 1947 to 1972 set the scene for Australia's defence capabilities and orientation in the late 1970s through to the present. For instance, the Australian Defence Force has recently introduced a secure, nationwide, strategic communications network which owes its origins to the decisions taken in the early

1 T.J. Barker, *Signals : A History of the Royal Australian Corps of Signals 1788-1947*, Royal Australian Corps of Signals Committee, Canberra, 1987.

post-war years to establish, for the first time, a peace-time, nationwide, strategic communications network. The structure of today's 'combat' and specialist signal units also owes much to the events during this period. For instance, field communications technology is now being introduced into the Army, the need for which was first realised as a result of experiences detailed in these pages.

This work also serves as a record of the Corps in the past, which has intrinsic significance in helping the Corps of today to understand how it has emerged the way it has and why it happened this way. There is also value in appreciating the significance of lessons learned along the way. This will perhaps help to shed some light on the way ahead - while also serving to preserve part of the Corps' corporate memory.

Martin Van Creveld, in his book *Technology and War*, correctly asserted that 'war is impacted by technology in all its forms' and 'military technology affects warfare like waves spreading from a stone thrown in a pond.'[2] Likewise, William H. McNeill in his book The *Pursuit of Power*, asserted that: 'one can detect in the historic record a series of important changes in weapons-systems resulting from sporadic technological discoveries and inventions that sufficed to change preexisting conditions of warfare and army organisation.'[3] This had become so critical by the late 1930s that all the belligerents realised by the time fighting began that some new secret weapon might tip the balance decisively. Accordingly, scientists, technologists, design engineers, and efficiency experts were summoned to the task of improving existing weapons, as well as communications and cryptographic equipment, and inventing new ones on a scale far greater than ever before.[4]

---

2 M.L. Van Creveld, *Technology and War: From 2000 B.C. to the Present,* MacMillan, New York, 1989. pp. 1-2 & 311.

3 W.H. McNeill, *The Pursuit of Power,* Basil Blackwell, Oxford, 1983, p. 9.

4 Ibid., p. 357.

With this in mind, I have argued that developments in the field of communications technology in the period from 1947 to 1972 have significantly affected the life of the Corps and, in particular, the way the Corps has been structured and has operated. In turn, these changes have had a direct and significant impact on the Army and its warfighting capability (including its strategy, tactics and administration), as illustrated by the varied, dispersed, mobile and flexible operations of the Vietnam War that were made possible by the communications systems then available. The Signal Corps and the wider Army have been forced to be responsive to changes caused by technological advances. Communications systems have increased in range, capacity and reliability. This, in turn, has improved the level of control that commanders can exercise. Better communications have given commanders a greater chance to seize fleeting opportunities as information collection and dissemination have improved. Also, increasingly, commanders at the highest level have been granted access to those involved in operations at the lowest levels. On the other hand, these improvements have also increased the complexity of command as commanders have become vulnerable to being overwhelmed with too much information.[5] They also presented members of the Corps with a seemingly endless supply of challenges to improvise and adapt. The ramifications of these changes have been profound.

The 'father' of Australian military history, C.E.W. Bean, established the Australian tradition of 'democratic history', where emphasis was placed on the actions of the Australians without paying too much attention to larger-scale aspects of

5 Martin Van Creveld has argued in *Command in War,* that in Vietnam, for instance, where the US Army established the most extensive, expensive and sophisticated signals network in history, it proved in the end incapable of dealing with this 'bottomless pit' of information. See M. Van Creveld, *Command in War,* Harvard University Press, Harvard, 1985, p. 258.

politics and command.[6] His style has been sharply contrasted in recent years by D.M. Horner, who has published various works that have focused on the upper echelons of military decision makers, and, as Michael McKernan pointed out, Australian military history is the richer for it.[7] In this work, however, I have tried to both explain some of the decisions made affecting the Corps and to provide an anecdotal expression of what it was like to be a Signalman in Melbourne or Korea or Vietnam.

This work is intended to focus more narrowly on the specialist contribution of the Royal Australian Corps of Signals to the Australian Army and Australia's involvement in military operations from 1947 to 1972 than the already published official histories of the Korean War, and the seven volume official history of Australia's involvement in Malaya, Borneo and Vietnam (of which four volumes have now been published).[8] Unlike general military histories, it is a narrative of a particular Corps, and unlike unit or regimental histories, one can not list simply the achievements or events chronologically, nor can one go into the amount of detail on the life of individual units because of the breadth and disparate nature of the components. On the same token, I have tried to avoid merely presenting a string of unit histories tied together without cohesion or without an overarching sense of meaning. I have done this by relating the narratives of the units serving in Australia and overseas to how communications technology affected them and how they, in turn, affected the units they supported.

6 M. McKernan, 'Writing About War', in M. McKernan & M Browne (eds), *Australia: Two Centuries of War and Peace,* Australian War Memorial and Allen & Unwin, Sydney, 1988, p. 13.

7 See for instance D.M. Horner, *High Command, Australia and Allied Strategy 1939-1945,* Allen & Unwin and Australian War Memorial, Sydney, 1982. He has, however, also written Corps and Regimental histories that have a more 'democratic' focus than *High Command*. See For instance *SAS Phantoms of the Jungle,* Allen & Unwin, Sydney, 1989; and *The Gunners: A History of Australian Artillery,* Allen & Unwin, 1995.

8 At the time of completion of the manuscript P. Edwards' *Crises and Commitments,* I. McNeill's *To Long Tan,* C.D. Coulthard-Clark's *The RAAF In Vietnam* and B. O'Keefe's *Medicine at War* had been published.

Writing a book about a part of the Army, such as the Royal Australian Corps of Signals, presents some difficulties for any historian trying to analyse data while also providing a narrative. In this work, I have tried to give a factual account of developments with sufficient detail to make the reader gain a clear understanding of how and why they came about and how the Corps operated while also keeping in focus the significance of how it contributed to shaping the Army and Australia's method of conducting military operations. It is a challenge to adequately place developments within the Royal Australian Corps of Signals in their wider military, strategic and technological context. I have endeavoured to do so by including references to the state of the wider Army, the directions in strategic and foreign policy, and observations about the developments in communications technology affecting the wider community.

My access to government-sourced information concerning Signals Intelligence, particularly at the level of the Defence Signals Bureau (or Defence Signals Directorate as it has become known) has been limited due to its sensitivity. Consequently, I have kept the analysis focused scrupulously on activities conducted by the Army's Signals Corps and have avoided the Signals Intelligence activities controlled at the Defence Headquarters level. There remains substantial valuable material at this level to consider, particularly concerning operations in Borneo and Vietnam. I have also sought corroboration from open source material wherever possible. My coverage of the Corps' Signals Intelligence capability and activities has also been subject to review by the Archives and Historical Studies Section of the Department of Defence. Fortunately, this has resulted in only a couple of small portions being removed, and this has not detracted from the observations that I have been able to draw.

Because of the multifaceted nature of the Australian Army's communications requirements, this book is divided into fields that equate with the chapter divisions. The first chapter explains the antecedents of the Corps and provides an overview of the key lessons learnt by the Corps by the end of World War II and the plans then being laid for the establishment of the Post War Army. The Chapter culminates with an overview of the Corps' situation by 1947 and the conferring of the 'Royal' title in recognition of its service in World War II.

Chapter Two deals with the Army's Field Force and specialist communications elements within Australia and its territories. In particular, it considers how and why they were established and the contribution they made. The Divisional Signal Regiments are looked at first, focusing on their role in the National Service Training Scheme and their metamorphosis through the various reorganisations in the late 1950s and 1960s. Next is the 1st Independent Infantry Brigade Signal Squadron which evolved into the 1st Division Signal Regiment and this is followed by brief overviews of other specialist Signal units including 3 Line of Communication Signal Regiment (later 2 Signal Regiment), as well as project and construction Signal units, the Command Signal units around Australia and in Papua and New Guinea, the role of the Women's Royal Australian Army Corps (WRAAC) in the life of the Corps and an overview of the place of Signals Intelligence in the Corps.

Chapter Three covers the various overseas deployments from 1947 to 1961. These include Japan, Korea and Malaya. Australia's deployments on operations in these countries generated unprecedented requirements for military communications support as well as opportunities for adventure for members of the Corps. Never before had Australia sought to maintain an international military communications network to support deployed forces in peacetime, and the demands for communications support in-country for the units deployed presented their supporting Signals units with unique challenges. They, in turn, led the Corps to search for and find improved ways of 'getting through'.

Chapter Four concerns Australia's deployments to Borneo, Singapore, Malaysia and Indonesia. Australia's deployments to Borneo were a result of the Confrontation between Indonesia and the newly formed Malaysia, and the experiences with Signals Intelligence here set a precedent which was to be capitalised upon during the Vietnam War. In Singapore and Malaysia, the drawdown of British forces in the Far East placed added pressure on the Royal Australian Corps of Signals personnel providing support to the British Commonwealth Far Eastern Strategic Reserve (BCFESR). The commencement of deployments in support of survey operations in Indonesia in 1969 presented further challenges, which the Corps took in its stride.

Chapter Five deals with the Army's strategic communications network that was established during World War II and maintained thereafter in light of both Australia's military deployments overseas commencing with the BCOF in Japan and the perceived need to maintain a nation-wide emergency backup to the civil inter-state communications network. The introduction of new technology on this network was to give the Chiefs of Staff in Australia unprecedented access to the operations of even the smallest units overseas. By providing this service, the Corps enabled the Government to ensure maximum national control of even the smallest of operational deployments alongside allied armed forces overseas.

Chapter Six concerns all aspects of training and aid to the civil community. It also deals with the Directorate, as well as with issues faced by the Signals Corps as a whole. Compatibility with allied armed forces was a key strategic concern of the Chiefs of Staff throughout the period, and the ill-fated Project Mallard bore testimony to the difficulties of developing capabilities on a shared basis.

Chapter Seven is the first of five chapters covering Australia's commitment to the Vietnam War. It covers the initial deployment of advisers in 1962, the deployment of a battalion group in Bien Hoa Province in 1965 with the support of 709 Signal Troop, leading up to the deployment of a Task Force to Phuoc Tuy Province in 1966. Several problems were experienced and lessons learnt during this period, which helped set the direction for the remainder of the time that Australian troops served on operations in Vietnam. The provision of a national rear link for the infantry battalion helped to ensure the effective maintenance of national sovereignty while it operated as part of a much larger force driven by different imperatives.

Chapter Eight concerns the provision of communications support for the deployment of the First Australian Task Force in Phuoc Tuy Province in 1966 and the subsequent build-up of the force through to the end of 1967, when the Task Force expanded to a complement of three Australian infantry battalions. In 1966, 103 Signal Squadron deployed as the Task Force Signal Squadron and 145 Signal Squadron as the theatre-wide and national rear link communications element. 103 Signal Squadron was supported by 547 Signal Troop for Signals Intelligence tasks, and the SAS Squadron was supported by a detachment from 152 Signal Squadron. In mid-1967, 103 and 145 Signal Squadrons were replaced by 104 and 110 Signal Squadrons, respectively. However, the difficulties experienced with the rotation of entire Signal Squadrons resulted in that being the last time an entire Squadron was rotated through Vietnam. From then on, individual replacements ensured a steady maintenance of the required communications support for the force. New communications technologies also gave commanders increased flexibility and the ability to rapidly concentrate combat power over an increasingly dispersed area of operations. This, in turn, gave them the added confidence to attempt more with less in circumstances that in earlier conflicts would have been suicidal.

Chapter Nine covers the events of 1968 when the Tet Offensive, which commenced at the end of January, and the subsequent May offensives presented the Corps' elements in Vietnam with the greatest challenges they were to experience there. For 110 Signal Squadron, attacks on their facilities in Saigon resulted in a relocation of the main components to the logistic support base at Vung Tau. For 104 Signal Squadron, the period was marked by repeated deployments away from the Task

Force base to adjacent provinces. Further technical innovations were introduced to meet the demands of the Task Force when it was stretched to the limit.

The following chapter examines operations in the final years from 1969, leading to the drawdown and eventual withdrawal of the Task Force in late 1971. The period was marked by further deployments of the Task Force away from its base at Nui Dat, the introduction of voice encryption equipment to help overcome a significant problem with radio communications security violations and the continued provision of reliable communications despite the drawdown of the Force.

Chapter Eleven, the final chapter on the Vietnam War, relates to the use of the electronic warfare unit - 547 Signal Troop and the 152 Signal Squadron Detachment in support of the SAS Squadron operating from Nui Dat. The adaptation of in-service equipment and the introduction of new technologies and concepts served to make these two elements remarkable in terms of the significance of their contribution to the overall effectiveness of the Task Force.

The final Chapter provides an overview of the impact of communications technology on the Royal Australian Corps of Signals, the Australian Army, and the conduct of its military operations from 1947 to 1972.

I have also included, as an Appendix, a narrative description of some of the main equipment in use by the Corps during this period. In essence, this is designed to assist the layman in understanding what the various equipment types were capable of doing and how they were used, as well as providing a window on how they evolved. For the readers with an understanding of the Corps' equipment, this may make interesting but less crucial reading to aid in the understanding of the intervening chapters.

CHAPTER ONE

THE ROYAL AUSTRALIAN CORPS OF SIGNALS 1947-1972

IN SEEKING TO understand the impact of communications technology on the conduct of Australian military operations in the period from 1947 to 1972, it is appropriate to provide some background concerning the Australian Army's communications arm - the Royal Australian Corps of Signals. This chapter explains how the Signal Corps came into being and how it had evolved by the end of World War II. It also outlines the structure of the Signal Corps and its roles, explaining why a Signal Corps was required in a peacetime Army and how it contributed to the Army as a whole. In so doing, this chapter provides a yardstick by which to assess the impact of communications technology on the conduct of Australian military operations.

Origins

Military communications can be traced back centuries. However, military and scientific influences in the nineteenth century set in motion the gradual evolution of telecommunications to the point where they have become not only an elaborate and integral part of modern armed forces but also a key military force-structure determinant. The great increase in size of armies during the Napoleonic wars and the development of more powerful and faster-firing weapons meant that a greater dispersal of forces was required to ensure survivability. So the need arose for greater and more sophisticated methods of telecommunication. Various technological developments throughout the nineteenth and early twentieth centuries all combined to stimulate and meet the requirement for faster, more sophisticated and more reliable telecommunications.[1]

In Australia, the Royal Australian Corps of Signals can be traced back to 4 November 1870, when a small 'Torpedo and Signalling Corps' was formed in Victoria, and three years later in New South Wales. These Corps continued until Federation as part of the Corps of Engineers. In 1885, a Signalling Corps, consisting of an officer and twelve other ranks, existed in South Australia and also remained active until 1901.[2] Following Federation, an Australian Corps of Signallers was formed on 12 January 1906. In 1912, it merged with the Australian Corps of Engineers and its members became known as Signal Engineers.[3]

The Signal Engineers, with their limited and unsophisticated equipment, including Begbie lamps and Heliographs, provided the Divisional and Corps Signal Companies for the Australian Imperial Force that served throughout World War I. By the War's end, wireless and line telegraphy, field telephones and motorcycle dispatch riders had become the most common means of telecommunication, and the Signal Engineers tailored their organisation for the improved circumstances.[4] Following the return to Australia, severe financial constraints were applied to the Army as no threat seemed imminent. It was during this period, on 14 February 1925, that the Australian Corps of Signals was officially born.[5]

---

1 Major General R.F.H. Nalder, *The Royal Corps of Signals: A History of its Antecedents and Development,* Royal Signals Institution, London, 1958, pp. 1-4; and J.C. Blaxland, *Organising an Army: The Australian Experience 1957-1965,* Strategic and Defence Studies Centre, Canberra, 1989, pp. 6-7.

2 T.J. Barker, Signals: *A History of the Royal Australian Corps of Signals 1788-1947,* Royal Australian Corps of Signals Committee, Canberra, 1987, pp. 11-25; and 'A Short History of the Royal Australian Corps of Signals', (RASignals Museum, Macleod, Victoria - RASCM) pp. 1-2.

3 Barker, *Signals,* pp. 29 & 33.

4 Barker, *Signals,* pp. 54-65; and RASCM, 'A Short History of the Royal Australian Corps of Signals', pp. 2-5.

5 Barker, *Signals,* p. 112; and RASCM, 'A Short History of the Royal Australian Corps of Signals', pp. 5-6.

In the years leading up to 1939, the Australian Corps of Signals consisted solely of field force militia units. After the outbreak of hostilities in 1939, divisional signal units were raised for the forces sent to the Middle East, Malaya and the South West Pacific. These units initially provided the same type of wireless, line and dispatch rider services for the field units that their World War I counterparts had done before them. The more fluid nature of the war led to innovations and improvements. In turn, the range, durability, reliability and portability of equipment were improved. Also, by 1945, new radio relay equipment was being introduced. By the final stages of World War II, efforts were being directed towards finding military applications for the latest developments in communications technology. Examples included the integration of wireless and line equipment (to provide a streamlined communication system) and the miniaturisation of equipment (to increase portability).[6] Improved communications equipment gave armies flexibility, mobility and added range to match the increased tempo and mobility of war. In so doing, the improvements enabled military commanders to capitalise on other military hardware that would otherwise have been of only limited value, such as close air support by fighter and bomber aircraft, as well as artillery and naval gunfire support.

Corps Structure

By 1947, following the rundown of the wartime forces, the only Permanent Military Force (PMF) Field Force units remaining were those stationed in Japan as part of the BCOF.[7] Nevertheless, by March 1947, the Government had decided that the Army would be raised entirely by voluntary recruitment.[8] The role of the Army

6 'Modern Concepts of Intercommunication', in *Australian Army Journal,* No 11, Feb-Apr 1950, p. 29; 'AHQ Administrative Symposium 'Sympex' - Administrative Communications', 23 October 1958, part 4; *Signals Bulletin,* Vol II, No III, October 1953, p. 19; and RASCM, Brigadier A.D. Molloy, 'Technical Developments in Recent Years', pp. 6-7.

7 AWM 123, Box 95/4, pp. 2-3.

8 AWM 123, Box 95/4, Army Post War Plan, March 1947, p.1-4.

was now to provide forces for cooperation in Empire Defence and regional security and to provide the basic organisation for expansion in time of war.[9]

On 1 July 1948, the new Australian Military Forces (AMF) came into existence. These included an Australian Regular Army (ARA) infantry brigade, three CMF infantry divisions, and an assortment of support area units.[10] The signals elements of the post-war Regular Army, consisting of about 1180 personnel[11], were all technically responsible to the Directorate of Signals.

The technical direction and control of all Signals personnel was vested in the Director of Signals. He acted as Head of Corps and Signals Adviser to the Chief of the General Staff and members of the Military Board. Under him at the Signals Directorate were a limited staff who were responsible for the implementation of signals policy. This responsibility included the detailed work in relation to signals training, equipment, organisation, and 'staff duties', security, liaison with overseas headquarters and other governmental departments. Each command also had a Signals Staff Officer attached to the General Staff. These officers acted as Signals advisers to the General Officer Commanding and staff of his respective command. In addition, they dealt directly with Signals units in the Command on all technical matters, thus providing a direct link between the units and the Signals Directorate.[12]

---

9 AWM 123, Box 95/6, 'Australian Memorandum on Machinery for Co-operation in British Commonwealth Defence', 1951, p. 1.; and J.J. Dedman, in *Commonwealth Parliamentary Debates,* (Hansard), H of R, II Geo VI, Vol 192, 4 June 1947, p. 3336.

10 For a summary of the Army Program see J.J. Dedman, in *Commonwealth Parliamentary Debates,* (Hansard), H of R, II Geo VI, Vol 192, 4 June 1947, p. 3340.

11 CRS A2653, Item 1949, Vol 1, Department signalof the Army Minute 53/431/24, Appendix A, Comparison of Strengths to Establishments: Signals Units ARA. This figure is approximately one 20th of the UK Army's Signal Corps strength at the time (22,115 personnel). See WO244/81 Appendix A to BM/288 (Sigs 5) Post War Orbat R. Sigs.

12 J.E. Murphy, 'History of the Post War Army', unpublished manuscript, 1955, p. 241.

The execution of these responsibilities was divided between various commands within the Army, but the technical oversight and control lay with the Directorate of Signals at Army Headquarters. The Director of Signals (along with the various Signal staff officers throughout the Army at various headquarters) had oversight over the following areas:

- the operation and maintenance of existing telecommunications;
- the development of communications policy and procedures;
- the provision of advice on communications requirements to commanders at all levels;
- the training of Signals personnel;
- liaison with other authorities concerned with telecommunications;
- the provision of assistance in the maintenance of communications security; and
- the sponsoring, development and purchasing of telecommunications equipment required to fulfil the Corps' responsibilities.[13]

The oversight role of the Directorate of Signals was a crucial one, given the breadth of tasks faced by the Corps and the number of challenges to be overcome. The strategic communications network had to be maintained and improved to keep up with developments in new technology and procedures overseas if it was to remain interoperable with its major allies. Considerable effort was also expended liaising with other authorities concerned with telecommunications, as there was a clear need to define and delineate the military's requirements for use of the

13 See *Signals Bulletin,* Vol 1, No 1, September 1951, p. 1; and 'The Royal Australian Corps of Signals', Handout to Students at RMC, Duntroon 1970; and *Signal Training, Vol 1, Signal Organisation and Tactics,* Pamphlet No 1, The Higher Organization of Signals, War office, 1950, p. 1.

frequency spectrum. Effort was also put into collaborating with allies in the development of new technologies.

In 1948 the Signals Corps consisted of the following units:

- the School of Signals, at Balcombe, Victoria;
- Army Headquarters Signal Regiment in Melbourne;
- Headquarters Signal Regiment, BCOF, Japan;
- 101 Wireless Regiment, based at Cabarlah in Queensland (for signals intelligence);
- 1st Independent Infantry Brigade Signal Squadron (raised on paper but not manned until 1953);
- 1 Base Signal Park, located at Kingswood in New South Wales;
- 1 Line Construction Project Squadron, based at Woomera until its disbandment in 1952;
- 1 Signal Equipment Troop, based in Melbourne;
- six communications troops, allotted one per military Command, to operate the command links to Army Headquarters (AHQ) in Melbourne; and
- training cadres of varying sizes for allotment to Citizen Military Force (CMF) signals units.[14]

The key ARA elements of the Corps at the time were the AHQ Signal Regiment, the Headquarters Signal Regiment, BCOF (later Britcom Base Signal Regiment) and the 101 Wireless Regiment. The ARA field elements of the Corps did not start expanding until the Britcom Base Signals Regiment returned to Australia in 1956.[15]

A number of additional elements were added, and some were deleted, as requirements changed in the 1950s and 1960s. These elements included a contribution to the signal unit of the British

14 RASCM, 'Post War Developments in R Aust Sigs', p.1; Woollard, letter, September 1988; RASCM, Colonel R. Clark, 'A Short History of the Royal Australian Corps of Signals', 1975; RASCM, Barker, unpublished manuscript, p. 370 & Appendix IV; Directorate of Signals, 'Signal Information Bulletin', Nos 32-33, Aug-Sep 1948; and Murphy, op. cit., p. 236.

15 Major Bernie Saunders, letter, August 1990.

Commonwealth Far Eastern Strategic Reserve (BCFESR based in Malaya), the raising of Special Action Forces Signal units for the Commando Regiment and the Special Air Service Regiment, and the provision of instructors for the National Service Training Scheme.[16]

On 1 July 1948, the CMF was re-activated as the basis for expansion and mobilisation in time of war - a prospect which had become more likely with the onset of the Cold War. The CMF Signal Corps units raised at this time consisted of the following:

- 1 Corps Signals (raised in South Melbourne in 1948 but disbanded in July 1950),[17]
- 2 and 3 Infantry Divisional Signal Regiments (in Sydney and Melbourne respectively);
- 1 and 2 Independent Armoured Brigade Signal Squadrons (also in Sydney and Melbourne, respectively),
- 1 Air Support Signal Unit (1 ASSU) based at Moore Park in Sydney,
- six command signal squadrons (one in each capital city),
- two Headquarters Army Group Royal Artillery (HQ AGRA) signal squadrons, and
- other miscellaneous troops for attachment to certain CMF units.[18]

Each of these units incorporated a cadre of ARA personnel responsible for instructional and administrative duties, and the care and maintenance of unit equipment, vehicles and drill halls.[19]

Relations with the UK and the USA

16 RASCM, 'Post War Developments in R Aust Sigs', p.2.
17 Major George F. Powell, letter, August 1990.
18 Murphy, op. cit., p. 236; and RASCM, Corps History File-I September '53 - December '55 'History of the Post War Army', p. 2.
19 AWM 123 Box 95/4 Army Post War Plan March 1947, p. 6.

The Army structure which emerged in 1948 reflected the prevailing economic, strategic and military orientation of the nation toward Britain and the Empire. This was largely as a result of Australia's membership of the British pound sterling area, whereby Australia's currency remained fixed in value to Britain's, and Australia's principal trade links were with Britain. Despite strong ties that had developed with the United States during World War II, Australia was still tied economically to the UK. Therefore, to assure Australia's economic security, it had no alternative at the time but to 'sink or swim' with the UK.[20] Consequently, in the early post-war years, as the Corps endeavoured to standardise its training and organisation, links with the UK remained strong.[21] These links were particularly illustrated in Japan and Korea, and in the integrated Signals units stationed in Malaya and Singapore in the 1950s, 1960s and early 1970s.[22]

Close liaison was maintained by the Corps with the UK to ensure that standardisation of components and manufacture was achieved wherever possible.[23] In some instances, the British and Australian industry was unable to meet the requirements. This sometimes resulted in the purchase of substitutes from the United States (US). In other cases, items initiated and developed in Australia were adopted for use by both Australia and the UK. In most cases, however, UK equipment was obtained complete and ready for use. This policy allowed the Corps to derive benefit from economies of scale, the greater British industrial capac-

20 For a discussion on the significance of the currency and trade links for Australian defence and foreign policy see David Lee, *Search For Security: The Political Economy of Australia's Postwar Foreign and Defence Policy,* Allen & Unwin and RSPAS, ANU, Canberra, 1995.

21 In 1947 it was decided that the organisation and training of the post war Army would be standardised with that of the UK so as to facilitate cooperation with the forces of other Commonwealth countries. See AWM 123, Box 95/4, p.4.

22 Lee, *Search For Security,* p. 160.

23 Various Army Headquarters Policy Statements were drafted by successive Directors of Signals, issued and revised each 2 to 3 years. They contained the Army Policy on Communications Equipment. Major General R.P. Woollard, letter, August 1990.

ity and from the experience and proven facilities open to the UK through its various forces stationed in overseas theatres.[24]

The tendency towards US as opposed to UK equipment in Field Force Signals and Strategic Communications units in the 1950s and early 1960s reflected wider changes in Australia's outlook. Restructuring of Australia's economic relations with the rest of the world was a major factor in shaping the nation's international political and military interests during these years. The growth of trade with Japan through the 1950s and 1960s focused attention on the north of Australia.[25] Growing trade links with the United States,[26] stimulated further interest in the purchase of US equipment for use in the Corps. These changes were hastened by the decline in economic significance to Australia of the UK during this period.[27]

In 1957, the Australian and US Governments signed a Military Standardisation Agreement, beginning a programme of integration and standardisation of equipment and procedures with the US and thereby overturning Australia's traditional policy of standardisation with Britain.[28] As Prime Minister Robert Menzies pointed out at the time, in the case of a major war, Australia would have depended upon the US for equipment and

---

24 RASCM, 'History of the Post War Army - R Aust Sigs' p. 6; MP 729/8, 67/431/74, 'Proposed Development of New Telecommunications Equipment for Field Army Use' - submission to Cabinet, 1952; and RASCM, Molloy, 'Technical Deve,lopments in Recent Years', p. 9.

25 P. Drysdale, 'The Transformation in Australia's Foreign Trade', in H.G. Gelber (ed), *Problems of Australian Defence,* OUP, Melbourne, 1970, pp. 188 & 195.

26 Trade links grew particularly following the abolition of restrictions on the importing of US dollars in February 1960. See J.G. Crawford, 'Partnership in Trade', in N. Harper (ed), *Pacific Orbit, Australian-American Relations Since 1942,* Cheshire, Melbourne, 1968, p. 50.

27 In 1959-60 British products represented 36 percent (by value) of Australia's imports, whereas in 1948-49 this figure had been 50 percent. In exports to Britain from Australia the figure had dropped from 42 percent in 1948-49 down to 26 per cent in 1959-60. See T.B. Millar, *Australia in Peace and War: External Relations 1788-1977,* ANU Press, Canberra, 1978, p. 189.

28 NSC 5713/2, 'Long Range US policy Interests in Australia and New Zealand, 23 August 1957, cited in G. Pemberton, *All The Way: Australia's Road to Vietnam,* Allen & Unwin, Sydney, 1987, p. 67.

main war supplies. Australia had also adopted a broad strategic alignment with the US and there was a clear implication that future operations in South East Asia would rely heavily on the US for transport and other logistic support. Furthermore, the Australian Government believed that in the field of sophisticated weaponry, the best cost-effectiveness deals came with procurement from the US.[29] The shift toward this procurement was a significant factor affecting the deployment of signal units overseas and those signal units remaining in Australia.

Divisional and Brigade Signals

For those units in Australia, the shift towards the US also resulted in the introduction of a new organisational structure. In its Review of Strategic Guidance of 1959, the Defence Department assessed that there was an increased requirement for Australia to develop independent priorities and structures because, in certain circumstances, 'Australia might have to rely completely on her own defensive and economic capacity for an indeterminate period.'[30] The organisation that appeared to best meet these requirements became known as the 'pentropic' organisation. The new pentagonal divisional organisation was an adapted version of the US 'pentomic' model, designed to increase flexibility and be compatible with the US organisation.[31]

The re-organisation resulted in a substantial restructuring of the Divisional Signal Regiments sub-units with the absorption of brigade signal troops, given that the pentropic organisation introduced the notion of Task Forces in lieu of the supposedly less flexible structure of brigade headquarters (see Diagrams 2 & 3). The reorganisation also coincided with the introduction of

29 H.G. Gelber, *The Australian-American Alliance: Costs and Benefits,* Penguin, 1968, pp. 34-37.

30 Department of Defence, 'Key Elements in the Triennial Reviews of Strategic Guidance Since 1945', April 1986, p. 7.

31 CRS A2653, Vol 4, 1959, Military Board Agendum, 47/1959, 'Adoption of the Pentropic Division Organization'.

a substantial amount of new radio and line terminal equipment. £30 million was planned for the three-year programme for the purchase of Army equipment associated with the introduction of the pentropic divisional organisation, commencing in 1960.[32] For the Signals Corps, this was to include the B70 radio relay set as well as the C11/R210 vehicular HF radio stations, AN/MSC-29 mobile field telegraph terminals and shelter equipment.[33]

The introduction of new equipment paralleled developments in the wider community. In 1959, for instance, the first broadband telecommunications link in Australia was opened for traffic. This was a microwave radio link between Melbourne and Bendigo, Victoria.

Although the restructured Divisional Signal Regiments were never tested in war, the introduction of radio relay equipment and the concomitant reduction in reliance on line made a significant difference to the way they operated and the level of support they could provide to commanders. Initially, the value of radio relay was limited due to technical difficulties experienced with the first sets introduced, but eventually these limitations were overcome, allowing the Corps to improve the level of service it provided to commanders. The provision of machine telegraph links down to the infantry battalions, while also being able to provide simultaneous telephone circuits, provided a further significant enhancement to commanders' ability to exercise command and control over subordinate units and formations, even with the increased span of command found in the five-sided pentropic organisation.

With the demise of the pentropic organisation throughout the Australian Army in December 1964, unit organisations were amended to adjust to the triangular concept of three battalions, and three 'task forces' (or brigades) within a division (see Diagram 4). The pentropic organisation was abandoned as it

32 *Commonwealth Parliamentary Debates,* House of Representatives, 26, 29 March 1960, p. 652; and *Signals Bulletin,* Vol 9, No 1, April 1960, p. 1.

33 *Australian Army Journal,* No 134, July 1960, pp. 47-50; and *Signals Bulletin,* Vol 10, No 2, December 1961, p. 4.03.

was no longer seen as being compatible with the organisation then in use with the United States' Army, which had changed to a different structure in 1961. It had also proven difficult to control and supply in its original design.[34]

From the time that Australia first deployed Signal elements in support of the occupation of Japan through to the withdrawal of forces from Vietnam, the structure of Signal units supporting brigades had undergone substantial change. This change was driven in part by wider force restructuring, such as resulted from the introduction and hasty abandonment of the pentropic organisation, which saw the temporary elimination of the brigade level command structure. The change was also driven by technological developments that resulted in brigade Signal units having to provide an increased number and an increased variety of types of communications links to the commanders they were supporting. Machine telegraphy and radio relay, and the subsequent reduction in reliance on line communications, altered the structure and tasks of the brigade Signal units and at the same time, dramatically increased the degree of control that commanders could exercise over their formations and units. This then allowed commanders to deploy their forces on exercises and on operations with increased flexibility and mobility.

Apart from the signal units directly tied to the infantry divisions, a number of additional units provided other necessary communications services to the Army. These include special action forces units.

Special Action Forces Signals

Special Forces techniques had initially been developed in Australia during World War II for special tasks behind enemy lines. In the late 1950s and early 1960s, there was a resurgence of 'special action forces' which resulted in the creation of CMF Commando units, as well as an ARA Special Air Service (SAS)

34 Blaxland, *Organising an Army*, p. 116.

unit. The Commando Signal Unit was raised in January 1960 as 301 Signal Squadron, at Lidcombe, New South Wales. Its role was to provide long-range communications for operations involving 1 and 2 Commando Companies (CMF), which were raised in Sydney and Melbourne, respectively, in the late 1950s.

In 1963, a Regular Army element was added to the unit to meet the demanding communications requirements of the Commando companies. This troop consisted of a troop commander, 10 operators and three technicians. In 1966, the squadron was relocated from Lidcombe, near Sydney, to interim accommodation at Albert Park in South Melbourne, Victoria, and was renamed 126 Signal Squadron (Special Forces).[35] In 1972, the unit moved to its new home at Watsonia Barracks, Victoria.[36]

In the Regular Army's SAS Company, raised in 1957, members of the Corps of Signals served as part of the Company's Signal Platoon.[37] By mid-1960, the Signal Platoon consisted of one officer and 38 other ranks and played an important role in enabling the SAS to overcome a communications weakness in the use of ground waves, Morse code and one-time letter pads. This training was to pay off on operations in Borneo. The SAS was expanded to become a Regiment in 1964, and on 12 September 1966, the Royal Australian Corps of Signals component of the Regiment was expanded to become 152 Signal Squadron, with Captain Ross Bishop as its first commander. Reliable, long-range and discrete communications links were critical for the types of operations for which the special action forces were trained, and these were increasingly available for use. The Squadron went on to serve with the SAS Regiment on exercises in Australia and overseas as

35 Major C. Lawson-Baker, 'Unit in Focus: 126 Signal Squadron, 1 Commando Regiment - A Different Unit', in *Signalman,* No 9, 1982, p. 13; and RASCM, Sergeant K.E. Larner, Unit History: 126 Signal squadron (unpublished).

36 Lawson-Baker, 'Unit in Focus', p. 13.

37 For a detailed history of the SAS see David Horner's SAS: *Phantoms of the Jungle,* Allen & Unwin, Sydney, 1989.

well as on its deployments to Borneo and Vietnam.[38] The details of these episodes are discussed in subsequent chapters.

Signals in Papua New Guinea

In the 1960s, the Australian Government committed itself to raising a viable military force for the defence of Papua New Guinea. Australia had been given the responsibility by the United Nations to foster its development into an independent nation. As part of this effort, the Pacific Islands Regiment (infantry) was developed along with supporting units. Considerable effort was also expended during the mid-1960s on the construction of accommodation, barracks and messes for use, initially, by Headquarters Papua New Guinea Command of the Australian Army but eventually by the Papua New Guinea Government when independence was gained.[39]

By 1970, the Signal Squadron in Papua New Guinea operated a 14-channel torn tape relay station located in a modern building within Murray Barracks, Port Moresby. This station was supported by a transmitting station, also within Murray Barracks, and a receiving station at Taurama Barracks, both of which were in modern buildings connected to Port Moresby by multi-channel radio circuits. Combined transmitting and receiving stations and communication centre installations operated at Vanimo, Wewak and Lae. These were installed by 127 Signal Squadron and located in modern permanent buildings.[40]

With the establishment of the multi-channel radio circuits and the torn tape relay facility, the Corps had established a modern communications network around the territory to provide for the Army's communications requirements in difficult terrain. By

38 Horner, SAS, pp. 48, 63 & 210-211; and 'Unit in Focus: 152 Signal Squadron Special Air Service Regiment (SASR)', in *Signalman,* Vol 13, 1984, p. 9.

39 CRS A2653, Military Board Proceedings, 'Reorganization of Forces in Papua/New Guinea: Works Progress', 18 Feb 1966, No 62/1966; and *Defence Report 1968,* p. 27.

40 AWM 121, 6/M/1, 'Rationalisation of Communications Army Fixed Communications Network', 22 May 1970.

doing this, the Corps had made a significant contribution to the development of the Papua New Guinea Defence Force, which was to emerge at independence in 1975.

Women in Signals[41]

The raising of the Signals units described earlier occurred after the cessation of the Australian Women's Army Service, or AWAS in June 1947. The AWAS signal women's positions were supposed to then be filled by male Army volunteers.[42] However, as a result of this ruling, a deficiency of 360 persons was experienced throughout the army's signal stations.[43] The cessation of the AWAS was in contrast to the position adopted by the United Kingdom and New Zealand, which both retained Women's Services for all three armed services.[44]

In the realisation that the enlistment of women in the Army was the only certain means of overcoming the serious manpower deficiencies, on 1 December 1950, approval by Cabinet was given to the enlistment of 250 women.[45] However, only 118 of these were allocated to the Signals Corps as operators (keyboard and cipher) and drivers,[46] of whom 80 were allocated to AHQ Signal Regiment.[47]

41 The history of the WRAAC is detailed in Lorna Ollif's *Colonel Best and Her Soldiers: The Story of the 33 Years of the Womens Royal Australian Army Corps*, Ollif, Sydney, 1985.

42 'Signals Information Bulletin', No 18, 30 June 1947, p. 1.

43 CRS A2653, Vol 1, 1949, 'Manning of Services Signal Stations: Comparison of Strengths to Establishments'.

44 CRS A2031, Defence Committee Minute No 197/1948: 'Manning of Service Signal Stations.'

45 Murphy, 'History of the Post War Army', pp. 295-6.women's

46 CRS A2653, Vol 3, 1950, Agendum 35/1950, 'Women's Services: Allocation of Ranks, Trades and Duties - Australia Women's Army Corps." The following year, in October 1951 the ceiling strength of the WRAAC was lifted to 1000. The ceiling was lifted again in June 1952 to 1320 but the actual ceiling reached was only 950 all ranks. See Murphy, 'History of the Post War Army', p. 297.

47 MP 729/8, 53/431/27, 'Introduction of a Women's Service'.

The members of the Women's Royal Australian Army Corps (WRAAC), as the new Corps was named, quickly attained a high level of loyalty to their shifts, which they carried into their social lives. Additionally, they had an amazing sense of esprit de corps, not so much as to their own Corps but to the Corps of Signals.[48] By July 1955, the involvement of the WRAAC in the Signals Corps had progressed to such an extent that approval was granted for eight officers and 93 other ranks WRAAC to be integrated into the establishments of certain Signals units. This did not mean that they had become full Signal Corps members, but that a suitable career structure for WRAAC members had been accepted. They could only be employed in mainland units and were not eligible for overseas service.[49]

The demands of the strategic communications network with its complex equipment (including cryptography), the wider demands of the Army at a time of manpower shortages and the social imperatives resulting from the changes to society brought about during World War II had resulted in the re-establishment of the Army's womens' corps. Once the decision had been made, members of the WRAAC were quickly integrated into the respective units and played a significant role in the history of the Royal Australian Corps of Signals. Their mastery of the skills required allowed for a redistribution of manpower to meet the other operational and training requirements of the Signals Corps and the Army. One area that required attention, and in which members of the WRAAC were to make a substantial contribution, was the field of signals intelligence.

48 Major B.G. Saunders, letter, August 1990.
49 *Signals Bulletin,* Vol 5, No 1, June 1956, pp. 19-20.

Signals Intelligence and Electronic Warfare[50]

The merits of electronic interception, or signals intelligence, had been so soundly proven in World War II that the Australian Army realised there was a need for an electronic intercept capability to be maintained. It was recognised that the most important means of gathering information about an enemy would be by intercepting and deciphering his signals.[51] There was close coordination at the highest level between the UK and US, which was extended to include Australia and Canada's code-breaking facilities.[52]

Cabarlah, approximately 12 miles north of Toowoomba in Queensland, was to become the permanent home for 'special wireless' in the Australian Army. This site was chosen because it was technically suitable, close to Brisbane and Toowoomba and (most significantly) already had an existing camp (economy measures were paramount at the time).[53] On 3 February 1947, the special wireless unit commenced operations at Cabarlah, under the Command of Major T.R. (Dick) Warren, and it was re-designated as 101 Wireless Regiment as from 1 November 1947.[54] Eleven days later, on 12 December 1947, the Chifley Government

---

50 Electronic warfare is divided into electronic counter-measures (ECM), electronic counter-counter-measures (ECCM), and electronic support measures (ESM). See AWM 121, 12/B/27, 'AHQ Weapons and Equipment Policy Statement No 27/2 - Electronic Warfare, June 1966.'

51 CRS A816, 43/302/18, Defence Committee Minute No 10/1940, 15 February 1940 cited in D.M. Horner, *High Command,* AWM, Canberra, 1982, p. 224. The British had commenced intercepting enemy wireless transmissions in World War I. See F.W. Winterbotham, *The Ultra Secret,* Weidenfeld and Nicholson, London, 1976, pp. 7-8.

52 Ibid., p. 275.

53 RASCM, S.W. Foley, 7th Signal Regiment Unit History'. During World War II the site had been occupied by the 11th Light Horse followed by Headquarters Line of Communication Signal Regiment until 1944 and from 1944 to 1946 by the Army's Staff School until it moved to Queenscliff, Victoria (and eventually became the Army's Command and Staff College).

54 Barker, Signals, p. 166; and MP 742, 240/7/388, 'Australian Corps of *Signals*: Unit Nomenclature', 22 October 1947. This coincided with the change of terms throughout the Signals Corps to 'Regiment,' 'Squadron' and 'Troop'.

approved the integration of the Australian signals intelligence agencies into the US and UK (or UKUSA) system.[55]

101 Wireless Regiment's operational tasking was determined by Joint Services machinery. Operational (as opposed to administrative) control was exercised by the Defence Signals Bureau in Melbourne.[56] This control was also exercised over the RAAF No 3 Telecommunications Unit at Pearce Air Force Base, in Western Australia, and the Royal Australian Navy's base at HMAS *Harman*, in Canberra.[57] During this period, 101 Wireless Regiment's main role was strategic intercept of telecommunications from and within certain Asian targets.[58]

The Defence Signals Bureau is described as an organisation within the Department of Defence 'responsible for defence signals and communications security'[59] and concerned with radio, radar and other electronic emissions from the standpoint both of the information and the intelligence that they can provide and of the security of our own Government communications and electronic emissions. It is an agency that serves wide national requirements in response to national priorities.[60]

---

55 Canada also became a co-signator to the UKUSA Agreement at this time. See Richelson & Ball, *The Ties That Bind,* pp. 142-143.

56 MP 729/8, 41/432/59, '101 WS Regt-Employment of Pers at Darwin'.

57 Whyte, interview; and Ball, *Australia's Secret Space Programs,* pp. 2-3.

58 Whyte, interview, August 1989.

59 *Commonwealth Government Directory,* cited in Richelson, J.T., and Ball, D., *The Ties That Bind,* Unwin Hyman, Boston, 1990, p. 36. Richelson and Ball provide a detailed description of the Australian security and intelligence community.

60 The Hon Mr Malcolm Fraser, cited in Richelson and Ball, *The Ties That Bind,* pp. 36-37. Richelson and Ball, assert that the organisation has two missions. The first is the protection of Australian defence, diplomatic and intelligence communications from foreign intelligence exploitation and from unauthorised disclosure - the communications security (COMSEC) mission. The second is to exploit foreign signals, communications and other electronic emissions to provide intelligence for other Australian agencies - the SIGINT mission. This mission involves the interception, processing, analysis, and dissemination of information derived from foreign electronic communications and other signals.

The Regiment's radio operators were highly trained and could also type and copy Morse quickly. The operators were trained in overseas radio procedures, after receiving two years of training at the School of Signals in Victoria, plus a further 18 months of training before being fully proficient. Operators were also required to learn signal procedures and characteristics.[61]

As part of the process of gathering the information received, traffic was passed to two sections, namely traffic analysis and cryptography. The traffic analysis personnel examined the network system, establishing who was talking to whom, building 'net' (user network) diagrams and establishing frequencies, call signs and other technical information from which could be drawn an 'order of battle' (a list of units involved in the activity). On the cryptography side, encoded messages would arrive, and an attempt would be made to break it. Once this was accomplished and the message was looked at, it would be sent to the linguists in plain language. They would then assess the document and see if it had any valuable information.[62]

The Corps' signals intelligence elements had a significant role to play in raising awareness of the potential problems concerning communications security that Australian soldiers would experience on operations in South East Asia. They also provided the Army with a substantial force-multiplying capability, which was to be demonstrated repeatedly on military operations, particularly in Borneo and Vietnam. Maintaining this capability meant that the Corps had to purchase the right equipment, keep abreast of developments and use imagination and ingenuity in designing ways to maximise the benefit of the technology available. Maintaining that capability would require the allocation of substantial resources and the training of the personnel selected to operate the systems.

61 Ibid.
62 Ibid.

The Army's Strategic Communications Network

The dramatic reduction in size of the Australian Army, after World War II, meant that decisions had to be made as to whether or not the Army's strategic telegraph message network that had developed during the war years should be maintained thereafter. By war's end, the network included extensive overseas and internal links. The overseas HF radio links consisted of transmitter and receiver sites, connecting the relay stations and signal centres. These formed part of the British Empire's Army Wireless Chain, which had links connecting the Army Headquarters of the participating nations. The main Australian station was at Army Headquarters Signals in Melbourne.[63]

The internal Australian links consisted of inter-state telephone, teleprinter and radio links as well as intra-state and district telephone links (at the time long long-distance telephony was in its infancy, and there were no civil or military methods of providing effective worldwide telephony). In Melbourne, wartime HF transmitters (including the Marconi SWB 8 and SWB 11) were used at Diggers Rest, while a mixture of HF receivers (including some dual diversity and single channel equipment) was used at nearby Rockbank (see Diagrams 5 & 6). Antennae were 'rhombics' and 'horizontal dipoles'. For 'dual diversity' reception, two antennae were used to overcome fading.[64]

Wartime experience had shown that Australia should never again allow itself the 'technological nakedness' that had left it so vulnerable in 1942 and which had required such effort to overcome. As a result, the PMG was active in developing a civil communications infrastructure in Australia in the early post-war

63 RASCM, L. Moore, 'Outline and Resume of the Development of Army Fixed Communication Network', p. 1; RASCM, Directorate of Signals 'A Short History of the Royal Australian Corps of Signals (809); RASCM, Royal Australian Signals, 'An Introduction to 402 Signal Regiment', p. 3; and AWM 121, 9/C/5, 'Manning Problems in the AUSTCAN Network'.

64 RASCM, Moore, op.cit., p. 1.

years. However, the role of the Army's strategic communications network was not so clear.[65]

One of the most remarkable features of World War II was the growth in the scope and quality of military communications. The scale of essential communications was greatly increased, as was the demand on the traffic-carrying capacity of the system. With the increased distances involved and the greater mobility of forces, it was normally not possible to provide line channels. Consequently, radio had become the primary and often the only means of communication.[66] In the immediate post-war years, long before the advent of satellites, long-haul 'wireless' communications worldwide were still by HF skywave, which, when reflected off the ionosphere, could achieve worldwide coverage given sufficient transmitter power and sophisticated antenna design.[67]

Increased traffic demands meant that there were more wireless circuits and congestion in the usable frequency bands. So a logical step was the development of high-speed multi-channel radio circuits, which enabled more traffic to be handled on a given frequency in a given period. The high-speed technique employed in the early stages of the war was usually the high-speed Morse system, but eventually the introduction of the teleprinter or teletypewriter was perfected, and this conferred on a radio circuit much the same traffic handling capacity as a line circuit employing similar terminal equipment.[68] The trend during World War II, in the sphere of the higher-powered radio sets serving the rear areas, was toward high-speed machine telegraph, either manually-operated or employing a tape relay system for both wireless and line.[69]

---

65 Moyal, *Clear Across Australia,* pp. 174-179.

66 RASCM, Address by the Director of Signals at Albert Park on 14 May 1947 - Appendix B to the Directorate of Signals' 'Monthly Information Bulletin', No 17.

67 Warner, *The Vital Link,* p. 239,.

68 Teleprinters had first been introduced on line circuits in Australia in the mid-1920s. See Moyal, *Clear Across Australia,* pp. 187-188.

69 RASCM, Address by the Director of Signals at Albert Park on 14 May 1947.

The US-developed teletypewriters were considered the best type of machine telegraph for use with radios and considerable interest in them was shown by the Australian Army's Directorate of Signals. They provided greater ease and efficiency of operation than the old High Speed Morse (Wheatstone) system. The system developed by the US Army Signal Corps, which employed teletypewriters and other similar instruments, using a '5-unit' (Baudot) code in their operation, was termed the 'Semi-Automatic Tape Relay System'. In this method, the information to be transmitted was converted initially into the form of perforations in a paper tape, using combinations of a '5-unit' code for each character. The message remained in tape form until it reached its ultimate destination. At all intermediate offices, it was necessary only to transfer the tape from one automatic instrument to another.[70] This process eliminated the slow and costly letter-by-letter, manual, reprocessing of traffic at intermediate offices that had been carried out up to that point, although it still retained the use of operators to relay the tape between message circuits (hence its description as 'semi' automatic).[71]

This new equipment had a number of advantages over Morse. These advantages included: a reduced risk of failure or transmission interference, a greater transmission accuracy, a simplified message handling procedure for the teletypewriter operators, a manpower saving as a result of the development of automatic or 'on line' encrypting facilities, greater speed and accuracy of transmission, and greater flexibility of use due to its ability to be integrated into other military or civilian communication systems.[72]

As this technology was developing, the British Army Wireless Chain had come into operation, using HF skywave as its mode of operation. In 1939, the Army Wireless Chain had connected with

70 Directorate of Signals, 'Modern Concepts of Intercommunication', in *Australian Army Journal,* No 11, Feb-Apr 1950, p. 28.

71 'Equipment for the AMF Semi-Automatic Telegraph Tape Relay System', in *Signals Bulletin,* Vol 1, No 2., 14 December 1951, p. 37.

72 MP 742/1, 323/6/425, 'AMF Communication System'.

Egypt, India and Hong Kong. Australia was connected shortly thereafter, and by 1947, the Chain reached its peak when East Africa, West Africa, Austria, Burma, Ceylon, Egypt, Germany, Greece, Hong Kong, Delhi, Iraq, Italy, Japan, Malaya, Australia, New Zealand and Canada were connected (see Diagrams 6 & 7).[73] A similar worldwide semi-automatic network had been established by the Americans stretching eastward and westward from the Pentagon in Washington, DC, known as the Army Command and Administration Network (ACAN).[74]

The British War Office proposed, in June 1945, that the Army Wireless Chain be converted from a chain of high-speed Morse stations to a chain of radio 'teleprinter' stations, using the 'semi-automatic tape relay system'. This would link the British War Office with the various British overseas commands and with the military headquarters in the various dominions and colonies. As part of this conversion, it was envisaged that the long-distance radio links would be converted to either Carrier Frequency Shift (CFS) or Single Side Band (SSB) operation, employing modern high-powered 'space diversity' reception.[75] CFS applied to circuits where one teletypewriter alone could cope with the traffic offering and was usually not a 'multi-channel' facility. CFS was also known as carrier shift working, frequency shift working or frequency shift keying. This technique enabled teleprinters to be used over radio circuits with greater efficiency.[76] Unlike CFS, SSB carrier operation applied to circuits where there was a heavy volume of traffic to be passed. It allowed up to six teletypewriter circuits to be operated simultaneously over the one radio circuit, thereby achieving economy in the use of both equipment

---

73 Colonel R.M. Adams, *Through to 1970,* Royal Signals Institution, London, 1970, p. 107; and Nalder, *The Royal Corps of Signals,* p. 479.

74 Coker & Rios, *A Concise History of the US Army Signal Corps,* pp. 23-24.

75 MP 729/8, 67/431/48, 'Conversion of the Army Wireless Chain to Radio Teleprinter Working'. This method involved the use of two separate or spaced antennas for the one receiver so as to overcome poor reception of radio signals.

76 MP 729/8, 67/431/48, 'Conversion of the Army Wireless Chain to Radio Teleprinter Working'.

and frequencies.[77] These upgrades in the US and UK demanded a response from the Australian Army. Would the Signal Corps update as well to enable it to continue to be interoperable internationally, or would it cut its losses and abolish the wartime network?

For strategic reasons, Australia was committed to operating the Australian terminals of the Empire Army Wireless Chain circuits in the post-war period, and parallel action was at least temporarily required to maintain compatibility. The channel to Kure was established when the force was sent to Japan in early 1946. It existed as a complementary requirement for both the Empire Wireless Chain and the Australian Army.[78] Australia provided and maintained this channel for military and higher-level traffic of the countries involved in the BCOF. The channels to the islands of Morotai and to Rabaul were only maintained until 1947, whilst Australian Army commitments in those areas continued to exist. Circuits internal to Australia were also maintained so that efficient command and control could be maintained over the elements of the Army retained for the post-war period.[79]

Radio links were maintained until May 1947 on the Australian Army's communications network from the Army Headquarters Signals in Melbourne to Adelaide, Hobart, Perth, Darwin, Duntroon in Canberra and Rabaul in New Britain. The links to Sydney and Brisbane included both teleprinter and telephone channels, but voice channels on strategic HF circuits were difficult to provide and rarely used. The maintenance of the BCOF commitment in Japan centred on Sydney, and there was a significant traffic load arising from the commitment. Overseas radio links, maintained in Australia as part of the Empire Army Wireless Chain at the time, were from Melbourne to Ceylon,

77 MP 742/1, 323/6/425, 'AMF Communication System'.
78 RASCM, 'The History of the Army Wireless Chain, the COMCAN and AUSTCAN' manuscript by Colonel R.G. Lawrence, 1977, p. 3.
79 CRS A2653, Vol 2 1947, - 32/1946 'Acquisition of "Grosvenor" as a Permanent Signal Office to Serve AHQ'.

Singapore, Delhi, Wellington and Kure in Japan.[80] Direct links to London were found to be spasmodic and unreliable, so messages for London and for further relay via London were relayed via the Colombo or Delhi stations.[81] With cross connection, it was possible to route messages to any part of the British Empire.[82] All overseas links were connected from Army Headquarters Signals, Melbourne, and traffic from all the other states to each other had to be relayed through the Melbourne complex.[83]

Army Headquarters Signals (renamed 403 Signal Regiment upon being relocated to Watsonia in January 1961, and then 6 Signal Regiment in 1965)[84] was formed on 28 May 1946 when the two wartime units known as Land Headquarters Heavy Wireless Group and Land Force Headquarters Signals were amalgamated.[85]

The Army Wireless Chain underwent significant development as the Korean War and the Malayan Emergency emphasised the need for the Army to acquire new equipment.[86] In June 1952, the term 'Army Wireless Chain' was replaced by 'Commonwealth Communication Army Network', abbreviated to 'COMCAN'.[87] In order to standardise the nomenclature with the UK-based international network and the United States' Army Command and Administration Network (ACAN), the Australian Army's interstate network became known as the Australian Communications Army Network or AUSTCAN. This term applied to the Army's radio and line network linking Army Headquarters and the British

---

80 CRS A2653, Vol 2 1947, - 59/1947: 'Summary of Proceedings of the Military Board' 19 May 1947 'AMF Communication System'; 32/1946 'AMF Communication System'; and *Signals Bulletin*, Vol 1, No 3, March 1952, Annex G.

81 RASCM, 'The History of the Army Wireless Chain, the COMCAN and AUSTCAN' manuscript by Colonel R.G. Lawrence, 1977, p. 2.

82 Ibid., p. 1.

83 Moore, op.cit., p. 3.

84 Barker, *Signals*, p. 141; and RASCM, 6 Signal Regiment Unit History, prepared in 1987 by Lieutenant M. Oliver-Weymouth, p. 4.

85 Barker, *Signals*, p. 141; and RASCM, 6 Signal Regiment Unit History, p. 2.

86 Woollard, letter, September 1988.

87 *Signals Bulletin*, Vol 1, No 4, 21 July 1952, p. 30.

Commonwealth Force in Korea.[88] As agreed to in the late 1940s, it was intended that Australian practice continue to follow as closely as possible that laid down by the War Office in Britain concerning principles, engineering techniques and procedure.[89]

The stations on the network played an important role in providing the communications necessary for the effective command of the diverse elements of the post-war Army. As demands continued to grow, the imperative for the network to grow became stronger.

Changes were also occurring in the area of codes and cryptographic equipment. Off-line encryption, using systems adapted from the UK War Office, Inter-Service and Army manuals, was in use, including 'Typex' and 'Rockex'. These were labour-intensive, slow and often prone to error.[90] A new cipher machine, called 'Derby', was introduced in early 1959 for use within Australia on selected AUSTCAN circuits. The Derby was an 'electronic teleprinter cryptographic regenerative repeater mixer' that could be used in connection with all standard 5-unit code teleprinter or teletype equipment to provide a teleprinter secrecy system.[91] Plain text or automatically encrypted text could be transmitted to or received from a distant station. Incoming text in cipher could be automatically or manually deciphered and made to appear on the associated teleprinter in plain text.[92] Later developments in the early 1960s saw the change to US online and offline equipment. This move reflected the growing ties with the US, marked by the Military Standardisation Agreement signed by Australia and the US in 1957.[93]

---

88 *Signals Bulletin,* Vol 3, No 1, January 1954, p. 36; and Letter from Lieutenant Colonel G.J. Lawrence.

89 *Signals Bulletin,* Vol 3, No 1, January 1954, p. 38.

90 'Signals Information Bulletin', No 21, 15 September 1947, p. 2. CD&R Troop is now 700 Signal Troop.

91 Derby was a compact, electrically operated unit, weighing 60 pounds and it used 'one time' key tape for the code. It operated at 45 bauds or 61 words per minute (normal US speed), or 50 bauds, that is, 67 words per minute (normal European speed).

92 *Signals Bulletin,* Vol 3, No 2, July 1959, p. 20.

93 NSC 7513/2 'Long Range US Policy Interests in Australia and New Zealand', 23 August 1957, cited in G. Pemberton, *All The Way,* p. 67.

The contrast in procedures between 1947 to 1972 was stark. In 1947, for instance, the process of sending a security-classified message to London involved a number of manually processed steps. Firstly, the message would be handwritten and passed to the signal centre, where it would be manually encrypted offline and then transmitted using high-speed Morse over single-channel HF skywave radios to Ceylon. There it would be relayed manually by an operator (ie re-keyed in Morse code), again using HF skywave, to the receiver station in London. There, the message would be received in Morse, decrypted manually (off-line) and typed out for reading by the intended recipient. By 1972, the message could be sent more reliably and quickly. It would be typed into a teleprinter at the communication centre, encoded online and transmitted using a single-sideband, carrier frequency shift multi-channel HF transmitter. The relay would occur automatically, probably in Singapore, and would be automatically forwarded to the receiving station in London, where it would be automatically decrypted and printed out for onward forwarding to the intended recipient.

These improvements altered both the method of operations and the actual network configuration to the point where integration with the Navy and Air Force networks became not only possible, but in their best interests. The British Services had introduced a rationalised Defence Communications Network in November 1965.[94] In Australia, the three Services' network controllers also saw that, corporately, they stood a better chance of keeping up with technological developments, given the financial constraints under which they had to operate.

The Defence Communications Network (DEFCOMNET), that was to emerge in 1975, and its successors in the late 1980s and 1990s, owed much to the work that had gone into the operation of the Army Wireless Chain and its successors the COMCAN and AUSTCAN. The development and introduction into service

94 DEFE 4/192 Defence Communications Network: UK Chiefs of Staff Meeting 64 of 1965, 30/11/65

of improved communications technology resulted in significant economies and improvements in the service provided. Messages could be online encrypted and transmitted at a far greater rate and with substantially improved reliability than had been possible in 1947. These improvements helped ease the personnel requirements necessary to maintain the network.

Australia's peace-time Field Force, strategic communications and specialist signal units faced significant strains in the period from 1947 to 1972. Financial constraints arising out of Australia's economic links with the British economy and manpower constraints imposed by the postwar shortage of labour, as well as obligations that arose from the national service scheme (itself a product of a perceived strategic threat to global peace), combined to make the job of the Signalman a challenging one. These factors, along with technological advances, resulted in significant changes in the way the Army's Field Force, strategic communications and specialist Signal units were organised and operated. Their efficiency, effectiveness, capacity and flexibility were progressing rapidly, given the changing nature of the equipment and methods in use, but false expectations and assumptions can be made from peacetime soldiering and testing of equipment on only military exercises. During this period, however, the Signal Corps was also faced with a number of obligations overseas, which served as proving grounds helping to mould not only the Signal Corps Field Force components, but its fixed strategic communications and other specialist units as well. The initial commitment to Japan, followed by the involvement in the Korean War and the Malayan Emergency, Borneo and the Vietnam War, all played significant roles in creating demands for further technological and operational innovations as well as in shaping the Corps from 1947 to 1972. These deployments also provide valuable insights into the impact of military communications technology on the conduct of Australian military operations.

CHAPTER TWO

OPERATIONAL DEPLOYMENTS 1947 to 1961

THE ROYAL AUSTRALIAN Corps of Signals had made significant contributions to Australia's deployment of military forces overseas during the two world wars. Wherever the Army went there was a need to 'get through', to maintain the vital communications link with higher headquarters and with Australia. Following World War II, Australia's foreign policy returned to a primary association with Britain and the Commonwealth, particularly given Australia's strong economic ties there, while looking to the United States as secondary support for this alignment. It was not until the 1960s that this alignment became effectively altered with the shift away from membership of the sterling economic area and the commitment of Australian troops to Vietnam.[1] It has been argued that, in the post-war period, Australian forces were deployed more for their diplomatic than their military effect. Amongst those deployed forces, Signalmen could be found in the Occupation Force in Japan from 1946, in Korea from 1950, and in Malaya from 1950. They could also be found in Singapore, Borneo and Indonesia, as well as in Vietnam as discussed in later chapters. These deployments were much smaller contributions to multi-national forces provided mainly by

1 Edwards, 'The Post-1945 conflicts: Korea, Malaya, Borneo and Vietnam,' in McKernan & Browne, *Australia Two Centuries of War and Peace,* pp. 296-302,

the UK or the US which had their own extensive communications networks.

This Chapter looks at why the Royal Australian Corps of Signals contributed to Australia's operational deployments overseas in the period after World War II and how it did so. It also looks at the effect of their contribution on the conduct of Australia's military operations in Japan, Korea and Malaya. The overall aim is to show how developments in the field of military communications technology had an impact on the way Australians fought in the period from 1947 to 1961.

Japan and Korea

The Australian Government had initially decided in 1945 that its contribution to the military occupation of Japan should be as a separate entity under an Australian commander who would be subject only to the Supreme Allied Commander, General Douglas MacArthur. Participation in a British Commonwealth Force was later agreed to on the basis that the Commander-in-Chief would be an Australian officer and that Australia would provide the bulk of the Force Headquarters. The force was commanded by two Australian Lieutenant Generals: Sir John Northcott in 1945 and 1946, and Sir Horace Robertson from April 1946 until June 1951.[2]

The original participants in the British Commonwealth Occupation Force (BCOF) were from the UK, India, Australia and New Zealand. The Force consisted of the following units:

2 For a discussion on the BCOF and the career of Lieutenant General Sir Horace Robertson, see Jeffrey Grey, *Australian Brass: The Career of Lieutenant General Sir Horace Robertson,* Cambridge University Press, Cambridge, 1992. See also AWM 123, Box 57/5, "British Commonwealth Occupation Force in Japan", pp. 1 & 4; D.M. Horner, *High Command, Australia and Allied Strategy 1939-1945,* Allen & Unwin, Sydney, 1982, pp. 419-425; and R.N.L. Hopkins, "Lieutenant General Sir Horace Robertson", in D.M. Horner (ed), *The Commanders,* Allen & Unwin, Sydney, 1984, p. 296.

- the British Indian Division (BRINDIV) consisting of the 5th British Infantry Brigade, the 268th Indian Infantry Brigade and the 7th Indian Light Cavalry Regiment;
- the 9th New Zealand Infantry Brigade;
- the 34th Australian Infantry Brigade;
- shore-based Naval personnel;
- an Air Group (which included squadrons from the British, New Zealand, Indian and Australian Air Forces); and
- elements for the BCOF Headquarters.[3]

These units were distributed in Southern Honshu (see Maps 1 and 2). BCOF Headquarters were at Kure and the 34th Australian Infantry Brigade was stationed in the Hiroshima Prefecture. Each Prefecture had a Governor and a civil administration and boundaries were closely adhered to by the occupation forces for supervision and liaison purposes.[4]

The role of the occupation force under General MacArthur was to supervise the demilitarisation and disposal of Japanese installations and armaments, temporarily safeguard them, protect Allied installations, and exercise military control over the country. Occupation duties for BCOF included: supervising the destruction of Japanese war materials and potential, supervising repatriation and demobilization centres catering for 700,000 incoming Japanese servicemen and 65,000 departing non-Japanese, assisting the Japanese Government agencies through liaison with the United States Military Government section, supervising elections, and setting an example to the people of Japan.

---

3 See Grey, *Australian Brass,* p. 131; and Major J.W. Willis, "The British Commonwealth Occupation Force", in *Australian Army Journal,* No 81, February 1956, p. 13.

4 Southern Honshu included the Prefectures of Hiroshima, Okayama, Tottori, Shimane, Yamaguchi, and the Island of Shikoku. See Hopkins, *History of the Australian Occupation of Japan,* p. 98; and A.W. John, *Uneasy Lies The Head That Wears A Crown: A History of the British Commonwealth Occupation Force, Japan,* Gen Publishers, Cheltenham, 1987, p. 8.

BCOF had no significant role to play in the conduct of the war trials. In the Australian sector a major task involved supervising the disposal of the massive main naval ammunition and chemical warfare depot deep in the hills beyond Hiroshima.[5] Units also took it in turns to mount guard at the Imperial Palace in Tokyo.[6]

Apart from these responsibilities, the constraints under which BCOF operated had some negative results. Many considered that Australia's part was futile because military government in the BCOF area remained in American hands.[7] Japanese were reported as saying that 'while the British parade to impress us, the Americans rebuild our country'. This was largely because General MacArthur had decided that the BCOF should not participate in the military government - an exclusively American responsibility. As a concession, even in the BCOF zone, this would have given grounds for the Soviet Union to make similar claims in the north of Japan.[8]

Most of the BCOF Commanders found this difficult to accept. In addition, it was observed that BCOF saw itself as a conquering force - an army of occupation with little of the reformist zeal that the Americans displayed under General MacArthur.[9] On the other hand, there was some balance to the ill feelings towards the Japanese, as evidence of the ravages of war could not be ignored.[10]

So it was in this context that the first Australian Signal Corps unit in Japan, a detachment of 88 Aust High Speed Wireless Section, arrived and established a communications link between Tokyo and Melbourne on 30 January 1946. The involvement of

5 Hopkins, *History of the Australian Occupation of Japan.*, pp. 98 & 111-115; Buckley, *Occupation Diplomacy,* pp 93-98; and John, *Uneasy Lies The Head That Wears A Crown,* pp. 9 & 43. See also Horner: *Duty First,* Chapters 2 & 3.

6 MP 729/8, 41/431/55, "Manpower Requirements in Tokyo to Meet BCOF and BC Agencies Commitments" - Memo, 8 May '47.

7 Captain F.C. Hutley, "Memoirs of the Occupation of Japan", in *United Service,* Vol 37, No 3, April 1984, pp. 17-19.

8 John, *Uneasy Lies the Head That Wears A Crown,* p. 59; and Horner, *High Command,* pp. 424-5.

9 Hutley, "Memoirs of the Occupation of Japan", pp. 17-19.

10 John, *Uneasy Lies the Head That Wears A Crown,* p. 42.

the Signal Corps in Japan from 1945 to 1947 is covered in detail in Barker's *Signals*.[11]

Within a short period, many of the tasks for which the force had been raised were nearing completion and forces began withdrawing. Britain faced severe financial and manpower constraints and withdrew the bulk of its contingent by April 1947. India had been granted independence and withdrew by October 1947. By December 1947 the force had been reduced from 37,000 (of which 13,500 were Australians) to only 16,000. The New Zealand component was withdrawn by November 1948. The Australian Government decided that by December 1948 the Australian contingent of BCOF would be reduced to one infantry battalion, one RAAF squadron and a naval support unit of one ship and the necessary support units, such as the Australian-commanded BCOF Signal Regiment, for the maintenance of an approximate strength of 2,750 personnel.[12]

Although life in Japan had its difficulties particularly given the lack of good accommodation along with temperature extremes and the constricting mandate for the BCOF laid down by the controlling US administration, the experience of the Signal units in Japan proved valuable for the Corps. BCOF served to justify, among other things, the continuation of both ARA Field Force Signals units and the Corps' participation in the Army Wireless Chain at a time when the Government was looking to

---

11 See Barker, *Signals*, pp. 278-280.

12 See also the following for general descriptions of BCOF: Grey, *Australian Brass,* Part 2; Murphy, op. cit., pp. 22-26; R. Buckley, *Occupation Diplomacy: Britain, The United States and Japan 1945 - 1952,* Cambridge, 1982, pp. 92-105; Rajendra Singh, *Official History of the Indian Armed Forces in the Second World War, 1939-1945: Post War occupation Forces - Japan and South East Asia,* Combined Inter Services Historical Section, Kanpur, 1958; Hopkins, "History of the Australian Occupation in Japan, 1946-50", pp. 93-116; Hutley, "Memoirs of the Occupation of Japan", pp. 5-19; RASCM Letter from Colonel Greville to Major General Simpson, March 1958; Horner, *High Command,* chapter 18; John, *Uneasy Lies The Head That Wears a Crown* and "British Commonwealth Occupation Force-Japan (1946-1951)", *Sabretache: Journal of the Military Historical Society of Australia,* Vol 17, No 1, July 1975, pp. 34-35

reduce Defence expenditure. Also, the occupation proved a challenge as the Signal units were given demanding assignments with inadequate resources and insufficiently trained personnel. To offset the shortage of skilled Army technicians, Japan Bureau of Communications personnel had to be trained by the units to perform key roles and in order to do this the difficult language barrier had to be overcome.[13]

In 1946 the Australian Signal units in BCOF included Headquarters BCOF Signals and 34 Brigade Signals (see Diagram 8). Headquarters BCOF Signals, using equipment shipped from Australia, was based at the old Japanese Naval Academy, Eta Jima, until late 1948 when it was relocated at Kure. The 34th Infantry Brigade Signal Troop was initially based with 34 Brigade at Kaitaichi and then in the Hiroshima Prefecture from July 1946 until its withdrawal with the brigade in 1948. 34 Brigade Signal Troop had carried out its normal role of providing communications to the three battalions and support units, mainly by using systems of telephone, despatch riders and HF radio.[14]

The telephone system used by the Signal units was the Japanese civil line switched through exclusive Australian and local Japanese exchanges. The two main exchanges in the BCOF area were the 'Dover' exchange at Kure and the 'Anzac' exchange at Eta Jima. Dover, with a manual and an automatic exchange, was the main switching centre. The automatic exchange handled trunk facilities and had tie lines to the manual exchange. This was a five-digit number type with a capacity for 1400 subscribers. By mid-1949 this exchange was the main switching centre for the surrounding 9 Prefectures and handled an average of 40,000 long-distance telephone calls per month. From May 1947 onwards, the exchange was operated by Japanese operators under the control of the Headquarters BCOF Signals. The

13 Buckley, *Occupation Diplomacy*, p. 95.

14 Weir, letter, July 1986; RASCM "Draft RA Sigs Historical Developments to 1953", pp. 4-5, in Corps History File; Murphy, "History of the Post War Army", p. 238; RASCM "BCOF Progress Reports - Signals", 1946-1949; and Hopkins, *History of the Australian Occupation of Japan.*, pp. 99.

Anzac exchange was an automatic exchange connected to Dover by 25 tie lines and with a capacity for 500 subscribers. By 1948 virtually all manual telephone exchanges had been replaced by automatic ones.[15]

The Signal Despatch Service (SDS) consisted of a motorcycle and jeep in the local area and civilian rail and ferry services to remote units. To a young signalman, some of those SDS courier runs by rail, sea and air within Japan and later Korea were 'the stuff that dreams are made of'.[16]

Using the Australian made WS 122 transceiver, HF radio links were maintained to units not co-located with the Brigade Headquarters. These provided duplication of the line links for emergencies and an internal security measure. Line equipment progressively became the responsibility of the Japanese Bureau of Communications.

Radios were also used for other purposes. In one instance, a link using WS 22 was employed in anti black market activities. The shore station was located at 34 Brigade in Hiro and outstations included launches and Auster aircraft. Results were 'satisfactory.'[17]

Headquarters BCOF Signals tasks included technical support for the main wireless transmitter station located at Kure, and the wireless receiver site, where 34 Brigade was based, on the outskirts of Kure - a few kilometres away in Hiro. The British Commonwealth (Brit Com) Base Signal office in Kure was the main distributing centre for BCOF. Signal centres were also located at

- the RAAF bases at Iwakuni and Bofu,
- Ebisu, where the BCOF guard battalion and leave personnel were accommodated,

---

15 Captain Belcher, letter, September 1987, pp. 2-3; and RASCM "History of BCOF Signals", p. 3.

16 Belcher, letter, September 1987, p. 2; and Weir, letter, September 1989.

17 Belcher, letter, September 1987, pp. 2-3; RASCM "BCOF Progress Report - Signals", May '48, p. 8; and RASCM "History of BCOF Signals", p. 6.

- Kobe, where the publishing house for the *British Commonwealth Occupation Newspaper* (BCON) was located,
- Okayama, where the 67th Infantry Battalion was located until 1948 when it moved to Hiro,
- Osaka, for the natural disaster network in southern Japan, and
- Tokyo, servicing the British Commonwealth diplomatic and military elements at the Marunouchi Hotel.[18]

Headquarters BCOF Signals also provided the rear link communications to Melbourne and Singapore for relay to Delhi and London on behalf of the Indian and British units until their withdrawal from Japan. The systems used for the rear link were radio teletype on HF radio using mainly SWB 8 transmitters (see Diagram 9). Communications were good except for minor difficulty experienced in the small hours of the morning due to interference resulting from ionospheric conditions. This necessitated urgent messages being cleared by hand speed morse.[19] This was the first time that the Australian Army had provided an international strategic communications link in peacetime. The merits of the service would serve to justify the maintenance of the strategic communications network for some time.

Headquarters BCOF Signals was renamed BCOF Signal Regiment in February 1949 and, after installation and exhaustive testing, 'Frequency Shift Keying' (FSK) equipment was introduced by mid 1949 - a definite improvement on the old system and one that brought with it personnel savings. FSK equipment was also found to be simple to operate and messages received were

18 Weir, letter, July 1986, p. 3; RASCM "Draft RA Sigs Historical Developments to 1953", p. 5, in Corps History File; Murphy, "History of the Post War Army", p. 238; RASCM "BCOF Progress Reports - Signals", 1946-1949; and Hopkins, 'History of the Australian Occupation of Japan'., pp. 99.

19 Belcher, letter, September 1987, pp. 2-3

printed accurately even when transmitted through channels with radio interference.[20]

Broadcasting radio stations were established in the BCOF area in 1946 and 1947, including a 200 watt and two 1000 watt stations operated by Australian signalmen at Kure. An auxiliary station was also set up at Iwakuni. The 1000-watt stations provided full day and night coverage on short wave to the whole BCOF area. Programmes included BBC news, the hit parade from Radio Australia and news from the United States for the benefit of American listeners in the area. They were also popular with many Japanese who were 'radio hungry'. The Chief Signals Officer of BCOF was responsible for the control of all Japanese radio stations in the BCOF area. These were visited and checked by signalmen who passed intercepted material to the Intelligence Branch.[21]

One of the major problems of Signals was to train incoming personnel to handle the Japanese and US equipment which BCOF installed and operated until the Japanese Bureau of Communications could take over. This was accentuated by the lack of fluent Japanese speakers in the Occupation Force. Moreover, as the strength of BCOF declined, the communications commitments did not decline so the problem became acute. Consequently, until the commencement of the Korean War, Australian Signals personnel attended 8 to 14 weeks-long technical courses at the Eighth US Army Signal Corps School at Yokohama. These courses covered subjects such as cable splicing, automatic teletype system maintenance, VHF radio, radio teletype, and teletypewriter repair.

20 Units reorganised as part of BCOF Signal Regiment included the following: HQ Sig Regt BCOF, and elements of 34 Line Troop, 21 Line Maintenance Troop, 22 Line Maintenance Troop, 23 Tech Maintenance Troop, and 5 Sig Equip Troop, as well as 11-14 Mobile Carrier Troop, 98 Hi Speed WS Troop, 112 TG Op Troop, 133 Tele Switchboard Op Troop, 115 Tele Switchboard Op Troop, 33 DR Troop, 88 Hi Speed WS Troop, 99 Heavy WS Troop, a proportion of 2 Broadcasting Station Mt Troop and 113 Cipher Troop. See MP 742, 240/1/3068. See also RASCM "BCOF Progress Reports - Signals", Jun '49, p. 6.

21 RASCM "History of BCOF Signals", p. 7; Weir, letter, July 1988, p. 1; CRS A816, 19/304/388; and RASCM "BCOF Progress Reports - Signals", 1946-1950.

In order to train morse operators on teleprinters and cipher operators, from volunteers in the Force, courses were conducted by BCOF Signal Regiment. Training was also carried out by BCOF for the operators supplied by the Japanese Bureau of Communications. Regimental training was conducted for all available personnel on Saturday mornings to ensure maintenance of the basic infantry and signals skills.[22]

Another measure proposed to increase the number of operators, given the acute shortage of qualified and available men, was that of including women on the establishment of the BCOF Signal Regiment. In Australia, consideration was given to the matter as early as 1945, but it was repeatedly rejected at Ministerial level with the exception of nursing and medical personnel. Lieutenant General Northcott, the Commander-in-Chief BCOF, had approached the Prime Minister directly on this matter when he visited Japan in 1946. Eventually Cabinet allowed women to work in the canteens and welfare organisations, but not in signal units because the Prime Minister objected to the idea.[23]

In May 1950 the Australian Government announced plans for the withdrawal of BCOF troops from Japan, and Australian servicemen were in the process of packing for this move when the North Koreans crossed the 38th Parallel into South Korea on 25 June 1950.[24] The origins of this war are complex but stemmed back largely to the partitioning of Korea in 1945, after decades of Japanese colonial rule, and the subsequent power struggle between Communist North Korea, with its Soviet-trained leader

22 RASCM "BCOF Progress Reports", 1947-1950; "History of BCOF Signals", p. 8; Colonel Greville, letter; and Weir, letter, July, 1988.

23 CRS A816, 52/301/247, "Women in BCOF"; and CRS A2031, Defence Committee Minutes, No 210/1946 &223/1946, June 1946.

24 *Sabretache,* Vol 17, No 1, July 1975, p. 35.

Kim-Il-Sung, and the US backed South Korea with its anti-Communist leader, the US emigré, Syngman Rhee.[25]

The forces which formed BCOF in June 1950 played an important role as readily available forces for deployment alongside US units in Korea and as close-at-hand administrative backup to those forces. In particular, on 29 June 1950, Menzies announced that two Australian naval ships in Japanese waters (the destroyer *Bataan* and the frigate *Shoalhaven*) had been 'placed at the disposal of the United Nations through the United States authorities in support of the Republic of Korea.' Also, RAAF No 77 Squadron was the first air force unit from any UN member state to give support to US ground forces there on 30 June 1950. Thus Australia became the first nation, apart from the US, to give active support to South Korea.[26] Uppermost in the minds of the Menzies Government's decision makers was the concern to foster better economic and political relations with the US. Loyalty to the United Nations and the strategic importance of Korea to Australia, while touted as the prime justifications for involvement, were not the prime considerations.[27]

Following the first weeks of the war, which saw some of the 'worst reverses to American arms in this century'[28], pressure was placed on countries such as Britain and Australia to supply ground

25 For a discussion covering some of the differing perspectives on the origins of the Korean War see R. O'Neill, *Australia in the Korean War, 1950-1953, Vol I, Strategy and Diplomacy,* Canberra, The Australian War Memorial and Australian Government Publishing Service, 1981. See also G. St J Barclay, *Friends in High Places,* Oxford University Press, Melbourne 1985; J. Grey, *British Commonwealth Armies & the Korean War: An Alliance Study,* Manchester University Press, Manchester, 1988; G. Greenwood & N Harper (eds), *Australia in World Affairs 1961-1965,* pp. 262-264; M. Carver, *War Since 1945,* Putnam's, New York, 1981, pp. 151-155; Fred Alexander, *From Curtin to Menzies and After,* Nelson, 1973, pp. 134-135; and Alan Watt, *The Evolution of Australian Foreign Policy, 1938-1965,* CUP, Cambridge, 1967, pp. 120-121.

26 O'Neill, *Australia in the Korean War 1950-53, Volume II:* p. 197; and *Sabretache,* Vol 17, No 1, 1975, pp. 35-7.

27 Lee, *Search For Security,* pp. 110-111.

28 Heller, C.E & Stofft, W.A. (eds), *America's First Battles 1776-1965,* cited in Grey, *Australian Brass,* p. 181.

forces. Thus, during his visit to the United States' Congress, which occurred shortly after the US call for further assistance, Prime Minister Menzies committed the Japan-based 3rd Battalion, The Royal Australian Regiment (3 RAR) on 1 August 1950. The battalion arrived in Korea from Japan on 28 September 1950.[29]

In the meantime, in August 1950, the Defence Committee had requested that Lieutenant General Sir Horace Robertson, the Commander-in-Chief BCOF, recommend the size of a forward base organisation for Australia's proposed military commitment to Korea. It was also agreed that the infantry battalion should work in conjunction with the British and New Zealand forces in Korea as the various units became available.[30]

Prior to the commitment, the first Australian troops in Korea were those sent as part of General Robertson's Advanced Headquarters for the British Commonwealth Forces at Pusan in southeast Korea (see Map 3). A small troop consisting of 1 officer and 6 other ranks of most trades, including Continuous Wave (CW) and Cipher Operators with a couple of Radio Mechanics, from BCOF Signal Regiment, went with him to set up the rear link to the BCOF Signal Regiment in Japan. Captain A. Findlay, the commander of the detachment, recalled:

> In September 1950 I was fortunate to be chosen by my Commanding Officer, Lieutenant Colonel S.J. Greville, to provide a Wireless detachment to Korea from Kure, Japan. RAAF flew me to Pusan and I found a site while I waited for the detachment to arrive by ship. Communications were established and most of the traffic was concerned with the servicing of the Australian Battalion [3 RAR] after their arrival. I was the

29 Grey, *Australian Brass,* p. 181; Barclay, *Friends in High Places,* pp. 39-43; and O'Neill, *Australia in the Korean War 1950-53,* Volume I:, p. 8.

30 CRS A2031, Defence Committee Minutes, No 137/1950. For a discussion on the role played by Lieutenant General Robertson during these deliberations see Grey, *Australian Brass,* pp. 179-192.

> first Australian Corps of Signals member to serve in the Korean conflict.[31]

The Force to which the detachment belonged was initially titled the Australian Force in Korea or AUSTFIK. The Signal Troop was initially named the AUSTFIK Maintenance Area (AUSTFIKMA) Signals Detachment. This arrangement lasted for approximately nine months. The troop was also affectionately known as 'Robbie's Sigs'.[32]

When 3 RAR moved forward from Pusan and joined 27 British Commonwealth Infantry Brigade on 1 October 1950, the need for a wireless link back to Kure from Pusan diminished, and the station was eventually moved to Taegu where it became the link from Headquarters Advance BCOF to Kure from 17 November 1950 (see Diagram 10). On 8 January 1951 Captain Findlay was replaced by Major J.A. Little who had arrived from Australia for the purpose. Although he was expected to be a liaison officer between the Chief Signals Officer (CSO) BCOF and the signals officers of the UK, composite and Canadian brigades as well as US 8th Army, he was hampered on all sides by the fact that there was little he could do with authority if any of them demurred on the grounds that they owed allegiances to other commands. It was belatedly recognised that a CSO should have been appointed for the Lines of Communication. At the same time, however, there appeared to be 'no lack of cordiality' in the relationship that existed between the officers of the various units. These ambiguities were not clarified until the arrival of 1 Commonwealth Divisional Signal Regiment on 25 June 1951.[33]

Apart from a couple of sub units, such as the AUSTFIKMA Signals Troop, no complete Australian Signals unit was deployed in Korea during the war, but members of the Corps served there

---

31 Captain A. Findlay (RL), letter.

32 Weir, letter, July 1986 and March 1988; and CRS A2107, K10.1, "BCOF Operational Instruction No 1", 23 September 1950.

33 WO 308/22 Headquarters BCFK, CSO's Branch Periodic Reports 1950-1953.

with other units. Because of a shortage of specialists within the infantry battalion support companies, and in recognition of the valuable training opportunity offered, personnel from Signals, Armour, Artillery and Engineers were posted to fill the specialist appointments in the signals, anti tank and assault pioneer platoons respectively. These men were transferred from their parent Corps to the Infantry Corps for the duration of their service with the battalions in Korea.[34]

The volunteer officers, non-commissioned officers and soldiers who participated in this exchange were given short infantry conversion courses to prepare them for their role with infantry. For signals platoons this training was conducted at the School of Signals at Balcombe, Victoria. A Battle School was also established later at Haramura in Japan and regular courses were conducted there for regimental signallers. These men were then taken on strength and served with all three battalions of the Royal Australian Regiment that served in Korea. In early 1952 the Regimental Signal Officers (RSOs) of both battalions stationed in Korea (1RAR and 3RAR) were Australian Signals officers posted from Army Headquarters Signal Regiment, Melbourne - Lieutenants Ken Taylor and Ray Clark (later Major General and Colonel respectively).[35]

Initially, a barrier existed between the Infantry signallers and the RA Signals members of the signals platoon, but within weeks this was broken down by shared experiences, and, before long, they were all working together harmoniously. During operations there were no distinctions at all, and a close relationship resulted

34 Murphy, "History of the Post War Army, p. 174; RASCM Barker, unpublished manuscript, p. 411(a); and MP 897, 106/8/1088, "Signals Officers - Infantry Battalions".

35 Murphy, "History of the Post War Army", p. 174; RASCM Barker, unpublished manuscript, p. 411(a); MP 897, 106/8/1088, "Signals Officers - Infantry Battalions"; CRS A816, 37/301/381, "BCFK Progress Reports"; Weir, letter, March 1988, p. 3; and RASCM "History of the Post War Army - Signals", pp. 5-6.

between the Corps involved in this programme.[36] It may well be argued that this bonding with Infantry helped the advancement of several Signals Corps officers to senior rank.

A close relationship was also fostered between the Commonwealth countries participating by emphasising the need for standardised procedures and equipment. The standardisation of equipment within the infantry brigades presented some initial difficulties for the Australians when 3 RAR joined the British battalions in 27 Commonwealth Brigade. In order to standardise the wireless sets used within the Brigade, and so ease the repair tasks of workshops, the Australian wireless sets were withdrawn in December 1950 and British WSs 31, 88A, 88B and 62 were issued. At Brigade Headquarters level, the Brigade Signal Troop used WSs 19, 62 and 53, 40-line F & F and 10-line magneto switchboards with 'J' telephones.[37] Once the war reached the static phase of trench warfare, procedures and equipment were sufficiently standardised to ensure the smooth conduct of operations.

Of all the Signals officers' jobs in the Division, under these circumstances of trench warfare, probably that of the Battalion Signal Officer in the line was the hardest. Duties of the Battalion Signals Officer included advising the battalion commander on all signal matters, supervising all communications in the battalion, and organising directly the communications at the battalion headquarters. The signal platoon had to provide:

- a signal office at battalion headquarters,
- a control set on the battalion command wireless net,
- four detachments to provide line communications from battalion headquarters down to company level,
- communications assistance to the companies,
- communications to supporting and neighbouring units,

---

36 RASCM Letter from Captain R. Bruce. Rogers to Colonel Bruton, 25 August 1952; and RASCM Barker, unpublished manuscript, p. 411(a).

37 RASCM Appendix A to CSO's Branch Historical Notes, Jun 50-Jul '53: "Report on Visit to Korea by Maj A.A. Mason 13 - 26 Dec '50"; and RASCM Barker, unpublished manuscript, p. 413.

- additional communications as required and
- signalmen to operate the rear link wireless set to brigade headquarters![38]

Captain Bruce (or 'Buck') Rogers, was the Signals Corps Regimental Signal Officer (RSO) for 1RAR in 1952, and he explained in a letter at the time some of the problems that were encountered in maintaining communications in the static warfare conditions that had become prevalent:

> It is normally desirable to site radios on hilltops in this very hilly country. However [Command Posts] are most popularly placed well on the reverse slopes, as shelling is more or less inevitable, our daily average here being 40 - 60 per day. Aerials or activity on hilltops always draws an extra ration of shelling and (worse) mortaring. [The best solution is] to remote the wireless... and use a horizontal aerial just below the crestline, concealed by scrub or... folds in the ground.
>
> Interference on all nets is a major problem. With radios on top of almost every hill this is well-nigh inevitable, however, and co-operation and goodwill are the only solutions. Only yesterday I changed the battalion command net channel with the Machine Gun internal channel (seldom used) at the request of a harassed Pronto [Signals Officer] in the US Marine Division four battalions across from me.
>
> Intercept is normal practice for both sides... and Charlie sends an astonishing amount of

38 "Infantry Training, Vol IV Tactics, The Infantry Battalion in Battle", UK War Office, 1952, pp. 47-49; and RASCM "1 Commonwealth Divisional Signal Regiment: Notes on Operations Dec '52 - Apr '53", p. 17.

> valuable information by radio in clear. The intercept is good enough to have it [translated by Chinese Nationalists and] passed to the patrol commander on the valley floor in front of the enemy in time for him to act on it. Such things as 'do not fire until the enemy is within 30 feet' and 'let the patrol through the outposts and then attack it from behind,' and so on.[39]

Communications to patrols required careful coordination of WS 88 and WS 31 frequencies owing to the hilly terrain and the range of the sets in the deep valleys between the opposing forces. In addition, a number of cases were reported in which the enemy came up on the WS 31 and 88 set nets and put over propaganda. At other times, when an attack was imminent, the nets were often jammed. To reduce the vulnerability to enemy interception, US codes were used by the Australian units as they were simple and universal. These were changed every third day.[40]

By 1952 the war had been static for so long that the positions had become like fortresses with fortress-like communications. In 1 RAR, four switchboards of an American crystal pattern were issued for use in companies, but they were inadequate and a ten-line magneto board or two six-line boards were required. Even then, the lines had to be 'partied'. In this static war, according to Rogers, the above was considered by everyone, including the Brigadier, as the minimum requirement.[41]

Rogers also followed a policy of having every line with an alternate, or two alternates in some cases, on completely separate routings. Together with a comprehensive system of later-

39 RASCM Rogers, letter to Bruton, 25 August 1952; and RASCM "1 Commonwealth Divisional Signal Regiment: Notes on Operations Dec '52 - Apr '53", p. 4.

40 RASCM "1 Commonwealth Divisional Signal Regiment: Notes on Operations Dec '52 - Apr '53", p. 17.

41 RASCM Rogers, letter to Colonel Bruton, 25 August 1952; and RASCM "1 Commonwealth Divisional Signal Regiment: Notes on Operations Dec '52 - Apr '53", pp. 4 & 17.

als, communication links were seldom lost entirely, even during heavy shelling, although field cable in the battalion areas was frequently cut by mortaring and shelling and the wastage rate of cable for battalions was fairly high. The longest company line was two miles, and two were over a mile. This was rendered necessary by the devious routing, as the forward infantry companies were located on forward slopes in full view of the enemy and any movement above ground during daylight courted enemy fire of all types - and the same was true of the enemy positions. Alternates were well separated, on opposite sides of features, and only met at the switchboard.[42]

In addition to the transfers to the Infantry Corps, arrangements were made for the semi-permanent attachment of officers of all Corps to appropriate British, New Zealand and Canadian units in Korea. As a result of this procedure, there was a continuous turnover of junior officers of all Corps and a heightened training opportunity which attachment to units on active service automatically provided. These tours of duty were of 18 months duration for married members and two years for single personnel.[43]

The provision of communications for effective artillery support had always been critical in battle and developments in radio communications technology enabled commanders to get better use from their 'indirect fire support' artillery units. According to Lieutenant General Sir Thomas Daly, who commanded the 28th Brigade in Korea in 1952-1953, improved artillery support was made possible by greatly improved radio communications, combined with the prolonged occupation of the defensive line known as the Jamestown Line. For instance, in an extreme case, an infantry section leader on patrol could bring down the divisional

42 RASCM Rogers, letter to Colonel Bruton, 25 August 1952; and RASCM "1 Commonwealth Divisional Signal Regiment: Notes on Operations Dec '52 - Apr '53", pp. 2-4.

43 Murphy, "History of the Post War Army", p. 174; and MP 897, 172/1/101, "Tours of Duty - Japan & Korea.".

artillery with remarkable accuracy if required.[44] This achievement produced a significant combat power multiplying effect for the isolated patrols and served to influence significantly the tactics employed as patrols were prepared to take greater risks than would have been contemplated otherwise.

On 28 July 1951 the First Commonwealth Division was formed in Korea, under the operational command of I US Corps. It was formed from the 25th Canadian Infantry Brigade, the 28th Commonwealth Infantry Brigade (consisting of Australian and British infantry battalions, with a New Zealand field artillery regiment), and the 29th British Infantry Brigade. A composite Divisional Signal unit was subsequently formed, under Lieutenant Colonel A.L. Atkinson, a Royal Signals officer, containing British, New Zealand and Canadian troops.[45] The British readily conceded the right of the nation which supplied the greater part of a force to also supply its commander and so, when a second Australian battalion (1 RAR) was despatched from Australia in March 1952, the command of the Commonwealth Brigade was handed over to the Australians who retained it for the rest of the war.[46]

Australian Signalmen served in various units on operations in Korea apart from those already mentioned. Signalmen were detached to 1 Australian Ancillary Unit (1 AAU) in Korea for duty with such units as 1 Commonwealth Divisional Signal Regiment, Communication Zone (CommZ) Signal Squadron in Seoul and Pusan, 25th Canadian Brigade Signal Squadron and 28th Commonwealth Brigade Signal Squadron (see Diagram 11). One exchange arrangement was a British Signals Captain from

---

44 Lieutenant General Sir Thomas Daly cited in Welburn, *The Development of Australian Army Doctrine 1945-1964*, p. 25.

45 Nalder, *The Royal Corps of Signals*, p. 484; and J. Grey, "A Military Alliance at Work? - Commonwealth Forces in the Korean War", *Journal of the Australian War Memorial*, No 9, October 1986, p. 44. For an account of Canadian Signals in Korea see J.S. Moir (ed), *History of the Royal Canadian Corps of Signals 1903-1961*, Corps Committee, Royal Canadian Corps of Signals, Ottawa, 1962, pp. 290-302.

46 Grey, "A Military Alliance at Work? ", p. 44; and CRS A816, 37/301/381, "BCFK Periodic Report, No 3 1 Oct '52 - 31 Mar '53".

the Commonwealth Divisional Signal Regiment and an Australian Signals Captain from the British Commonwealth (Brit Com) Base Signal Regiment in Japan exchanging positions. This gave Royal Signals representation in an otherwise wholly Australian unit, and Royal Australian Signals reciprocal representation in a unit predominantly manned by British and Canadian Signallers. At one point in 1952, when the Commonwealth Division was out of line, 34 (Australian) Line Maintenance Troop from the Brit Com Base Signal Regiment in Japan was sent to lay and maintain lines throughout the Division lines. This task lasted for approximately six months.[47]

The composite Division in the field maintained rear links with Headquarters I US Corps. These links consisted of six circuits - four telephone, one full duplex teletype and one duplex radio teletype for the command net. Rear links from Rear Divisional Headquarters (consisting of one telephone circuit and one hand speed telegraph circuit), were also maintained with Advanced Headquarters British Commonwealth Forces Korea (BCFK); and to Headquarters BCFK in Japan.[48]

Numerous difficulties were experienced with communications in Korea. Distance, speed, the difficult terrain, and Korea's road nets restricted the use of wire. Telephone circuits were not practical. The rugged hills hampered radio relay teams from sending signals between stations. Relay trucks were targets of guerilla warfare and sabotage. The answer was very high frequency (VHF) radio which became more dependable than wire as the primary method of communication. Its role was so critical that, for the first time, VHF radio communications became the 'backbone' of the communications system in a theatre of military operations on land.[49]

47 Weir, letter, July 1986; and CRS A816, 37/301/381, "BCFK Periodic Report, No 3, 1 Oct '52 - 31 Mar '53".

48 RASCM "1 Commonwealth Divisional Signal Regiment: Notes on Operations Dec '52 - Apr '53", p. 12.

49 Coker & Rios, *A Concise History of the US Army Signal Corps,* p. 26.

The Commonwealth Division encountered many problems apart from those found when establishing links to the various component Home countries. The different types of equipment demanded a separate line of communication running parallel to the American one involving duplication of effort. Maintenance of signals stores was difficult and US assistance in the provision of common pattern or compatible equipment was often gratefully received.[50]

Liaison with the US Army Signal Corps elements in Korea remained on a well established footing throughout the war and, in general, fair and reasonable consideration was given to the requirements of the British Commonwealth Forces. It was important that these good relations were maintained as all long distance telephone and telegraph circuits in Korea were under the control of the US Signals Corps. Commonwealth local switchboards had access to the long lines network by connection to the appropriate US exchange in the area concerned. There were instances, however, when the impression was gained that the US Signals Corps regarded the handling of signal traffic for British Commonwealth forces as a favour. This was partly the result of an instruction, released by the US Commander-in-Chief, Far East, on the passing of administrative signal traffic of UN forces in the Korean theatre. The wording of the order was such that it could be interpreted that this was to be a favour rather than a requirement. Consequently, Commonwealth and other commanders felt that they were being regarded as supplicants and not equal partners when trying to use this service. In the light of this, a substantial proportion of BCFK signals traffic was sent overseas from Korea via the wireless link maintained between the Brit Com Base Signal Office in Kure, Japan and Rear Headquarters 1 Commonwealth Division.[51]

50 RASCM Letter from Colonel Greville to Major General Simpson, March 1958.

51 CRS A816, 37/301/381, "BCFK Periodic Report No 3: 1 Oct '52 - 31 Mar '53"; RASCM Appendix A to CSO's Branch Historical Notes, Jun 50-Jul '53: "Report on Visit to Korea by Maj A.A. Mason 13 - 26 Dec '50", pp. 6-10; and RASCM Barker, unpublished manuscript, p. 412.

The validity of having an independent communications link, not subject to the whims of commanders of other nations' forces was keenly felt at this time.

The British Commonwealth Communication Zone (CommZ) Signal Squadron was raised to provide communications links for Headquarters Brit Com Sub Area in Pusan, Advanced Headquarters BCFK in Seoul, and Brit Com Forward Maintenance area (FMA) in Seoul.[52]

Although many Australian Signals personnel saw action in Korea as members of Commonwealth units, the Royal Australian Signals, as already mentioned, mainly provided the line of communications and base signals unit in Japan. Australia had decided not to contribute a signals unit for the Commonwealth Division because of these pre-arranged commitments in Japan which stretched the resources of the Corps at a time when there were competing requirements at home, including the provision of ARA cadre staff for the National Service Training Scheme. The Royal Australian Corps of Signals had proportionately contributed its fair share of forces to the British Commonwealth's joint efforts with the manning of the Base Signal Regiment in Japan. The provision of these communications links was a formidable task for the Signal unit remaining at Kure, but it was simplified by cooperation with the US Army Signal Corps as Kure Base Signal Office was made a minor relay station on the US 'American Command and Administration Network' (ACAN). This provided an easy and efficient communications link between Korea and Japan through US channels. It also provided a means whereby Canadian traffic could be cleared from Kure through Tokyo back to Canada. New Zealand traffic was cleared through the existing Australian link and a new radio teletype link was established between Singapore and Kure for the traffic generated by the British forces.[53]

52 CRS A 816, 37/301/381, "BCFK Periodic Report, No 3, 1 Oct '52 - 31 Mar '53".
53 RASCM Greville, letter to Major General Simpson, March 1958; and CRS A2031,93/1951, "Ground Force for Korea - Formation of a Commonwealth Division"; and interview with Major General D. Vincent, September 1988.

The arrangements that had existed prior to the outbreak of war in June 1950, in Japan, facilitated the use of the BCOF base as a support area for the forces stationed in Korea. The British Government had initially insisted on relying on US supply lines for their administrative support rather than being subject to Australian administrative control. Eventually, however, given that the US sources were already stretched and that a British Commonwealth facility was operating and available for use, the British forces were compelled to rely on the BCOF under its Australian Commander in Chief, Lieutenant General Sir Horace Robertson. Consequently, the non-operational control and the general administration of the British, Australian and New Zealand Army and Air Force units stationed in Korea and Japan was vested in General Robertson.[54]

In spite of the facilitating role of BCOF at the outbreak of war in Korea, there were in fact few international communications links working at the time. As at June 1950, there were five wireless nets in existence but only one of these was in active use. This was the Kure-Melbourne No 1 radio teletype duplex link working on a daily schedule from 0800 to 2330 hrs.[55] By November 1950 a new Kure -Singapore CW link was opened due to the sharp increase in the volume of traffic. It was not until 4 December 1952 that a voice wireless link was established (for unclassified conversations only) between Headquarters 1 Commonwealth Division and Headquarters BCFK.[56]

It was intended, in 1950, that BCOF would be scaled down to a size required only for the maintenance of the British

54 Grey, *Australian Brass*, pp. 181-182.

55 The other links were as follows:
Kure-Yokohama RTT (duplex) - standby command net, backing for line,
Kure-Iwakuni-Tokyo CW backup,
Yokohama-Kure-Eta Jima CW disaster net, and
Kure-Melbourne No 2 RTT (duplex) standby net.
See WO 308/22, Headquarters BCFK, CSO's Branch Periodical Reports, 1950-1953.

56 WO 308/22, Headquarters BCFK, CSO's Branch Periodical Reports, 1950-1953.

Commonwealth Force in Korea (BCFK), including the maintenance of 77 Squadron RAAF, and that the surplus, together with all other Army elements not required, would be progressively returned to Australia.[57] This was particularly important in light of the signing of the Japanese Peace Treaty which would abolish any legal grounds for the maintenance of an 'Occupation' Force in Japan. The scaling down occurred because the whole of BCOF could not be withdrawn while BCFK was relying on its facilities to avoid unnecessary and costly duplication. For example, the BCOF Signal Regiment and Cipher Troop handled all the British Commonwealth military signals traffic in and out of Japan and would, therefore, be required after the September 1951 signing of the Japanese Treaty. The same was the case for various ordnance, engineering and medical units. Consequently these units would be transferred to the command of BCFK and be integrated with elements from the other participating Commonwealth countries. In the case of the BCOF Signal Regiment, the unit continued to be operated mostly by Australians and under the command of an Australian officer.[58]

In the meantime, on the international scene, the Treaty of San Francisco and the United States-Japan Security Treaty were signed in San Francisco on 8 September 1951, seven days after the signing of the ANZUS Treaty, which provided for consultation between Australia, New Zealand and the United States 'whenever the territorial integrity, political independence or security of any of the Parties is threatened in the Pacific'.[59] At the end of October 1952 BCOF ceased to exist and BCFK remained in its place.[60] The communications task of the British Commonwealth (Brit Com) Base Signal Regiment and the British Commonwealth Base Cipher Troop, as the BCOF Signal Regiment and Cipher Troop came to be

57 CRS A2031, Defence Committee Minutes, 178/1950.
58 Grey, *British Commonwealth Armies & the Korean War,* pp. 219-220.
59 The origins of the ANZUS Treaty are discussed in works such as Pemberton's *All The Way,* pp. 1-34. The ANZUS Treaty is included as an Appendix to his book.
60 CRS A816, 19/304/503.

named, was now the provision of communications between BCFK Headquarters and Headquarters 1 Commonwealth Division in Korea, as well as the maintenance of channels between BCFK and Australia. This led to only minor changes in the roles of the units remaining in Japan, as they had been progressively reorganising in anticipation of the name change.[61]

During this period the BCFK Signal Regiment was operationally overstretched. So the unit employed Japanese civilians in the non-sensitive areas as teletype operators, technicians, drivers and switchboard operators as well as in positions such as labourers, typists, storemen and gardeners. Unit members had to attend defensive training which included air raid precautions, manning of air mounted bren guns and even aircraft recognition practice.[62]

Operations at Kure were on a 24 hour basis so shift work was a part of life for unit members. Throughout this period the unit worked a Carrier Frequency Shift (CFS) radio teletype (RTT) circuit to US Army Headquarters Tokyo, using US procedure - this circuit also carried traffic to and from Canada, United Nations Headquarters and the Brit Com Sub Area in Tokyo. Other links maintained included the following:

- RTT circuits to Army Headquarters Melbourne, and the UK Far Eastern Land Forces (FARELF) Headquarters in Singapore;
- a morse backup circuit to British Commonwealth Communications Zone (CommZ) and field formations in Korea, using either the US telegraph relay network or a hand speed morse circuit;
- a landline teletype circuit to Ebisu (in Tokyo), Iwakuni, Hiro and Kure;

61 RASCM "History of The Post War Army - Signals", p. 6; and *Signals Bulletin*, Vol 1, No 3, March 1952, p.4.

62 Weir, letter, March 1988, p. 3; B.G. Saunders, letter, August 1990; and Letter from Major General R.P. Woollard, AO, September 1988.

- a Cipher office in Kure; and
- an SDS service between BCFK Japan and BCFK Signals elements in Korea.

Despatches were handled through the RAAF. The courier boarded the RAAF DC3 courier aircraft and travelled to Kimpo Airport Seoul. Despatches were exchanged with the British or New Zealand Despatch Rider from the Communication Zone Signal Squadron in Korea and the Australian Signals courier then returned to Iwakuni, Japan, and back to the unit.[63]

After two years of prolonged negotiations, there was a cessation of hostilities in Korea, at 11 pm on 27 July 1953[64] and Signals commitments increased to facilitate the withdrawal of Commonwealth units and to provide the communications links for Operation 'Homeward Bound' - the repatriation of Commonwealth ex-Prisoners of War. The communications plan was arranged by 1 Commonwealth Division after consultation with Headquarters BCFK. The initial prisoner exchange in March 1953 relied on a morse link established between Pan-Mun-Jom and Kure, Japan. The radio link was established after the armistice.[65]

On arrival at 'Freedom Village', the ex-prisoners were met by a senior officer, given a quick medical inspection, identified on the rolls and clothed in British uniform. They were then moved by road about 24 miles to a specially prepared camp called 'Britannia' in the rear of the Divisional area. Here they were properly inspected, kitted out and documented. They stayed the night at Britannia Camp and were then moved to an airfield near Seoul,

63 CRS A816, 37/301/381, "BCFK Activity Report - Signals" No 3; Weir, letter, March 1988, p. 3; Woollard, letter, September 1988; RASCM "British Commonwealth Base Signal Regiment: Ebisu Relay Station, Tokyo Japan", p. 1; and RASCM Barker, unpublished manuscript, p. 412.

64 For a discussion on the reasons for the prolongation of the negotiations see Carver, *War Since 1945*, p. 168; and McCormack, *Cold War Hot War*, pp. 118-127.

65 Weir, letter, July 1990.

from where they were flown to Japan. A total of 1005 prisoners passed through the organisation which lasted for five weeks.

There was an obvious need for speedy and reliable communications in passing on particulars of each returning prisoner to the Commonwealth country concerned - if possible beating the Press releases. The Divisional Signal Regiment also provided communications to facilitate the control of movement of the ex-prisoners from Britannia Camp to Seoul and then to Kure, Japan. In addition, the internal communications for Britannia Camp were installed and based around a 40 line switchboard.[66]

Shortly afterwards, the US Army withdrew the automatic link into the US Army's ACAN from Kure in Japan (despite the fact that no appreciable decrease in the total traffic volume handled had been observed). As a result, the existing small Signal Office in Ebisu, Tokyo, was transformed into a minor relay station with circuits to Brit Com Kure and ACAN Primary Relay Station in Tokyo. A land line Voice Frequency (VF) Telegraph circuit from Ebisu to Kure was hired from the Japanese Post Office and a duplex RTT circuit was installed using on-off keying techniques as an alternative. An automatic telegraph tape machine was also acquired to handle the additional traffic load. By June 1954 the Relay Station was fully functional.[67] It was proposed that the hand speed circuit from Kure in Japan to the 1st Commonwealth Division in Korea would be upgraded to a radio teletype circuit at the end of 1954 when it was used to provide a radio teletype conference facility on a schedule basis but the unavailability of suitable equipment and frequencies prevented this from occurring.[68] Radio teletype con-

---

66 RASCM "1 Commonwealth Divisional Signal Regiment: Notes on Operations Dec '52 - Apr '53", p. 12; and CRS A816, 37/301/381, "Operation 'Homeward Bound'".

67 CRS A816, 37/301/381, "BCFK Periodic Reports - Operation Homeward Bound"; and RASCM "British Commonwealth Base Signal Regiment: Ebisu Relay Station, Tokyo Japan", pp. 1-2.

68 WO 308/2, HQ BCFK 6 Monthly Report, 11 November 1954, Appendix B; and WO 308/3 HQ BCFK 6 Monthly Report March 1955, Appendix D.

ference facilities, although of limited application in Korea, were to be more extensively used in Vietnam.

Reductions of units under the auspices of BCFK had commenced soon after the cessation of hostilities, in 1954, when the withdrawal of US units also commenced. By August 1955 the Australian Government, along with the British Government, was pushing for further reductions to the forces stationed in Korea and Japan and the US eventually agreed. Operation 'Greygoose' was the plan for the withdrawal of the Brit Com Base from Japan. The main body of Australian troops sailed from Kure on 4 November 1956 and arrived in Sydney on 18 November 1956. Headquarters BCFK Kure closed on 23 November 1956 and the Commanding Officer of BCFK Kure departed on 25 November 1956.[69]

After having been reduced in 1954, the Commonwealth Division ceased functioning on 15 March 1956, along with the Divisional Signal Regiment, leaving behind a residual contingent.[70] The remaining Commonwealth Forces stationed in Korea were regrouped as the Commonwealth Contingent Korea (CCK and with the closure of the Britcom Base (BCFK) Signal Regiment in Japan, personnel were available to operate the new unit -the CCK Signal Squadron. This was the first complete Australian Signal unit to be deployed to Korea. The unit was based at Inchon with detachments in Seoul and with the Cameron Highlanders on the De-Militarized Zone on the Imjin River.[71]

In order for the CCK Signal Squadron to achieve its tasks, new equipment had to be procured in Australia, shipped to Korea

69 CRS A816, 19/323/176, "Reduction of BCFK"; CRS A816, 19/304/551, "Reduction of BCFK"; and CRS A816, 19/323/189, "Summary of Progress Made on Withdrawal of Brit Com Base from Japan."

70 CRS A816, 19/323/176, "Reduction of BCFK"; Warner, *The Vital Link,* pp. 115-116; Nalder, *The Royal Corps of Signals,* p. 486; and *Commonwealth Parliamentary Debates,* 5 Eliz II, Vol H of R 9, p. 96.

71 Commonwealth of Australia, *Parliamentary Papers,* Session 56-57, Vol 3, 'Commonwealth Forces in Korea'; RASCM "A Short History of the Royal Australian Corps of Signals" (written for publication in the RA Sigs Newsletter Vol 10 No 11, November 1981), p. 11; *Signals Bulletin,* Vol 5, NO 2, September 1956; and CRS A816, 19/323/172, "Reduction of Commonwealth Forces in Korea".

and installed. This included 5 Kilowatt CFS transmitters, VHF and channelling equipment, installation stores, aerial masts, generators, cable, telephone poles and wire. The planning, procurement of stores, engineering design, loading and despatch of stores were all completed within 14 days. The installation Troop of 1 Officer and 12 other ranks from 1 Signal Project Squadron arrived at Inchon by early April 1956. They began pouring the concrete slabs for the building and antenna bases in preparation for the arrival of the remaining stores from Australia. CCK Signal Squadron personnel began arriving on 26 April 1956 and were accommodated at 'Commonwealth' Camp.

On 15 May 1956 the Squadron took over the residual Commonwealth communications tasks previously performed by the Divisional Signal Regiment and the Communication Zone Signal Squadron. The unit remained partly based on UK Royal Signals personnel until the shortfall could be rectified. By June the Australian component had risen to four officers and 52 other ranks. The installation was finished three weeks after the unit became operational - great co-operation being an important factor during this interval.[72]

After the radio circuits were established, a system of land lines for both telephones and telegraphs was installed in the CCK-Commonwealth Camp area. This gave some trouble as some of the local Koreans stole lengths of cable. In order to prevent this problem from recurring, 240 volts were fed into the lines when they were not being used. Soon afterwards the problem ceased. Operations by CCK Signal Squadron continued under Major Redgwell until November 1957 when the last link to Singapore was closed down.[73]

The experience of the Australians in the BCOF in Japan had been, in some ways, a frustrating one, particularly because of the constraints of the mandate given by the US occupying powers.

72 *"Signals Bulletin"*, Vol 5, No 2, September 1956, pp. 49-50; "CCK Signal Squadron Inchon", June 1956; and interview with Major General D. Vincent.

73 Barker, unpublished manuscript, p. 416; and *Signals Bulletin*, Vol 6, No 1, December 1957, pp. 19-24.

Nevertheless, Australia's participation in the Force served initially to justify the maintenance of an ARA Field Force and later, to provide readily available forces from all three Services to support the US forces early in the Korean War. From the Signal Corps point of view, the contribution of forces to BCOF helped to justify the maintenance of a Field Force Signal unit and to justify the maintenance of the international links with the British Commonwealth Army Wireless Chain and their eventual upgrade to ensure standardisation with the modernising services of the UK and US networks. This was not just because of the requirements to communicate with Australia, but also because of the communications support tasks for the attached British, New Zealand and Indian (and later, Canadian) troops stationed there and later in Korea. The introduction of Carrier Frequency Shift or Frequency Shift Keying was a consequence of this effort to standardise. This had the effect of improving the speed and accuracy of message handling.

Operations in Korea provided a further catalyst for change. Early on, the Australian infantry battalions were reissued with radios that were compatible with those of their UK and US allies to overcome communications difficulties. US codes were also adopted to further enhance security and intercommunication. Greater reliance on VHF radio communications was a hallmark of the Korean War due to the improvements in VHF radios as well as both the terrain and the nature of the conflict. The greater flexibility inherent in increasingly reliable and portable communications equipment served to add to the combat power available at very short notice. Consequently, artillery support could be called for with unprecedented accuracy and timeliness by even section sized patrols equipped with a radio. This gave commanders greater confidence and more options in the way they could conduct military operations, despite the higher strategic constraints that dictated the way the war was conducted. By the time the CCK Signal Squadron was established in Korea, the level of service provided was markedly superior to that which had been provided by the High Speed Wireless Section for the BCOF when the occupation force first arrived in Japan.

Malaya

Having reduced its forces in Korea after the ceasefire in 1953, the Australian Government was in a better position to deploy more of its forces in other places. One place which was calling for further support was Malaya. This segment looks at why Australia committed elements of the Royal Australian Corps of Signals to the conflict in Malaya. It also considers what they hoped to achieve and the contribution that they made to the way in which the campaign was conducted.

The outbreak of the Malayan Emergency had its origins in the Malayan Communist Party and the Anti-Japanese movement in the 1920s which received direct British military assistance during the Japanese occupation of the country during World War II. Following the restoration of British authority in 1945 - an event lacking the aura of victory and liberation that accompanied allied conquests in Europe - the British authorities had proposed to form a Malayan Union. This proposal met with support from the Chinese but strong opposition from the ethnic Malays, who forced Britain to back down on the matter in 1948 and thus managed to alienate the Chinese element of the population. The end result was that an insurrection was sparked by the Malayan Communist Party in early June 1948 and a State of Emergency was declared on 18 June.[74]

---

74 For a discussion concerning the origins of the Malayan Emergency see P. Edwards with G. Pemberton, *Crises and Commitments,* chapters 1 to 4. See also J. Coates, *Suppressing Insurgency: An Analysis of the Malayan Emergency 1948-1954,* Westview Press, Boulder, 1992, Chapter 1; R. Clutterbuck, *The Long Long War: The Emergency in Malaya 1948-1960,* Cassell, London, 1966; A. Short, *The Communist Insurrection in Malaya 1948-1960,* London, 1975; R. Thompson, 'Emergency in Malaya', in R. Thompson (ed), *War in Peace: An Analysis of Warfare Since 1945,* Orbis, London, 1983, pp. 81-89; Carver, *War Since 1945,* pp. 12-15; Director of Operations, Malaya, *The Conduct of Anti-Terrorist Operations in Malaya,* (3rd edition), 1958, pp. 112-6; and E.D. Smith, *Counter Insurgency Operations: 1- Malaya and Borneo,* Ian Allan, London, 1985, pp. 5-10.

Following the decision in May 1948 to authorise strategic planning with Britain and New Zealand for the defence of the south-west Pacific area, including Malaya, Australia offered to supply small arms and even military equipment but did not consider sending any forces.[75] The May 1948 decision also marked the beginning of ANZAM. The acronym ANZAM referred to Australia, New Zealand and the Malayan area.[76] However, the Labor Government was reluctant to support any substantial effort in Malaya.[77]

By 1950 Robert Menzies had been elected to office with the outlook on military and economic commitments that reflected concerns about the apparent southward advance of communism. Australian involvement in South East Asia was thus seen as helping to halt this advance.[78] So in mid 1950, the Australian Government sent across some Royal Australian Air Force transport aircraft. They also sent a military mission to Malaya in order hopefully to gain access to planning and decision making in the

75 House of Representatives Debates Vol 198, p. 1027, cited in T.B. Millar, "Anglo-Australian Partnership in Defence of the Malaysian Area", in A.F. Madden & W.H. Morris Jones (eds), *Australia and Britain: Studies in a Changing Relationship,* SUP, Sydney, 1980, p. 73.

76 Chin Kin Wah, *The Defence of Malaysia and Singapore: The transformation of a security system 1957-1971,* CUP, Cambridge, 1983, pp. 8-9; and T.B. Millar, *Australia's Defence,* MUP, (2nd edition) 1969, p.57.

77 For a detailed discussion of Australia's military involvement in the Malayan Emergency see Peter Dennis and Jeffrey Grey, *Emergency and Confrontation: Australian Military Operations in Malaya and Borneo 1950-1966,* Allen & Unwin and the Australian War Memorial, 1996. See also P.G. Edwards, *Crises & Commitments,* Chapter 6, and 'The Australian Commitment to the Malayan Emergency', in *Australian Historical Studies,* Vol 22, No 89, October 1987, pp. 604-610 for a discussion of the political machinations behind the decision to commit RAAF and later Army forces to Malaya.

78 See Edwards, *Crises and Commitments,* Chapters 6 & 10. See also P. Firkins, *The Australians in Nine Wars,* Rigby, 1971, p. 425; Chin Kin Wah, *The Defence of Malaysia and Singapore,* CUP, Cambridge, 1983, pp. 10-11; and D. Hawkins, *The Defence of Malaysia and Singapore: From AMDA to ANZUK,* RUSI, London, 1972, p. 17. See also T.B. Millar, *Australia in Peace and War: External relations 1788-1977,* ANU Press, Canberra, 1978, pp. 238-241.

Emergency and to confer with the British military authorities on the method of conducting the campaign.[79]

The Australian Government remained reticent in its committal of forces in Malaya as it believed that political stability and economic development could best be achieved by placing more emphasis on the long term economic approach. One plan that emerged from the Commonwealth Foreign Ministers Conference in Colombo, Ceylon, in January 1950 was the 'Colombo Plan', designed to provide economic co-operation for developing countries to reduce the factors which helped the spread of communism.[80]

By May 1951, military measures were adopted with Australia providing RAAF aircraft and crews to assist in operations in Malaya and in servicing certain RAF aircraft, but an offer of Australian land forces was not forthcoming for another four years, following the withdrawal of the infantry battalions from Korea and the French defeat at Dien Bien Phu in Vietnam.[81] The Australian Government considered that a line now had to be drawn to prevent further communist expansion. Thus an increase in Australia's commitment to the collective defence of South East Asia would be necessary.[82]

In the meantime, the British Administration set up a combined civil, military and police structure (on all levels down to districts) responsible for the setting of priorities and responsibilities both by tasks and areas. By 1955, the situation had stabilised and elections were held in anticipation of the granting of independence

---

79 Edwards, *Crises and Commitments,* p. 98. See also Dennis and Grey, *Emergency and Confrontation,* Chapter 2.

80 'Australia and the Colombo Plan', *Current Notes on International Affairs,* Vol 42, 1971, Department of Foreign Affairs, Canberra, pp. 116-117. See also Edwards & Pemberton, *Crises and Commitments,* pp. 68-71.

81 RAAF No 1 Squadron, flying Lincoln bombers, was operational for eight years from July 1950 to July 1958. The Squadron flew 4,000 sorties from its base at Tengah, Singapore. It delivered 85% of the bomb tonnage dropped during the entire campaign. No 38 Squadron, flying Dakotas, was employed in supplying the British and Gurkha forces with food, ammunition and equipment. See P. Firkins, *The Australians in Nine Wars,* p. 425; Millar, *Australia in Peace and War,* p. 238; and Edwards, 'The Australian Commitment to the Malayan Emergency', p. 611.

82 Edwards with Pemberton, *Crises and Commitments,* pp 147-148.

in 1957. Tunku Abdul Rahman led the United Malay Nationalist Organisation (UMNO) to victory and became the country's first Prime Minister upon the declaration of independence.[83]

Prior to these developments, shortly after the declaration of the 'Emergency', the 1 Australian Observer Unit was raised and sent to Malaya. 1 Australian Observer Unit never functioned as a normal unit as the Australian Government was keen to have elements in Malaya but were not prepared to commit complete combat units for a land battle. It was formed to overcome difficulties that would otherwise arise in obtaining certain benefits and re-establishment rights resulting from service in Malaya. Its organisation was 'completely elastic' and could grow or contract with the arrival and departure of personnel. It had no Commanding Officer, no administrative or quartermaster personnel, no transport of its own; in fact none of the normal things a unit is expected to have.[84]

From 1951 to 1959, 101 Wireless Regiment also maintained a detachment in Singapore and Malaya as part of 1 Australian Observer Unit. The detachment was a Wireless Troop type 'F' and included one officer and 15 other ranks from the Regiment.[85] According to one participant, the detachment 'roamed around the jungle doing tactical intercept of Communist Terrorist communications (of which there were not a great deal)'.[86]

Although the 1 Australian Observer Unit's personnel worked under the British Army, there was still a deal of liaison and administration to be carried out on matters peculiar to Australian conditions of service such as pay and leave arrangements. These administrative duties were performed by Major G. Sadler, followed by Captain Cattanach and then Captain (later

83 Thompson, 'Emergency in Malaya', pp. 83-87; and Clutterbuck, *The Long Long War*, pp. 57-64 & 135-149.

84 OW 84/5 'Australian Observer Unit War Diary', Annex B to Nov '52 report. See also Dennis and Grey, *Emergency and Confrontation*, pp. 54-56.

85 RASCM 101 WS Regt, Monthly Report to D Sigs, October 1953; and Minutes of D Sigs Staff Conference: Held at 'Grosvenor' 6 September 1955.

86 Interview with Lieutenant Colonel K. Whyte, August 1989.

Major General) J.I. (John) Williamson - a technically well qualified Duntroon graduate of 1946. It was convenient for the 101 Wireless Regiment detachment to perform these duties as the detachment commander had the majority of Australian personnel under his direct command. The strength of the Australian Observer Unit in December 1952 was 5 officers and 25 other ranks, of which one officer, Captain Williamson, and 15 other ranks were attached to the British General Headquarters Signal Regiment, which served the UK's General Headquarters, Far Eastern Land Forces (GHQ FARELF) in Singapore (see Map 4). The other members of 1 Australian Observer Unit included an officer serving in Hong Kong, two officers serving at GHQ FARELF, one in the Intelligence Corps and the other in Engineers. Another officer and three non-commissioned officers (all infantry) instructed at the FARELF Training Centre, Kota Tinggi, Johore, (see Map 5) and a few electrical and mechanical engineer corps members were employed with British and Gurkha workshops in Malaya.[87]

The Signals Corps became more formally involved in Malaya for the second time (the first time being with the Australian 8th Division in World War II), when an Australian element was included in the 28th Commonwealth Independent Infantry Brigade at Butterworth on 1 October 1955. The Brigade had been originally disbanded in Korea in 1954 and reformed at Penang. Australia's contribution to the Brigade followed the reduction of forces in Korea and the establishment of the British Commonwealth Far Eastern Strategic Reserve (BCFESR).[88] This contribution initially consisted of 2nd Battalion Royal Australian Regiment, 105th Field Battery, Royal Australian Artillery, 4th Troop (Royal Australian Engineers) 11th Independent Field Squadron (Royal Engineers), and various brigade-level support elements. The Australian infantry battalion brought the number of battalions on operations in Malaya up to a total of 23, in addition to

87 OW 84/5 'Australian Observer Unit War Diary', Annex B to Nov '52 report and Annex B to Dec 52 report.

88 See Horner: Duty First, pp. 95-96.

the armoured, artillery, engineer and other combat support units. The Royal Australian Corps of Signals contribution to the 28th Commonwealth Brigade Signal Squadron was two officers and 24 other ranks.[89] By May 1956, the overall Australian contribution to the FARELF would be 1474 personnel of whom 69 were Royal Australian Corps of Signals personnel.[90]

Initial responses to the arrival of Australian troops in Malaya were mixed. One newspaper editorial in the Malay language daily *Utusan Melayu* wrote saying 'we do not want Australian troops to be stationed in Malaya.' On the other hand, the nationalist leader who was to become a warm friend of Prime Minister Menzies, Tunku Abdul Rahman, spoke more favourably saying 'we shall accept the help with thanks on condition that we do not bear their expenditure.'[91]

The components that arrived at this time were designated as being part of the Australian Army Force, Far Eastern Land Forces (AAF-FARELF) and formed part of the British Commonwealth Far Eastern Strategic Reserve (BCFESR). Australian Signals personnel served also in FARELF Headquarters and, as mentioned earlier, with One Wireless Troop (Type 'F') from 101 Wireless Regiment, which was in the AAF-FARELF, but not in the BCFESR. The primary function of Australian forces in the BCFESR was to provide forces readily available to help defend the SEATO area in 'a SEATO situation' and in so doing, 'provide a deterrent to further Communist aggression in South East Asia.' The second function

89 In November 1957, 2 RAR was relieved by 3 RAR, 105 Fd Battery was relieved by One Battery of 1 Fd Regt and 4 Troop RAE was relieved by 2 Troop RAE. See CRS A2653, 1957, Vol 1, "Relief of Aust Army Force, FARELF". See also RASCM "A Short History of the Royal Australian Corps of Signals", p. 13; RASCM "Post War Developments in R Aust Sigs", p. 7; CRS A2653, 1955, Vol 3, "Composition of the Australian Army Force, FARELF"; CRS A2653, 1956, Vol 2, "Personnel Establishment of AAF FARELF as at 30 May '56"; and Smith, Malaya and Borneo, p. 110.

90 CRS A2653, Item 1956, Vol 2, Personnel Establishment of Aust Army Force, FARELF As At 30 May 56.

91 Items found in file CRS A816, 19/321/35 "Australian Troops in Malaya - Local Press Reaction."

was to form part of the defence of Malaya against communist armed insurrection and external attack.[92]

The 28th Commonwealth Brigade Signal Squadron's establishment was that of a normal British Independent Infantry Brigade Signal Squadron, except that a portion of the strength of 92 was provided by the Royal Australian Corps of Signals. This portion included a Captain, appointed as Second in Command (2IC) of the unit, two sergeants, two corporals and twenty signalmen, most of whom were operators wireless and line (OWL) and linemen-field. Everyone serving in the unit was subject to the same rules and discipline as prescribed by the UK Army Act and Queen's Regulations. Although this was an unusual arrangement for Australian soldiers, the system worked successfully.[93]

From the arrival of the Australian contingent in mid-1955 until June 1957, the Signal Squadron was located astride the main coastal road about one mile north of the Penang Island Ferry terminus at Butterworth, and three miles south of the RAAF Air Base. After June 1957 Headquarters 28 Commonwealth Brigade and its associated supporting elements, including the signal squadron, moved about sixty miles south to Taiping, a small garrison town of approximately 20,000 people. The Brigade Headquarters took over the buildings previously occupied by 2 Federation Brigade (a Malayan Regular Army Brigade). The signal squadron occupied the buildings vacated by its Federation Brigade counterpart. From there the unit provided the Brigade Command net and linked into the Divisional Command net and the international Commonwealth Communications Army Network (COMCAN) link.

Participation in COMCAN was to meet the special needs of the Brigade Headquarters. Originally traffic from Australia and

---

92 Millar, *Australia's Defence,* p. 58; CRS A 816, 19/321/35; CRS A2653, 1955, Vol I, Military Board Minutes - "Composition of the Australia Army Component of the British Commonwealth Far Eastern Strategic Reserve"; Weir, letter, November 1988; and CRS A2653, 1955, Vol 3, "Directive to Commander Australian Army Force, FARELF".

93 Hanley, letter; "Old Sweat in a Slouch Hat" in *Sunday Tiger Standard* (local newspaper in Malaya); and Barker, unpublished manuscript, p. 417.

Headquarters AAF, FARELF in Singapore was passed through General Headquarters (GHQ) FARELF to Headquarters Malaya Command, then to Headquarters 1 Federation Division and so to its destination. Later, to expedite message traffic, three stations were joined on the COMCAN network. These included GHQ FARELF, Headquarters Malaya Command and Headquarters 28 Commonwealth Brigade.[94]

28 Commonwealth Brigade was responsible, among other things, for eliminating Communist Terrorist insurgency within its area of operations in Northern Malaya. In effect, this was the main day-to-day role of the brigade and the battalions. The primary role was ostensibly to act as a strategic reserve.[95] However, it was only every so often that units and sub-units of the brigade were taken off day-to-day operations and given periods of up to a month for 'major war training.'[96] Where the aim of eliminating Communist Terrorists was achieved, areas were designated as 'white'. Areas in which terrorists operated were known as 'black'. Control measures by police, such as spot checks on vehicles and identity cards, were mounted in black areas and in conjunction with battalion operations. These operations were based on intelligence provided by either informers or material taken from captured or killed terrorists.[97]

The lines used by the Squadron were mostly supplied by the Department of Communications, Malaya as it was more economical. Much use was made of Signals Despatch Service (SDS) which operated by train, road and air. Travelling by road was often dangerous and to ascertain how safe they were, it was necessary to check their security classifications. A 'black' road required the minimum combination of two soft skinned or one armoured vehicle and one soft-skinned vehicle or one armoured vehicle for protection, whereas on a 'white' road, vehicles, whether soft-skinned

94 Hanley, letter; and Barker, unpublished manuscript, p. 417.
95 Lieutenant Colonel R.E.P. Cowley, letter.
96 Major General B.H. Hockney, letter, August 1990.
97 Hanley, letter; and Director of Operations, Malaya, *The Conduct of Anti-Terrorist Operations in Malaya,* (3rd edition), 1958, p. XIII. 4-5.

or armoured, could travel singly. In all cases an armed man(Shotgun) was carried in addition to the driver.[98]

In general, the role of the Brigade dictated an almost static life for the Signal Squadron. The Squadron possessed a number of radio vehicles for use if they were required in a mobile role but only on one occasion, in mid 1958, was it found necessary for these vehicles to be deployed. The radio vehicles (Austin vans), or 'Gin Palaces' as they were known in the British Army, provided communications for an advanced Brigade Command Post and various supporting elements.[99]

In 1957, the decision was made, by the British, Australian and New Zealand Governments, to house the three battalions of 28 Commonwealth Brigade in a large new purpose built camp at Bukit Terendak, 12 miles from Malacca. Approximately 30 line laying personnel from the Squadron, combined with Gurkha Signallers, were employed to carry out the installation of both underground and poled telephone routes in the camp area. This vast camp (which could eventually house 4000 troops and which was largely self contained with amenities) was not completed until 1964.[100]

The Brigade Command net included links from Butterworth (and from Taiping after June 1957) to:

- the Brigade Tactical Headquarters at Sungei Siput and, after 1958, at Ipoh;
- the battalion of the Royal Australian Regiment with its headquarters at Kuala Kangsar and some smaller elements at Penang;
- the 2/6 Gurkha Regiment at Ipoh;

98 Barker, unpublished manuscript, p. 417; and Cowley, letter, July 1991.

99 Ibid.

100 *Sunday Tiger Standard,* 28 December 1958; Hawkins, *The Defence of Malaysia and Singapore,* p 17; and Hanley, letter. The Brigade Force remained in Malaya and then Malaysia until 1970 when the Brigade was disbanded. The tasks and size of the Brigade Signal Squadron remained essentially unchanged during this period.

- 1 Royal Lincolnshire Regiment (later replaced by 1 Royal New Zealand Regiment) at Taiping; and
- two other British battalions at Ipoh.[101]

Where possible, Squadron personnel at each of the unit locations were provided on the basis of one Australian Signals to one UK Signals operator. This system of integrated staff worked well. Staff were detached for a period of three months and had to be 'dragged back' to the joys of a regular squadron existence. They were often loath to leave and on a number of occasions one of the two went out on operations with battalion patrols. The Battalion Signals Platoon would provide a CW (morse) operator to stand in during the absence of the RA Signals operators as often as possible. Both the Infantry and Signals personnel involved gained in experience from the arrangement.[102] It was always apparent, however, from the 'rough' procedures and slower CW speeds, when these unofficial change overs had taken place. The effect on the efficiency of the net was only marginal and commanders were prepared to turn a blind eye to these activities.[103]

Regimental infantry signallers were required to have a minimum morse speed of eight words per minute and the communications from all platoons back to company or battalion headquarters were usually by morse. They were not as fast or as fluent as Australian or British Signal Corps operators but they were undoubtedly capable operators and took pride in getting through in difficult conditions and in achieving proper morse standards.[104] Each of the Brigade Signal Squadron detachments were operationally self contained, complete with battery charging facilities, spare radios, etc. The battalion to which they were attached provided accommodation and rations.[105]

101 Cowley, letter; and Hanley, letter.
102 Hanley, letter ; and interview with Hockney, August 1989.
103 Hanley, letter.
104 Cowley, letter; and Interview with Hockney.
105 Hanley, letter ; and Interview with Hockney, August 1989.

As indicated earlier, the battalion communications were not provided solely by infantry signallers. In fact, the then Colonel D. (Tim) Vincent, the Director of Signals from 1954 to 1958 and later Major General, had a standing arrangement with the Director of Infantry, which dated back to the time when Brigadier A.D. Molloy was the Director of Signals from 1947 to 1952. The arrangement allowed for two Signal Corps officers to be attached to the overseas infantry battalions. Consequently, as had been the case in the Korean War, Signal Corps officers served with the battalions in Malaya from 1955 onwards.[106] This was important for the Corps of Signals because in the Malayan Emergency, it was the battalion communications which were critical for co-ordinating and controlling the operations of platoons dispersed over an area of over 600 square miles. As a consequence, there were about as many Royal Australian Signal Corps officers in the infantry battalion as there were in the 28 Commonwealth Brigade Signal Squadron.[107]

The Regular Army infantry battalions of the Royal Australian Regiment (RAR) were rotated through Malaya on two year tours, commencing with 2 RAR in October 1955, followed by 3 RAR in October 1957, and 1 RAR in October 1959. In the following years (up to 1969) 4 RAR and 8 RAR also served there, and 2 RAR and 3 RAR returned for their second tours. 1RAR and 6 RAR subsequently served in Singapore from 1969 through to 1973.[108]

Whilst on operations in Malaya, the Australian infantry battalions were organised into three rifle companies. The support company also acted as a rifle company and one rifle company was organised into a training company with reduced manning.

---

106 Vincent, interview 1988.

107 Cowley, letter, September 1989.

108 'Royal Australian Regiment: Standing Orders', pp. 26,36-37 and Annex A to Chapter 2; and 'The History of 1 RAR: From Morotai to Townsville', p. 7. For an account of the RAR in Malaya see D.M. Horner (Ed), *Duty First, the Royal Australian Regiment in War & Peace,* Allen & Unwin, Sydney, 1990, Chapters 5 & 7. See also Dennis and Grey, *Emergency and Confrontation,* Chapters 4 to 8.

The unit had armoured cars and armoured 3 ton vehicles allotted to the companies.[109]

Battalion headquarters was at Kuala Kangsar in the central-north-west of Malaya, located in a former sultan's palace.[110] The palace was, 'a flighty, airy, building with cream lattice work and fluted columns'. The various companies were dispersed. 'A' Company was at Lasah, an exotic place perched on high ground and surrounded on three sides by thick jungle, with the remaining side facing a broad river - the Sungei Plus. The camp site itself consisted of a jumble of tin huts and damp stained tents, along with water purification equipment and a diesel powered cookhouse.[111]

From these bases, fighting patrols would go out with the aim of destroying CTs or their food supplies. They would gather information which was combined with information provided by the Malayan police network. Plans were then formulated for ambush operations by soldiers in company or battalion strength. The ambushes were timed to coincide with the CTs' exit from the jungle to replenish stores. At times 'Junior Civil Liaison Officers' (interpreters) and 'SEPs' ('surrendered enemy personnel' who had been turned to support the government) would accompany patrols.[112] A British Signals Officer, J.C. Clinch, recalled that

> having been captured by us [they] then led our patrols back into the jungle where their comrades, who had been operating with them for eight or nine years, were still living. They would give us details of where their comrades were coming in to collect food, make proposals on how the ambush should be set up and

109 Captain R.E.P. Cowley, 'Talk on Anti-Terrorist Operations in Malaya, 17 Oct '62'. For an account of 3 RAR's experience from 1957 to 1959, see Colin Bannister, *An Inch of Bravery: 3 RAR in the Malayan Emergency, 1957-59*, Directorate of Army Public Affairs, Canberra, 1994.

110 Hockney, letter, August 1990; and Cowley, letter, July 1991.

111 J. Mossman, 'Australian Troops in Malaya', in *Sydney Morning Herald*, January 1958.

112 Hockney, interview, August 1989; and Cowley, interview, August 1989.

> wait for the 'comrades to walk into our trap. It was not therefore surprising that we had a high element of success in these mopping-up operations. [These men] were persuaded that self interest was best served by helping the Malay Government and that their former leaders in the jungle were now the main threat to them.[113]

The patrols were physically arduous and the terrain extremely difficult. The strain of continually laying all night ambushes with very little result required a high standard of soldiering. These patrols were usually sent out from a platoon base in the jungle in section strength (10 men) or less and once satisfied that the area was clear, the commander of the operation would relocate the patrols and repeat the performance somewhere else. Navigation was a critical matter and the ability to read the ground was paramount. There were times when patrols were not quite lost but at times did have considerable doubt as to exactly where they were.[114]

Watching the health of the men on patrol was a major problem - one soldier died of leptospyrosis (from drinking water contaminated with rats urine). Casualty evacuations for people who had allergic reactions to bites of various sorts caused problems. Casualty evacuation was usually by means of stretcher carried by other members of the patrol to a vehicle pick-up point, although in the case of very seriously injured people, helicopters were available and could be called for on the patrol's radio set.[115]

The usual method of communications for these patrols was with the WS A510 HF set, discussed in the Appendix, and the normal method of operation was with morse code. Once operators became familiar with the set it was extremely reliable. In the meantime, the British Army was using a much more cumbersome wireless, the WS 68, which had a more limited frequency range

113 Cited in Warner, *The Vital Link*, p. 81.
114 Cowley, interview, August 1989.
115 Ibid.

and used larger batteries. The frequency range for the A510 was 2 -10 MHz on four crystals, using dry batteries, of which there was one each for the transmitter and the receiver components. It was used with an 'end fed' antenna - that is, a single wire with one end connected to the set. Operators would try to get it up in the trees at a 45 degree angle away from the station being transmitted to. The hours of operation were from about 7 am to 4pm, after which time the changed ionospheric conditions made it virtually impossible to communicate with the WS A510.[116]

Occasionally HF relay stations were required for distant patrols out of radio range from base headquarters. These were usually located on high features equi-distant for all patrols working in a specific area of operations.[117] In instances where patrols had not made contact with their headquarters in 48 hours, contact would be made through a contact reconnaissance aircraft sent to fly over the area. Contact could be made over the HF wireless set carried in the light aircraft but a signalman's pride made him feel that it wasn't quite right to have to use an aeroplane.[118]

The battalion's signal platoon of 3RAR was commanded, from 1957 to 1958, by the then Captain B.H. (Barry) Hockney, a physically small, but well built, Duntroon graduate of the Class of 1953. Besides being an Australian Rules footballer of note (1951 premiers) he was one of a batch of the 'new breed' of capable, intelligent and technically trained, officers fostered by the Corps to provide the expertise it needed to cope with the technological era that it faced. His job, as Signal Platoon Commander and Regimental Signals Officer (RSO), was to provide radio detachments for patrols and to provide telephone and switchboard facilities at each of the company bases. He was also to provide a

116 Cowley, interview, August 1989; and Hockney, interview, August 1989. See also Australian Military Forces, *User Handbook: A510 Wireless Station*, 1956.

117 Cowley, interview, August 1989.

118 Hockney, interview, August 1989.

signal centre facility at both the battalion headquarters and 'command post' which were located separately.[119]

Other responsibilities of the battalion RSO included the issuing of the relevant 'griddle' and 'slidex' codes to the rifle companies. Griddle was a simple code used for map grid references and slidex was a word replacement code. To use slidex, a message would be made up around the words on the code sheet and the operator would send the axis on the sheet. Thus a bi-gram (two letters) would identify a phrase or word that was standard on the sheet.[120]

Each battalion had a discreet area of operations and consequently, co-operation with the other battalions in the brigade, other than on boundaries, did not happen very often. However, when an operation required more than a single battalion, General Hockney recalled that:

> we had no problem as we co-operated mostly with the New Zealand battalion and the relationship was first class. This was brought about because the New Zealand and Australian battalion commanding officers, (Lieutenant Colonels Kim Morrison and John White respectively), were old friends who had served as instructors together at Duntroon. It was great fun to go to brigade conferences and see the Australian and New Zealand commanding officers really having a shot at each other with much fun and the UK commanding officers not understanding the humour.
>
> On the communications side, the New Zealand battalion signals officer, Captain Dan Rodda (a Royal New Zealand Signal Corps officer),

119 Ibid.
120 Hockney, interview, August 1989; and Cowley, interview, August 1989.

> and I were great friends having gone through Duntroon together.[121]

In 1957, as these events were occurring, Malaya was granted independence from Britain and as part of the continuing defence arrangements between Britain and Malaya, the Anglo Malayan Defence Agreement (AMDA) was signed. This covered the defence of Malaya and British territories in the Far East against external attack. Instead of negotiating a separate treaty, Australia and New Zealand were linked in by means of an exchange of letters. As a part of this agreement, Britain, Australia and New Zealand agreed to maintain a military presence in Malaya for the then foreseeable future, even after the Emergency was to end. Australian defence thinking was dominated by the 'domino theory' and the forces in Malaya were seen as being part of the forward defence of Australia itself.[122]

On 31 July 1960 the Malayan Emergency was brought to an end, after the expenditure of approximately £A211,000,000 by the Malayan Government. Nearly 11,000 terrorists were either killed, captured or surrendered for the loss of 1865 Security personnel killed and 2560 wounded, and 2400 civilians killed and over 2000 injured.[123]

In 1960 28 Commonwealth Brigade Signal Squadron became tri-national, with a New Zealand officer joining the unit. In 1961, the brigade was involved with the signal squadron in exercises in Thailand and the unit title changed to 208 Commonwealth (COMWEL) Signal Squadron. In December 1961, the brigade moved into the specially built barracks at Terendak camp near Malacca.[124]

---

121 Ibid.

122 David Hawkins, The Defence of *Malaysia and Singapore: from AMDA to ANZUK*, RUSI, London, 1972, pp. 16-17.

123 'Federation of Malaya: End of Emergency', in *Current Notes on International Affairs*, Vol 31, 1960, Department of External Affairs, Canberra, pp. 447-8.

124 Major S.R. Hinton, letter, September 1975.

Australia's deployment on operations in Malaya contributed to the maintenance of military skills for the infantry battalions that were later to serve on more intense operations in Borneo and Vietnam. The deployments also served to continue justifying the maintenance of the Australian Army's strategic communications network known as AUSTCAN, particularly with its links into the British Commonwealth-wide network known as COMCAN through the 28 Brigade Signal Squadron. The initial deployment of a detachment from 101 Wireless Regiment as part of the Australian Observer Unit helped in maintaining the skills of signals intelligence operators which were to prove so significant in later operations in Borneo and Vietnam. However, they made little contribution to the actual outcome of the campaign in Malaya from 1948 to 1960. The experience of the other Signals contributions was in contrast to that of the 101 Wireless Regiment detachment.

The introduction of manportable and reliable HF radios, particularly with equipment such as the Australian designed A510 radio, made a significant difference to the conduct of operations for the infantry patrols tasked with finding the CTs. In World War II, the distrust of wireless communications and reliance instead on vulnerable line communications by Australian and allied commanders had contributed to the difficulties experienced in fighting against the advancing Japanese.[125] By the mid 1950s, the benefits of HF radio communications were becoming well established, particularly for use in hilly and jungle terrain (where VHF communications were less effective because of the limited range that could be achieved). These benefits, combined with significant increases in reliability and portability, proved to be critical determinants of the manner in which military operations could be conducted against the CT forces in Malaya. Patrols would deploy confidently away from their bases in difficult and remote terrain carrying their portable HF morse communications equipment that could be used reliably to report in

125 See Barker, *Signals*, p. 240.

or call for assistance when required. Consequently, operations were planned around widely dispersed patrols operating in the remotest localities, thus making it impossible for the CT forces to operate with impunity.

Australia's contributions to operations in Japan, Korea and Malaya were relatively small ones compared to those of their allies alongside which they served and to the Australian forces deployed during the world wars. Given the difficulties experienced by Australian commanders in retaining and effectively exercising national control of much larger forces during World War II, there was a conviction that Australian commanders must be able to retain control of these smaller dispersed forces. Consequently, the imperative for effective and reliable communications both within the area of operations and back to Australia was great. The Royal Australian Corps of Signals provided that link effectively.

CHAPTER THREE

OPERATIONAL DEPLOYMENTS 1962 TO 1972

Borneo and Confrontation

THE DISPERSED AND low-intensity nature of military operations during the Malayan Emergency had prompted certain innovations in the provision of military communications. However, unlike in Malaya, where operations were fought with essentially conventional forces within sovereign territory, in Borneo, there was a marked reliance on covert cross-border operations. This chapter explains why there was a requirement for the Royal Australian Corps of Signals to be involved in operations in Borneo and what difference these changed circumstances made for the provision of military communications support. This chapter then looks at the deployments to Singapore and Malaysia as well as Indonesia.

As calm descended after the Malayan Emergency, the proposal was made in 1962 for the formation of a Federation of Malaysia in 1963, which would include Sabah, Sarawak, Brunei and Singapore (see Map 6).[1] The Indonesian President, Sukarno,

1 For a detailed discussion on the origins of Australia's involvement in Confrontation see Edwards with Pemberton, *Crises and Commitments,* Chapters 14 to 17. See also Dennis and Grey, *Emergency and Confrontation,* Chapter 10; and Carver, *War Since 1945,* p. 83.

initially appeared to accept the idea but later took the view that the proposal was an example of neo-colonialism.[2]

Meanwhile, the sensitive and, at times, difficult nature of Australia's relations with Indonesia was being reflected in the issue of the fate of Dutch-administered West New Guinea (Irian Jaya). In August 1962, Indonesia had managed to convince the United Nations (despite Australia's objections) to place West New Guinea under UN temporary authority, with the eventual transfer of authority to Indonesia to take place in May 1963.[3] The Australian Government tried vainly to prevent this from happening, and its response to the consequent diplomatic isolation was, among other things, to increase its own defence spending.[4]

On 8 December 1962, a locally inspired revolt broke out in Brunei, but this was quickly suppressed by the Royal Marine Commandos. Subsequently, Indonesia conducted several raids into Sabah and Sarawak by 'volunteers' seeking to stir up opposition to the formation of Malaysia, which took place in September 1963.[5] Australia was reluctant to support Britain and Malaysia in fighting the Indonesians for a number of reasons.[6]

---

2 D.M. Horner, 'The Australian Army and Indonesia's Confrontation with Malaysia', in *Australian Outlook,* April 1989; Hawkins, *The Defence of Malaysia and Singapore,* p. 22; and Carver, *War Since 1945,* p. 83.

3 For a discussion of the Dutch New Guinea issue see Pemberton's *All The Way,* pp. 70-106.

4 T.R. Reese, *Australia, New Zealand, and the United States: A Survey of International Relations 1941-1968, OUP, London, 1969,* pp. 216-7.

5 Carver, *War Since 1945,* pp. 86-88; and Smith, *Malaya and Borneo,* pp. 50-58.

6 Horner, 'The Australian Army and Indonesia's Confrontation with Malaysia', pp. 3-6 & 12. The reasons were: firstly, Australia was not tied to North Borneo in the way Britain was; secondly, only a limited military capability was available (Australia was acutely aware of its limitations, following the dispute with Indonesia over West New Guinea); thirdly, troops sent to Borneo would not be available for SEATO tasks if they arose; fourthly, Australia was also keen to maintain good relations with Indonesia now that it had a common land border in Papua New Guinea (then an Australian administered territory), and fifthly, such a commitment would mean that Australia faced the possibility of having to commit troops to Borneo, New Guinea and Vietnam simultaneously. For a discussion on the governmental discussions concerning this see Edwards with Pemberton, *Crises and Commitments,* pp. 285-292.

Prime Minister Menzies eventually supported the formation of Malaysia in September 1963 and served notice on where Australia stood, by stating '...we shall to the best of our powers... add our military assistance to the efforts of Malaysia and the United Kingdom in the defence of Malaysia's territorial integrity and political independence.'[7]

Through late 1963 and into 1964, Indonesian armed parties continued to cross into the newly created states of West Malaysia, incorporating Sabah and Sarawak, but not Brunei (Brunei had opted out of the Federation because of its wealth from oil).[8] In April 1964, 7 Field Squadron of the Australian Army Engineers was sent to Borneo for road and bridge construction. The Government was not yet ready to commit infantry battalions.[9]

In August 1964, approval was given for 693 Signal Troop to be deployed to perform 'communications support' in place of a UK Signal Troop.[10] 693 Signal Troop was deployed to the island of Labuan, off the coast of Sabah, where, in 1945, the Commander of the 9th Australian Division had received the Japanese surrender in North Borneo and Sarawak. The troop was commanded by Lieutenant T.J. (Trevor) Richards (a later Commanding Officer of 7 Signal Regiment and a brigadier) and was staffed by Operators Signals, communications centre personnel and a technician - all members of 201 Signal Squadron (which later became 121 Signal Squadron) from Singapore, on detachment for set periods.[11] The troop provided valuable communications support and made a significant contribution to the overall operations. The wireless station was prefabricated in a container at

7 Cited in T.B. Millar, 'Anglo-Australian Partnership in Defence of the Malaysian area', in A.F. Madden & W.H. Morris-Jones, (eds), *Australia and Britain: Studies in a Changing Relationship*, SUP, Sydney, 1980, p. 83. See also Hawkins, *The Defence of Malaysia and Singapore*, p. 24.

8 Smith, *Malaya and Borneo*, pp. 59-76.

9 Edwards and Pemberton, *Crises and Commitments*, pp. 285-292.

10 Millar, Anglo-Australian Partnership in Defence of the Malaysian area', p. 84; Horner, 'The Australian Army and Indonesia's Confrontation with Malaysia', p. 11; and interview with Staff Sergeant J.E. Danskin October 1989.

11 J.E. Danskin, letter, September 1989.

7 Signal Regiment, Cabarlah, Queensland and lifted into the prescribed site very rapidly by a Chinook helicopter. According to Lieutenant Colonel Ken Whyte, the deployment was 'a masterpiece of organisation'.[12]

At the same time as 693 Signal Troop was getting established, according to the then Corporal J.E. (Jim) Danskin, 651 Signal Troop (Royal Signals) was deployed to Kuching, staffed by personnel on special duty from 13 Signal Regiment (Royal Signals) in Germany. The two sub-units then initiated an exchange posting for selected personnel so that it was possible for both Australian and UK Signals members to serve in Singapore, Labuan, and Kuching. Both deployments were operationally successful. Following the cessation of confrontation, the personnel of 693 Signal troop were absorbed back into 201 Signal Squadron.[13] By the time the unit left, it was acknowledged that they had made a significant contribution to the overall effort.

In essence, signals intelligence functioned at several levels: it included a substantial radio direction-finding effort; simple tapping of telephone lines on the Indonesian side of the border; and intercepted radio traffic. All information from signals intelligence was kept highly circumscribed and, despite concerns about information sometimes being stale, the Commonwealth forces were able to gather high-grade intelligence that significantly contributed to the outcome of the conflict.[14]

In October 1964, 7 Field (Engineer) Squadron was replaced by 1 Field Squadron with a Signals detachment attached.[15] The detachment underwent five weeks of familiarisation training in Sydney before departing for Jesseltown (now Kota Kinabalu)

---

12 Interview with Lieutenant Colonel K. Whyte, August 1989.

13 Danskin, letter, September 1989; and interview with Ron Ratchford, August 1993.

14 Dennis and Grey, *Emergency and Confrontation*, pp. 248-249.

15 Millar, Anglo-Australian Partnership in Defence of the Malaysian Area', p. 84; Horner, 'The Australian Army and Indonesia's Confrontation with Malaysia', p. 11; Letter from Warrant Officer Class Two A.W. Lane; and RASCM Military Board, Designation of Signal Units, 22 Dec 1964, Brigadier Roseblade, Director Staff Duties.

in Sabah. From Jesseltown, the troop travelled by train to the railhead at Tenom and then by truck to Keningau, the original base camp of 7 Field Squadron. The then Signalman Bert Lane recalled the following:

> Our communications role was twofold. We operated a rear link for administrative, control and supply purposes to Keningau on a two-hourly schedule. We were also a sub-station on a 16-sub-station CW (Morse) [and] voice network with a control station in Brunei. This net was manned entirely by Ghurkhas, with ourselves being the only exception. My CW capabilities rapidly improved, although the Ghurkhas did send Morse with an accent laced liberally with strange idiosyncrasies.

As the road building progressed, the forward camp would advance to keep up with the roadhead and, for ease of administration, this would occur every four to five miles. Some local Dyaks would be employed around the camp, and at each relocation of the camp, their help would be enlisted to cut some large jungle bamboos to support the half-wave dipole antennas. The radios on both nets comprised British C11/R210 (medium-powered transmitter/receiver) combinations, and for encoding purposes, the BID 50 encryption device was used. The C11/R210s did not like the tropical conditions and were constantly breaking down.[16]

By late 1964, Major General Walter Walker, the Commander of British and Commonwealth forces stationed in Borneo, had 18 British and Malaysian battalions under his command with supporting regiments of armoured cars, artillery, engineers, signals and administrative support.[17]

---

16 Lane, letter; and *The Royal Australian Corps of Signals Reference Manual*, p. 3.40.

17 Carver, *War Since 1945*, p. 94; and Hawkins, *The Defence of Malaysia and Singapore*, p. 26.

Australia's attitude towards Indonesia had also hardened by this stage. In October 1964, Indonesia forbade the overflying of Indonesian territory by Australian aircraft. In November, Australia announced the introduction of compulsory military service, and this was partly due to concerns over the seriousness of the Indonesian threat.[18]

In late 1964, following a number of small-scale Indonesian attacks on West Malaysia, General Walker was granted permission by the Malaysian and British Governments to conduct Operation 'Claret'. This was part of a new strategy to stop Indonesian forces from being able to cause havoc in Malaysian territory by forcing them onto the defensive. It involved deploying troops on operations up to 5000 and later up to 20,000 yards across the Indonesian border. The ability to cross the border in 'hot pursuit', combined with the judicious use of helicopters to rapidly position blocking forces, made deep penetrations into Malaysian territory by Indonesian troops more harrowing and less likely to succeed.[19] Claret operations were based on information gathered from wireless intercept and interrogations, and were surrounded by the strictest secrecy. They proved to be highly effective in wresting the initiative from the hands of the Indonesians and in rapidly turning the tide very clearly in favour of the British and Commonwealth forces.[20]

On 3 February 1965, further assistance consisting of the Australian infantry battalion in the Commonwealth Strategic Reserve in West Malaysia (3 RAR), and a squadron of the Australian SAS was announced.[21] The 100-man, 1 SAS Squadron, commanded by Major Alf Garland (later promoted Brigadier and sub-

18 Hawkins, *The Defence of Malaysia and Singapore,* p. 24.
19 Raffi Gregorian, 'CLARET Operations and Confrontation, 1964-1966' in *Conflict Quarterly,* Vol XI, No 1, Winter 1991, pp. 46-53; and Carver, *War Since 1945,* p. 95.
20 Carver, War Since 1945, p. 95; and Smith, *Malaya and Borneo,* pp. 73 & 77-80.
21 'Malaysia: Further Australian Military Assistance', in *Current Notes,* Vol 36, 1965, p. 98; and Hawkins, *The Defence of Malaysia and Singapore,* pp. 23-24.

sequently elected as National President of the RSL), deployed to Borneo on 6 February 1965. This was the first operational deployment for the SAS. They trained with the British SAS until the end of March, familiarising themselves with the latest techniques in use by the British.[22]

A Signal Corps detachment from 126 Signal Squadron accompanied 1 SAS Squadron to Borneo, and its radio operators participated in the cross-border operations which commenced in May 1965. Lance Corporal P.H. (Paul) Denehey, an Australian Corps of Signals radio operator, was killed on one such operation on 3 June 1965.[23] Denehey died from wounds sustained when he was struck in the chest by a rogue elephant's tusk while on operations across the border, the details of which are described in David Horner's *SAS: Phantoms of the Jungle.*

Communications links with the patrols were maintained using British HF radios for off-line encoded Morse communications. These had to be kept dry and were more delicate than the Australian models they had trained with in Australia. Clearly, radio communications were crucial for the success of such sensitive deployments. Patrols were required to send daily encoded reports at pre-arranged times. Patrols were also equipped with search and rescue beacon (Sarbe) radios, which could be activated in an emergency for contact with overflying aircraft. Alert codewords were also maintained, which, if used, meant that the patrol was being chased and was headed for the border or exfiltration rendezvous point. In such instances, a helicopter would be tasked to extract the patrol, homing in on the Sarbe beacon.[24] The radios were heavy, required specialist skills for morse

---

22 For an account of the involvement of the Australian SAS in Borneo see D.M. Horner, *SAS: Phantoms of the Jungle,* Allen & Unwin, Sydney, 1989, Chapters 4 to 10; and Horner, 'The Australian Army and Indonesia's Confrontation with Malaysia', p. 21.

23 *Signals Bulletin,* Vol XIV, No 1, July 1965, p. 3.04; Ratchford, interview, August 1993; RASCM Larner, 'Unit History 126 Signal Squadron'; and Horner, 'The Australian Army and Indonesia's Confrontation with Malaysia', p. 21.

24 Horner, SAS: *Phantoms of the Jungle,* pp. 94, 103 & 111.

operations and encoding of messages, and added to an already substantial load for the SAS patrol signaller; however, they significantly affected the manner in which operations could be conducted by increasing the supportable range of deployments, enabling patrol sizes to be kept to a minimum, and providing them with access to emergency back up if and when required. They thus added to the capability of the unit.

Sergeant Ted Blacker received the British Empire Medal (BEM) for the work done in ensuring effective and smooth communications for up to twenty patrols. On many occasions, when communications from the field were difficult, he went forward to provide the link. On every occasion, the difficulties were quickly and efficiently remedied despite the often dangerous location of the forward communication posts. No communication problem at base or in the field seemed beyond him.[25]

3 RAR, commanded by Lieutenant Colonel B.A. (Bruce) McDonald (later Major General), arrived in Sarawak in the latter half of March 1965, accompanied by a troop of 2 Field Squadron, Royal Australian Engineers, and 589 Signal Troop Royal Australian Corps of Signals.[26] 589 Signal troop formed part of the tri-national 208 Commonwealth Signal Squadron based in Malaysia (referred to earlier). For ease of control, it was permanently attached to the Australian battalion stationed at Terendak Garrison, Malacca, Malaysia. 3 RAR's Battalion Headquarters was situated at Cambrai Camp near the small hamlet of Bau, thirty miles from Kuching, the capital of Sarawak, where Captain Lionel Matthews was executed by the Japanese during World War II. Matthews had been a gallant Australian Corps of Signals officer who, whilst in captivity during World War II, personally directed an underground intelligence organisation and arranged a radio link with the outside world.[27]

25 Citation for British Empire Medal cited in RASCM, Larner's, 'Unit History 126 Signal Squadron'.

26 For an account of Australian infantry operations in Borneo see Horner, Duty First, Chapter 7.

27 Barker, Signals, p. 246.

Sergeant Stan Perijmibida (later Warrant Officer Class One and Regimental Sergeant Major), affectionately known as 'pair of pyjamas', commanded the Signal troop which provided communications from battalion headquarters to West Brigade Headquarters. One of the Gurkha signal squadrons provided communications for West Brigade Headquarters at Kuching. 589 Signal Troop's other responsibilities included battery charging and minor repair of radio equipment within the battalion.[28] 3 RAR withdrew from Sarawak in July 1965 and was not immediately replaced by an Australian battalion. However, 4 RAR was dispatched to Sarawak from April 1966 until August 1966, and 589 Signal Troop accompanied the battalion for this tour.[29]

Peace came on 11 August 1966, ten months after the October 1965 coup in Jakarta, which culminated in Suharto taking power from Sukarno on 12 March 1966. The victory was first and foremost a military one, as logistically, Britain had a large base conveniently placed and readily accessible to Borneo. Moreover, Indonesia faced problems in deploying its numerically superior forces and these problems were added to by the political ferment in Jakarta.[30]

Claret operations had played a critical role in ensuring this military victory; for through them, Britain, Malaysia and Australia were able to avoid any unnecessary escalation of Confrontation. The very success of these operations could have encouraged further belligerence from Indonesia, but the fact that they were not showcased to the world encouraged de-escalation, as Indonesian prestige was not directly threatened. Claret operations were intended to be deniable with covers for operations suggesting that they were actually being conducted in Malaysian territory. For the Indonesian Generals, this proved handy. They

28 Letter from Warrant Officer Class One S. Perejmibida, November 1985; Horner, 'The Australian Army and Indonesia's Confrontation with Malaysia', pp. 26-8; and Carver, *War Since 1945*, p. 96.

29 'Royal Australian Regiment Standing Orders', Annex A to Chapter 2 ; and Perejmibida, letter, November 1985.

30 Carver, *War Since 1945*, p. 96; and Smith, *Malaya and Borneo*, pp. 99-101.

were keen not to disturb this fabrication as it made them out to be more successful than they really were. Thus, politically, the secrecy of Claret operations allowed the Indonesians to make a face-saving backdown once military aggression proved ineffective. Claret operations were successful both tactically and strategically, demonstrating a clear example of 'war as a continuation of policy by other means'.[31]

Perhaps ultimately, these operations were as successful as they were because of the information gathered from 'broad range intelligence' (which was surrounded with the strictest secrecy), combined with the effective application of that information by discrete patrols. The Director of Operations in Borneo was noted to have praised the signal intelligence effort as a significant force multiplier, which had played a critical role.[32] The patrols proved to be highly effective in wresting the initiative from the hands of the Indonesians and in rapidly turning the tide very clearly in favour of the British and Commonwealth forces.[33] These vital patrols were also made possible (at least in part) by more robust, portable and reliable communications technology. The combination of the HF manpack radio with the Sarbe radio proved critical in determining the kind of operations that the SAS patrols could conduct and thus contributed to the outcome of Confrontation with Indonesia.

The lessons were well learnt in Australia as the merits of signals intelligence had been demonstrated. Moreover, the experience of patrols on Claret operations served to further hone the skills and techniques of the SAS patrols as well as to refine their communications requirements. It was not surprising, therefore, that with the experiences in Borneo fresh in people's minds, Australia's deployment of a Task Force to South Vietnam in 1966 included a signals intelligence troop, 547 Signal Troop, and

---

31 Gregorian, 'CLARET Operations and Confrontation 1964-1966', pp. 63-65.

32 Ratchford, interview, August 1993.

33 Carver, *War Since 1945,* p. 95; and Smith, *Malaya and Borneo,* pp. 73 & 77-80.

a detachment of 152 Signal Squadron with the SAS Squadron. Meanwhile, the benefits of signals intelligence continued to be appreciated elsewhere, such as in Singapore.

Singapore and Malaysia

With the end of the Malayan Emergency, the military communications support requirements for all British Commonwealth forces stationed in Malaya had diminished. Confrontation with Indonesia loomed large for a while, but Australian forces remained there both before the onset of Confrontation and for several years after it. This section looks at why the Royal Australian Corps of Signals continued operating in Malaysia and Singapore until the mid-1970s and what they did there.

In Singapore, between 1959 and the 'Confrontation' in 1963, the strength of 101 Wireless Regiment detachment had been built up to squadron level. In 1964, the Regiment had been redesignated as 7 Signal Regiment, and the Squadron in Malaysia had been designated as 121 Signal Squadron. As indicated earlier, elements of this squadron also served in Borneo from 1964 to 1966. Females first served in 7 Signal Regiment in 1964, and from 1967, the first WRAAC detachment ever to serve outside Australia was integrated into 121 Signal Squadron in Singapore.[34] 208 Commonwealth Signal Squadron, which had been formed in 1961 as part of 28 Commonwealth Brigade, remained at Terendak Army camp, near Malacca, throughout the 1960s (see Map 5).

In April 1966, the signal squadron integrated with the brigade headquarters to become 28 Brigade Headquarters and Signal Squadron. This was in accordance with the organisational changes implemented in the British Army. Shortly thereafter, on 11 August 1966, the Confrontation with Indonesia ended officially, and it was

34 RASCM 7th Signal Regiment Unit History', prepared for DCOMMS by Captain S.W. Foley, June 1986.

not long before Australian personnel with experience from South Vietnam were being posted into the Squadron.[35]

According to Major Dick Twiss (later Lieutenant Colonel), who was the last commander of the Squadron from December 1968 to December 1969. This brigade signal squadron organisation was significantly different from the Australian one.

> As squadron commander, I commanded the whole brigade headquarters and therefore was responsible for conducting the reconnaissance and siting for the whole brigade headquarters [when it relocated on exercises]. As a major, I was at every conference run by the brigadier and, in effect, was treated as a lieutenant colonel - as if I was one rank level above the brigade general staff (apparently this was quite normal in the UK Army).[36]

The deployed squadron was approximately 350 all ranks, including a defence and employment platoon provided by the British Pioneer Corps, a transport platoon, a quartermaster troop and the Signal Corps component, which included a radio relay troop, a radio troop, a communications centre troop and headquarters troop. The brigade was there as a strategic deployment force in support of SEATO. It was trained to operate as part of a deploying force from the UK and as part of a UK infantry division. It was also deployed on exercises with the 17 Gurkha Division Signals.[37] 1969 saw the squadron testing air movement techniques in several major exercises before the rundown of the brigade, and the move of the remaining elements to Singapore commenced in line with the withdrawal of British troops from Malaysia.[38] One such

---

35 Smith, *Malaya and Borneo,* p. 99; and Hinton, letter, September 1975.

36 Interview with Lieutenant Colonel R.L.C. Twiss, August 1989.

37 Ibid.

38 Hinton, letter, September 1975; *Current Notes,* Vol 40, 1969, p. 19; and Department of *Defence, Defence Report, 1969,* Canberra, p. 27.

exercise was a brigade deployment called 'Exercise Crowning Glory'. The exercise was followed by a period of stocktakes, when 'the sins of predecessors really came to the fore'.[39]

In September 1969, the Squadron held a farewell parade and disbandment party, after which the details were finally handed over to Malaysian forces. The unit was reduced from a squadron to a troop and placed under the command of Captain B.D.A. (Bruce) Roach. Thereafter, it was officially disbanded on 28 March 1970.[40]

The rundown of Headquarters 28 Commonwealth Infantry Brigade at Terendak was followed by the build-up of Headquarters 28 Commonwealth Infantry Brigade in Singapore by March 1970, with a signals detachment remaining as part of the new organisation.[41] It was later revealed that the arrangement between Britain and Australia over Terendak had cost the Australian Government $5,600,000 in capital costs. However, the then Minister for Defence, David Fairbairn, noted in 1971 that 'Australia obtained good value for its capital outlay during almost a decade of use by some thousands of Australian soldiers and their families.'[42] The arrangement may well have continued had it not been for the changes in strategic outlook for Britain.

In 1968, the British Government announced its intention to withdraw all of Britain's forces from east of Suez by the end of 1971.[43] Consequently, following the British withdrawal from Malaysia by early 1970, and the reduction of its presence to a representative force in Singapore, AMDA was replaced (in November 1971) by the Five Power Defence Arrangements (FPDA) between Malaysia, Britain, Singapore, Australia and New Zealand. The expansion from two to five signatories reflected the reduced

---

39 Twiss, interview, August 1989.

40 Hinton, letter, September 1975; and Twiss, interview, August 1989.

41 AWM 200, R579/2/20, folio 27, 3 October 1969.

42 *Current Notes,* October 1971, pp. 653-4; and *Commonwealth Parliamentary Debates,* 20 Eliz II, Vol 1, H of R, 75, 1971, p. 3873.

43 T.B. Millar, *Australia in Peace and War: External Relations 1788-1977,* pp. 245-6; and Madden & Morris-Jones (eds), *Australia and Britain,* p. 87.

British contribution to the defence of the region. The Five Power Defence Arrangements provided for continued co-operation in the field of defence with scope for consultation in relation to an external attack or threat against Malaysia or Singapore.[44]

Before the British withdrawal, Australian and New Zealand planners had recommended that a single joint force communications unit be formed with Australian and New Zealand participation. The unit was to be called 9 Signal Regiment and was to consist of a headquarters, and three squadrons as follows: 146 Signal Squadron - a joint army-navy communications squadron responsible for the operation of the transmitting and receiving stations at Suara and Kranji, and for the operation of the communications facilities at the Naval Headquarters Building; 153 Signal Squadron - an administration squadron; and 121 Signal Squadron, under command for local administration only. 121 Signal Squadron had been in Singapore, in one form or another, since the time of the Malayan Emergency.[45] 121 Signal Squadron's role was such a well kept secret that even many military personnel did not know that it performed tasks for intelligence purposes.[46] 9 Signal Regiment was also responsible for provision of a detachment at Bangkok to take over a link from SEATO Headquarters to Singapore. This link had been operated by UK Army personnel and was taken over by the Australian 548 Signal Troop.[47]

---

44 *Defence Report, 1972,* p 5.

45 '9 ANZUK Signal Regiment', in *Signals Bulletin,* January 1973, p. 6.10; AWM 121, 209/C/4, 'Role of 9 Signal Regiment' Annex H to JS Report No 9/71; and AWM 207, 529/1/1, 'Raising/Reorganization Instruction 10/70: Raising 9th Signal Regiment.'

46 Interview with Brigadier K.P. Morel, August 1989. See also Richelson & Ball, *The Ties That Bind,* pp. 192-193.

47 AWM 121, 209/C/3, 'Service Communications Detachment for Bangkok, September 1970'; and AWM 207, 529/1/1, 'Raising/Reorganization Instruction 10/70: Raising 9th Signal Regiment.' This link remained opened until December 1973, when the detachment was withdrawn from Bangkok. See AWM 207, 256/F8/8, 'Withdrawal Detachment 9 ANZUK Sig Regt Bangkok', 31 December 1973.

The communications plan for the ANZ Force was nearing completion when the newly elected British Government committed itself to maintaining land forces in Singapore. Consequently, the need arose to amend and expand the communications requirement planned for the ANZ Force. The unit was completely reorganised and expanded by 50 per cent. Systems and equipment installation plans were also extensively re-planned and modified at this time.[48]

The advance party of 9 Signal Regiment had deployed to Singapore in June 1970 under the command of the quartermaster, Captain Bill Blair. In August 1970, Major Mick Rooney arrived with a large detachment of the Army's signal construction and project unit -127 Signal Squadron. The installation teams immediately commenced the extensive tasks of expanding the existing Royal Navy transmitter and receiver stations at Suara and Kranji and constructing a large tape relay centre in the old Commander Far Eastern Forces Naval Headquarters building.

Although the tape relay equipment was old technology by this time, financial constraints prevented the purchase of more modern equipment. Additional tasks included transferring 121 Signal Squadron from Amoy Quee to its new home at Kranji in 1971, establishing tributary communications centres throughout Singapore Island, and completing various specialist facilities that were required within the Joint Force Headquarters Building. During this period, the ANZUK Force had not officially come into existence, and all work had to be done in conjunction with the UK Forces that had been planning their withdrawal from areas which the Regiment was to occupy.[49]

Following staff discussions held in Canberra in October 1970 between Australia, New Zealand and the United Kingdom, it was agreed that Australia would be responsible for the development

---

48 '9 ANZUK Signal Regiment', in *Signals Bulletin,* January 1973, p. 6.10.

49 '9 ANZUK Signal Regiment', in *Signals Bulletin,* January 1973, pp. 6.10-6.11; and AWM 207, 193/F1/8, 'Handover of Buildings 121 Sig Sqn to RAF Amoy Quee', May 1971.

and operation of the ANZUK Combined Communications System in Singapore. This system was to provide, wherever possible, the national and single service communication requirements of the ANZUK Force in Singapore and Malaysia by the end of 1971. It was also to be operated and maintained by the Australian Army using manpower contributed from all nations.[50]

On 1 November 1971, the ANZUK Force officially came into being and the term 'ANZUK' was adopted for the remaining integrated force, which had a strength of about 9,000, signifying the tri-lateral composition. Australia provided almost half of the ground force, all of the air defence capability, and a proportion of the naval forces - a total of about 4,000.[51] These 4,000 troops were dispersed in various bases around Singapore Island, extending from Changi in the far eastern tip of the island, to Nee Soon in the centre and Suara near the old Naval base opposite Johore Bahru on the north side, to Tengah on the west of the island.[52]

The Signal Corps components of the ANZUK Force were also dispersed around the island. 28 ANZUK Brigade Signal Squadron was located at Nee Soon. Headquarters 9 ANZUK Signal Regiment was at Kranji, across the causeway, south of Johore Bahru, and 121 Signal Squadron, under the command of 9 ANZUK Signal Regiment, was divided between Amoy Quee, south of Nee Soon, and Kranji.[53]

9 ANZUK Signal Regiment was commanded by Australians: initially Lieutenant Colonel John McGreevy and later Lieutenant Colonel Keith Munro. Their commands included members of the

50 AWM 121, 209/C/4, 'Communication Facilities Provided or Served by ANZUK Combined Communications System'. Annex B to JSPD No 2/1971; and AWM 207, 571/F1/37, 'HQ ANZUK Force Communications Instruction No 5 - The ANZUK Force Communications System, March 1972.'

51 *Defence Report, 1972,* p 5; and Madden & Morris-Jones (eds), *Australia and Britain,* p. 87.

52 *Commonwealth Parliamentary Debates,* 20 Eliz II, Vol 1, H of R, 74, 1971, pp. 1598-9; and AWM 207, 5820/F1/3, 'Unit Locations ANZUK Force Singapore as at May 1972'.

53 AWM 121, 209/9/1, 'ANZ Force Singapore', May 1970; and Annex A to Brief for QMG - Real Estate Singapore'.

Army, Navy and Air Force of the UK, Australia and New Zealand. They also included Royal Malaysian Navy personnel, Australian and UK civilians, and locally employed civilians. Support was also provided by the Singapore Armed Forces.[54]

By November 1972, 9 ANZUK Signal Regiment had undergone a number of organisational changes, and what remained was a regimental headquarters, an administration squadron and an operations squadron. The Commanding Officer also had planning and project officers at his disposal in the Regimental Headquarters group. Administration Squadron, under Major Les Hubble (later Lieutenant Colonel), included the administrative and quartermaster elements of the Regiment, located at the various sites around Singapore Island. Operations Squadron, under Squadron Leader Harvey, exercised control over all the technical and operating elements of the Regiment. The day-to-day control of the squadron was vested in the Operational Control Group, which worked under Squadron Leader Harvey.[55]

It had been envisaged, in 1970, that the tactical land forces stationed in Singapore would be organised into a brigade-sized group. This group came to be known as 28 ANZUK Brigade, and the signal squadron raised to support the brigade was to be a nationally integrated tactical squadron, based on the Australian Army's brigade (or task force) signal squadron organisation, and equipped with Australian-provided communications equipment. The scale for provision of manpower was to be based on a division of 40 : 40 : 20 for United Kingdom : Australia : New Zealand.[56] This allowed for the position of squadron commander to be rotated between officers from the three countries.[57] This new signal squadron was given three basic tasks. It was to provide for:

---

54 Weir, letter, July 1990.

55 AWM 206, 30/57 Administration, '9 ANZUK Sig Regt Functional Organization, 31 Oct '72.'

56 AWM 207, 605/F2133/1, 'ANZUK Land Force Signal Squadron Preliminary Examination, December 1970'.

57 AWM 121, 209/C/3, 'Organization of HQ 28 Bde and Sig Sqn', March 1971.

- the administration of brigade headquarters and the signal squadron;
- communications for operational and training aspects of the brigade; and
- the operation and maintenance of the communications centre and despatch rider service for the brigade headquarters elements based near Sembawang airfield, near the former naval base on the north side of Singapore Island.

Furthermore, unlike standard Australian brigade (or task force) signal squadrons, 28 ANZUK Brigade Signal Squadron was not capable of providing protracted operational communication for the brigade, nor were provisions made for telecommunication with air force units acting in support of the brigade.[58]

Following the Labor Party victory in the Australian Federal election of 1972, the forward defence policy of pre-empting direct threats to Australia through force deployments in South East Asia was abandoned. The newly elected Prime Minister, Gough Whitlam, caused quite a stir at the time by publicly announcing the previously secret nature of the roles and tasks of 121 Signal Squadron. Shortly thereafter, the forces stationed in Singapore began to be withdrawn.[59] In line with the 1973 Review of Strategic Guidance issued by the Department of Defence, the Australian government no longer saw the need to maintain land forces in Singapore and so began to bring them back to Australia.[60] In December 1973, the withdrawal of 121 Signal Squadron was announced, and the departure of the remaining

58 AWM 121, 209/C/3, 'Organization of HQ 28 Bde and Sig Sqn', March 1971; and AWM 207, 605/F2133/1, 'ANZUK Land Force Signal Squadron Preliminary Examination, December 1970'.

59 R.J. O'Neill, 'Defence Policy', in W.J. Hudson (ed), *Australia in World Affairs 1971-75*, George Allen & Unwin, Sydney, 1980, p. 11.

60 Department of Defence, 'Key Elements in the Triennial Reviews of Strategic Guidance Since 1945', April 1986; and Madden & Morris-Jones (eds), *Australia and Britain*, p. 88.

Royal Australian Corps of Signals elements from Singapore was completed during 1974.[61]

Indonesia

Although Australia had long considered Malaysia and Singapore important for forward defence, Indonesia had always loomed large. Confrontation with Indonesia had served to heighten awareness of this. So how could it be that, having provided military communications support for operations against Indonesia, the Royal Australian Corps of Signals found itself providing communications support for operations in support of the Indonesian government? This section seeks to answer this question and to outline the nature of that support.

Until the period of confrontation subsided under the administration of the new President Suharto, few opportunities for defence co-operation had emerged. So it was a very encouraging sign to see Indonesian-Australian relations emerge from their all-time low ebb of the mid-1960s. By 1969, they had improved to such an extent that the Minister of Defence approved a request made to Australia by the Indonesian Government for survey assistance on a mapping project in Southern Sumatra.

The work was to be carried out in conjunction with Indonesian surveyors and the aim was to remap the whole of the island of Sumatra and, with the help of Australian Army surveyors, to produce large-scale maps. The Indonesian Government needed these maps to allow urgent planning to proceed for hydro-electric schemes, irrigation projects, transmigration schemes and other national objectives.[62] The Royal Australian Survey Corps, who were tasked to provide this assistance, requested signals support from the Directorate of Signals. 1 Signal Regiment was chosen to provide the signals support.[63] Bert Lane, then a Radio

61 AWM 206, 3/29/1, message: 'Withdrawal of 121 Signal Squadron From Singapore' 7 December 1973.
62 'Mapping Aid for Indonesia', in *Current Notes,* Vol 43, 1972, p. 226.
63 *Signals Bulletin,* Vol XVII, No 4, p. 84.

Troop Sergeant with 1 Signal Regiment, Ingleburn, New South Wales, was fortunate enough to be selected to command the Signals Detachment which was to accompany 1 Field Survey Squadron to Indonesia as the first unit to be involved in the mapping project. The detachment consisted of Lane, Corporal Roger Mayberry, and six radio operators.[64]

The group joined with the members of the Survey Squadron at Randwick in Sydney on 10 April 1970, where they were briefed on their role and took delivery of seven AN/GRC-106 radios and 13 of the newly introduced PRC/F1 HF radios. All the radios came brand new and still in the manufacturer's packing cases, so they spent several days thoroughly checking each piece of gear. They also refreshed themselves on 'one-time letter pads' used for classified message traffic over the radios.[65]

The detachment flew to Indonesia via Singapore, where Corporal Mayberry and two of the radio operators remained to establish a rear link facility in a building in Nee Soon Barracks. The remainder flew on to Pontianak, West Kalimantan and moved into a camp beside the airfield. There, using the larger AN/GRC-106 radios, they established the administrative rear link with Mayberry in Singapore, which was to operate 16 hours per day for the remainder of the tour.[66]

The PRC-F1 radios were used on a forward communication net to the Survey teams established on various high points around the countryside. These were normally manned by only one Surveyor who had been given extensive familiarisation training on the PRC-F1s, and they managed to operate them quite satisfactorily. Lane remembers that the PRC-F1 radios, which were developed and manufactured by AWA, 'performed beautifully throughout the four-month operation despite some rough handling at some of the outposts.' The same, however, could not be said about the AN/GRC-106s, which were notorious for over-

64 Letter from Warrant Officer Class II A.W. Lane.
65 Ibid.
66 Ibid.

heating and causing a heartache as they constantly broke down and had to be backloaded to Singapore for repair.

There were a number of aircraft supporting the operation, including one RAAF 'Caribou', a Pilatus 'Porter', five Army Bell helicopters and a chartered 'Beechcraft' aircraft. The Caribou would shuttle daily to Singapore and communicate to the forward detachment on the radio rear link. All other aircraft used the forward communication net when they were airborne.[67]

Over the next couple of years, Survey teams returned to Indonesia, accompanied by Signals Corps personnel, to continue the mapping project. The 1971 survey was conducted in the area around Palembang, with Sergeant W.B. Shearing and Lance Corporal Gill at Palembang and Lance Corporal D.J. (David) McCleen and Signalman D.J. Clarke as the rear base radio operators in Singapore.[68] Further surveys were also conducted in 1972, covering 70,000 square kilometres in four months.[69] The follow-up visits occurred at the same time each year, so as to take advantage of the months when the skies were clearest, to enable the survey equipment to function adequately.[70]

Australia's strategic deployment of forces to Singapore and Malaysia was intended to be in support of SEATO. Once the Confrontation with Indonesia had passed, only limited opportunities for gaining operational experience presented themselves, such as the deployments to Indonesia in support of survey operations. Training activities thus became central to the maintenance of operational skills for members of 28 Brigade Signal Squadron. In Singapore, the formation of 9 Signal Regiment, later renamed 9 ANZUK Regiment, filled the void left by the reduced British presence. 121 Signal Squadron continued fulfill-

67 Lane, letter; AWM 207, 788/F2/1, 'Military Survey Assistance to Indonesia, Report No 1 for Period 15 Mar-29 Mar 71', p. 3; and Report No 4 for Period 26 Apr-11 May 71', p. 2.

68 AWM 207, 788/F2/1, Annex A to Report No 1, Mar 71.

69 *Defence Report 1972*, pp. 6 & 34; and AWM 207, 788/F2/5, 'HQ ANZUK SPT GP After Action Report: Military Survey Assistance to Indonesia'.

70 Lane, letter.

ing its signals intelligence role until withdrawn by the Whitlam Government. Throughout the 1960s and into the early 1970s, the skills and techniques of the Royal Australian Corps of Signals had remained critical for the proper functioning of the forces stationed in Malaysia and Singapore.

Conclusion

The commitment of Australian troops on operations in South East Asia reflected Australia's strategic, economic and political priorities. Australia's response to this was to commit forces as part of its forward defence strategy. These priorities also served to shape the Royal Australian Corps of Signals in the post-World War II period. For instance, the deployments alongside British units (and often under British command) resulted in Australian unit organisations and equipment significantly conforming with, or at least being heavily influenced by, that of their British counterparts. A similar and perhaps stronger influence was to be exerted by the United States' Army on the Corps, and on the Australian Army as a whole, during the course of the Vietnam War, discussed in subsequent chapters.

In the case of the Royal Australian Corps of Signals, developing communications technology had to carry out specific functions determined by the geographic environment and the operational role being performed, but the technological advances themselves presented the Corps with a range of new options for how to carry them out. In the field of signals intelligence, the development of specialised equipment and skills served as a significant combat power multiplier, providing commanders with ready access to critical information that had a dramatic impact on the manner in which they conducted military operations. In the field of signals support for dispersed jungle operations, patrols could be deployed in a manner that would not even have been contemplated during World War II because lighter, more robust and more reliable radios provided patrols with ready access to additional combat power, reinforcement, resupply and evacuation when necessary. The advances in 'ruggedised', more efficient and more powerful equipment, in turn, had an impact on the remainder of the Army, giving commanders greater flexibility in the exercise of their command. Nevertheless, the job of the signalman remained essentially the same - to get 'through', but the manpower requirements and the methods employed in doing so varied as new technology was introduced. Experience on operations in Vietnam through most of the 1960s and into the 1970s would provide further evidence of this phenomenon, as the subsequent chapters demonstrate.

CHAPTER FOUR

VIETNAM TO 1966

Introduction

AUSTRALIA'S PARTICIPATION IN the Vietnam War produced the largest commitment of Australian military forces overseas since the 1940s and, at its peak strength of 8300 men (of whom about six per cent were members of the Royal Australian Corps of Signals), the Force was two-thirds larger than the one committed to the Korean War. Yet, the Australian contribution was small when compared with that of the US, as the peak US strength was over 65 times as many troops, with a total of 542,000.[1] Moreover, only 46,852 Australian military personnel served in Vietnam, including 17,424 National Servicemen with 496 killed, whereas 330,000 Australians served in the 1914-1918 war with 60,000 fatalities.[2] As much as $500 million was spent in its execution by the Australian Government[3] as compared to US$180 billion by the US Government.[4] So on the scale of

1 G. St J. Barclay, *A Very Small Insurance Policy: The Politics of Australian Involvement in Vietnam, 1954-1967,* University of Queensland Press, St Lucia, 1988, p. 165.

2 Edwards, 'The post-1945 conflicts', in McKernan & Browne (eds) *Australia Two Centuries of War and Peace,* p. 297. Over 56,000 US servicemen, 4208 Koreans, 169 Thais and 26 New Zealanders were also killed in action. See Barclay, *A Very Small Insurance Policy,* p. 165.

3 See F. Frost, *Australia's War in Vietnam,* Allen & Unwin, Sydney, 1987, p. 1.

4 T.C. Thayer, *War Without Fronts: The American Experience in Vietnam,* Westview Press, Boulder, 1985, p. 23.

Australia's involvement in the conflict alone, it could be argued that Australia's role was not significant. One of the features of the role, which was significant, however, was the way in which the developments in communications technology, in conjunction with the introduction of the helicopter, facilitated a whole new approach to the conduct of military operations. This chapter and the following chapters on the Vietnam War show the impact of developments in communications technology on the conduct of Australia's military operations in Vietnam.

Australian troops were initially committed, in 1962, as advisers. From the time of the first Australian commitment, the Royal Australian Corps of Signals was represented. This chapter covers the first commitment of troops in 1962 to the deployment of a battalion-sized force in 1965 and the deployment of 709 Signal Troop through to 1966.

Australia's commitment to Vietnam in 1962 was aimed at preventing the spread of communism, encouraging regional stability and economic growth, and ensuring the continued regional engagement of the US in continental South East Asia.[5] One such initiative was the South East Asia Collective Defence Treaty, or SEATO, mentioned earlier. SEATO was a predominantly military organisation, with its headquarters in Bangkok, Thailand. There it had a military planning office, including a communications and electronics division headed by Royal Australian Corps of Signals Officers, including Lieutenant Colonels Marsh, Taylor and MacLean. This office was concerned with the drafting and refining of plans for possible military operations in the

5 For more detail on the reasons for Australia's commitment in 1962 see P. Edwards with G. Pemberton, *Crises and Commitments: The Politics and Diplomacy of Australia's Involvement in South East Asian Conflicts 1948 to 1965*, Allen & Unwin, 1992, Chapter 13.

South East Asian Region.[6] SEATO served as a base for wider appeals for involvement in Vietnam and to give legitimacy to Australia's contribution.[7] Similarly, the ANZUS alliance facilitated a rapid integration of Australian troops as part of a larger US force, mainly as a result of the efforts aimed at standardisation of equipment and procedures that had been followed since the late 1950s.

Following the return, in 1945, and eventual collapse of the French colonial forces at Dien Bien Phu in the north-west of Vietnam in 1954, American support was committed to the newly-emerged South Vietnam regime of Ngo Dinh Diem. This support escalated after the commencement of military operations by the North Vietnamese in 1960 against the South Vietnamese government following the abrogation of the 1954 Geneva Accords by South Vietnam.[8]

Australia's initial military contribution to support the South Vietnam regime, announced on 24 May 1962, consisted of military instructors. This decision came five months before the Dutch evacuation of New Guinea and shortly before Indonesia actively began to oppose the formation of the new Federation of Malaysia, in 1963, by means of Confrontation, as discussed in Chapter Three.[9] Despite these preoccupations, the Australian Government decided to go ahead and support the US efforts in South Vietnam. As Australia's Ambassador to Washington, Sir

---

6 For a discussion on the creation of SEATO see Edwards with Pemberton, Crises and Commitments, Chapter 9. For a brief discussion on the Military Planning Office in Bangkok see I. McNeill, *To Long Tan: The Australian Army and the Vietnam War 1950 - 1966, pp 12-13 and p. 21. See also* L.B. Swifte (Lieutenant Colonel), 'SEATO Today', in *Australian Army Journal,* No 204, May 1966, pp. 41-43.

7 See Barclay, *A Very Small Insurance Policy,* pp. 61, 106-109 & 118.

8 Carver, *War Since 1945,* pp. 172-176.

9 *Current Notes,* Vol 33, 1962, p 36; Reese, *Australia, New Zealand, and the United States,* pp. 216-219. For a discussion of Australia's response to confrontation in 1964 and for a discussion of the political machinations behind these crises in Australia's external relations see Edwards with Pemberton, *Crises and Commitments, Chapters 13 to 18; and Pemberton's All The Way,* pp. 219-248.

Howard Beale, made clear on 5 December 1961, demonstrable Australian support for Vietnam would 'make a very favourable impression' on the United States administration.[10] As subsequent official publications would explain, it was clear...that the motivation for assisting was predominantly political and that the Services were unwilling, for manpower reasons, to make more than a token commitment to the Republic of Vietnam.[11]

In the meantime, the number of American servicemen in Vietnam increased from 948 in November 1961 to 5,500 by June 1962.[12] During this time, US military communications facilities were being installed to create an expandable communications system known as 'Black Porch' to meet the defence needs of the South Vietnamese in their counter-insurgency operations. This consisted of tropospheric scatter radio trunks, capable of providing numerous circuits between locations more than 200 miles apart, and line-of-sight microwave radio relay links to cover shorter segments of between twenty to thirty miles (see Diagram 12). The system was operated by the 39th US Signal Battalion, which arrived in South Vietnam in February 1962.[13]

Early Days - Signals and Vietnam 1962-1965

The original Australian advisory team, known as the Australian Army Training Team Vietnam (AATTV), consisted of 15 officers and 15 warrant officers and sergeants. They left for Vietnam at the end of July 1962, under the command of Colonel F.P.

10 Frost, *Australia's War in Vietnam,* pp. 14-15.

11 *Australia's Military Commitment to Vietnam,* Canberra, 1972, p. 5, cited in Frost, *Australia's War in Vietnam,* p. 15.

12 Carver, *War Since 1945,* p. 179.

13 T.M. Rienzi, *Vietnam Studies: Communications-Electronics 1962-1970,* Department of the Army, Washington D.C., 1972, pp. 6 -11; and Coker & Rios, *A Concise History of the US Army Signal Corps,* p. 30.

Serong.[14] There they were integrated with the US Military Assistance Advisory Group (USMAAG), later to be subsumed under the United States Military Assistance Command Vietnam (USMACV), and located in the I Corps (northernmost) military region (pronounced 'Eye' Corps) in South Vietnam (see Map 7). The USMAAG provided the maximum professional advice to as many sections of the governmental and military organisations as possible. Advisory Teams were sent down to and including the battalion level. Teams were also located with provincial governmental organisations.[15] So, given that the members of the AATTV were administratively provided for by the US armed forces and were involved in primarily infantry training, what role was there for members of the Royal Australian Corps of Signals? The initial strength of the Team would be about thirty personnel with five signal instructors. The initial role of the AATTV was strictly training. They were not to accompany Vietnamese forces on operations.[16] That would soon change.

On 3 August 1962, Captain B.R. (Barry) Tinkler (later Lieutenant Colonel and Commanding Officer of 6 Signal Regiment) arrived in South Vietnam as the first Royal Australian Corps of Signals member of the AATTV. Four other officers and twelve warrant officers of the Corps also served with the AATTV, with one, Warrant Officer Class 2 M.W.T. (Mike) Gill, being killed in action on 6 May 1969.[17]

---

14 B.R. Tinkler, 'Australian Army Training Team Vietnam', in *Signals Bulletin,* Vol XII, No 4, December 1963, pp. 2.05-7; and *Current Notes,* Vol 33, 1962, p. 52. For a comprehensive history of the AATTV see Ian McNeill, *The Team: Australian Army Advisers in Vietnam 1962-1972,* Australian War Memorial, Canberra, 1984.

15 RASCM, Tinkler, 'Australian Army Training Team Vietnam', pp. 2.05-7; and I. McNeill (Major), 'An Outline of the Australian Military Involvement in Vietnam July 1962-December 1972', *Defence Force Journal,* No 24, September-October 1980, pp. 42-53.

16 McNeill, *To Long Tan,* pp. 39-42.

17 RASCM, G.J. Lawrence (Lieutenant Colonel), 'Royal Australian Corps of Signals - Corps History: RASCM, The South Vietnam Campaign 1962-1972' unpublished manuscript held by the Directorate of Signals, pp. 1-2; and RASCM, D Sigs file 723-3-4, 'Employment of AATTV as at 30 Apr 72', Annex A to AAAGV Report dated 12 May 1972.

According to Tinkler, there were two major limitations imposed on an adviser. The more frustrating limitation was that he had no command function over the Vietnamese he was advising. They could either accept or reject his advice. The second one was that an adviser could only fire at the enemy if he was fired at first. The appointments with the AATTV were also purely of an infantry nature, and so Signals advisers did not get to advise using their Signal Corps specialist skills.[18] Not surprisingly, the first batch of Signals personnel came predominantly from the Corps' sergeant majors and line tradesmen.[19]

The AATTV provided Australia with the means of gaining additional intelligence on the war in Vietnam, which, in turn, led to the recognition of the serious state of the campaign. Consequently, in consultation with the United States, the AATTV was increased in June 1964 to 80 personnel while their role was extended to operational advising in field units.[20] The Royal Australian Corps of Signals continued to contribute members to the AATTV until the withdrawal of the force in 1972.

Back in Australia, on 10 November 1964, the Prime Minister Robert Menzies announced the introduction of selective national service and an expansion in the nation's military forces.[21] The Prime Minister's statement was aimed at meeting the perceived threat from Indonesia, but it also gave the Government the ability to deploy combat forces in Vietnam at the same time and to provide a reserve in Australia.[22] On 7 April 1965, the Foreign Affairs and Defence Committee of Cabinet gave approval in principle to the proposal to send an infantry battalion to Vietnam. An infantry battalion was the minimum identifiable national group that the

18 RASCM, B.R. Tinkler, 'Australian Army Training Team Vietnam', p. 2.05.

19 Woollard, Letter, February 1977.

20 McNeill, 'An Outline of the Australian Military Involvement in Vietnam', p. 47; and McNeill, *The Team.*

21 *Commonwealth Parliamentary Debates,* Vol H of R 44, pp. 2717-2718.

22 See Edwards with Pemberton, *Crises and Commitments,* Chapter 16 for a discussion on the lead up to the introduction of conscription in November 1964.

Australian Government agreed to provide.[23] Moreover, Australian military personnel, with experience in Korea and Malaya, knew that a small identifiable force under Australian command would have more political impact than the dispersed members of the AATTV, however expert they might be.[24]

709 Signal Troop

Within 25 days of the Government's announcement made on 29 April 1965, the 1st Battalion, The Royal Australian Regiment (1 RAR), together with a cavalry troop, and an Australian Logistic Support Company began moving to Vietnam. The move was made by charter aircraft and the troop carrier HMAS *Sydney*.[25] Communications support would also be required for the deploying elements, which could perform a broad range of tasks from tactical communications at the battalion level to strategic communications links back to Australia. Given the range of tasks and the significance of a national rear link for the effective exercise of Australian national sovereignty over forces deployed as part of a much larger force, what kind of communications support would be provided and would the planning for this support be adequate? The Army's initial response to the problem was to send a composite troop known as 709 Signal Troop.

In late April 1965, 2 Signal Regiment in Watsonia received a warning order to form the required composite troop of 50 all ranks and to move it into a concentration area at Ingleburn, where it was to get the necessary equipment into a serviceable

---

23 Australia's Military Commitment to Vietnam, Canberra, 1975, p. 15, and Memorandum of conversation, Rusk, Bundy, Waller and Renouf, 13 April 1965, cited in Pemberton, *All The Way*, pp. 274-275. For a discussion on the decision making process that led to the commitment of the infantry battalion see McNeill, *To Long Tan*, Chapter 3; Edwards with Pemberton, *Crises and Commitments* Chapter 18; M. Sexton, *War For the Asking*, Ringwood, 1981; and Barclay, *A Very Small Insurance Policy*, pp. 79-105.

24 D.M. Horner, *Australian Higher Command in the Vietnam War*, Canberra Papers on Strategy and Defence, No 40, ANU, Canberra, 1986, p. 3.

25 McNeill, *To Long Tan*, pp. 75-79.

state, carry out tropic proofing and acquire maintenance stores. The order was apparently typical of previous exercises, and the troop commander, Captain R.L.C. (Dick) Twiss, thought the task would be the provision of a high-frequency rear link within Australia. He was allowed to select the best men in the Regiment to form a heavy radio detachment, a medium radio detachment and a communications centre detachment. Unfortunately, the newly reorganised field force signals units did not allow for a troop designed to perform a composite task with a specifically prescribed total of 50 all ranks, nor had the necessary radios fully entered service. The expedient forced on the Corps for this particular Troop was to prove how difficult operations can be when an untried organisation is put to the test.[26] Twiss later recalled that the organisation looked fine on paper, but it needed a squadron headquarters to function properly.[27]

It was early May before Twiss was told that the role of 709 Signal Troop was to support Headquarters Australian Army Force Vietnam (HQAAFV) by providing a rear link to Australia and a forward link to a battalion group. He was not told where HQAAFV (later designated HQ AFV-Army Component), nor 1st Battalion, The Royal Australian Regiment, were to be located in Vietnam.[28] The problem was that the plans concerning where the contingent would go had yet to be finalised.

Twiss arrived on an Air Vietnam plane at Tan Son Nhut airport in Saigon on 26 May 1965 with two other soldiers assisting him with the reconnaissance. The next day, he contacted the US Military Assistance Command Vietnam (MACV) and the 39th US Signal Battalion to arrange for the initial Troop site and for clearing radio frequencies for use on the Australian links. From the outset, Twiss encountered a staff-level problem of liaison with the US MACV Signals section as he did not hold a suffi-

26 RASCM, Lawrence, 'The South Vietnam Campaign', p. 2.1; and Letter from Major General R.P. Woollard, February 1977.

27 Interview with Lieutenant Colonel R.L.C. Twiss, August 1989.

28 RASCM, Lawrence, 'The South Vietnam Campaign', p. 2.2.

ciently high rank to gain the favour of higher-ranking American staff officers on the MACV headquarters.[29] In contrast to this, there was the 39th Signal Battalion. This unit was the only US Signals unit then stationed in Vietnam at the time, and it was one with which the Royal Australian Signal Corps was to have much contact in the following seven and a half years.[30] Twiss noted that the Americans of the 39th Signal Battalion 'did all that they could for us, but [even that] wasn't much because they too were short of gear as they were expanding from a Signal Battalion to a Signal Brigade.'[31]

Whilst liaising with the MACV, he also discovered how spread out the Australian forces actually were. HQ AAFV was at that stage located in the Cholon area of Saigon, some distance from Tan Son Nhut. 709 Signal Troop was allocated an area within the confines of the 39th Signal Battalion area at Tan Son Nhut. The 1 RAR group was some 30 kilometres away from Tan Son Nhut at the Bien Hoa air base, along with the US 173rd Airborne Brigade.[32] Two sites were needed, one for transmitting and one for receiving the radio signals from Australia. The transmitter site was selected on one side of Tan Son Nhut airfield, but from the outset, the site was unsatisfactory from a security point of view, as it was beyond the airfield perimeter defences. It did not take long for them to request a move. The receivers were placed in a Kingstrand hut at another site around the airport, at a place called Ba Queo, adjacent to the US receiver station.[33]

The Troop's advance party of ten soldiers arrived a couple of days later at Tan Son Nhut airport, on 29 May 1965, by RAAF

29 Interview with Lieutenant Colonel R.L.C. Twiss, August 1989.
30 AWM 95, 709 Sig Troop War Diary, Jun 1965; Interview with Lieutenant Colonel R.L.C. Twiss, August 1989; and RASCM, Lawrence, 'The South Vietnam Campaign', p. 2.2.
31 Interview with Lieutenant Colonel R.L.C. Twiss, August 1989.
32 RASCM, Lawrence, 'The South Vietnam Campaign', p. 2.2.
33 Letter from Major General B.H. Hockney, September 1989; RASCM, Lawrence, 'The South Vietnam Campaign', p. 2.4; RASCM, D Sigs 723/3/4, R.P. Woollard (Colonel), 'D Sigs Notes: South Vietnam - Malaysia visit Dec 66', p. 3; and AWM 95, CD - 709 Sig Troop War Diary, June & July 1965.

C-130 Hercules transport. They brought with them one AN/TRC-75 HF radio terminal (Australian modified version, working duplex instead of simplex as it was designed to do) and this was put in position and its antenna was erected on the day that they arrived. From there they laid D10 cable into HQ AAFV several miles away in Cholon, Saigon. The next morning, Morse (CW) contact was established with 6 Signal Regiment in Melbourne by using the AN/TRC-75. The men were jubilant when their first call was answered. Soon after, on 2 June 1965, the main body arrived in Saigon, and they were accommodated with the 39th US Signal Battalion.[34]

The AN/TRC-75, as mentioned earlier, was successfully used by the US armed forces in its originally designed simplex mode. The Australian sets were modified for duplex operation and came equipped with a different generator, the infamous Bucknell. The Bucknell generators had a 400 hertz output, necessary for the AN/TRC-75, and this made it impossible to operate the radios from the mains supply at Tan Son Nhut. The problem, however, was that it was often also impossible to use the Bucknell generators.[35]

Unfortunately, it was not long before the Troop began to encounter difficulties. The first reported problem occurred on 4 June 1965, five days after the first message had been sent, when all the Bucknell generators broke down.[36] Maintenance of the generators was not helped by the fact that all four generators were different prototypes.[37]

Not surprisingly, therefore, for the first three weeks from 30 May, operators had to rely on Morse code telegraphy to communicate with Australia.[38] The situation was so bad that it caused

34 RASCM, D Sigs Files - Cable, 'Communications to Aust Army Force Vietnam 30 May 1965; AWM 95, CD - 709 Sig Troop War Diary, June 1965; and RASCM, Lawrence, 'The South Vietnam Campaign', p. 2.2.

35 *Signals Bulletin,* Vol X, No 1, October 1961, pp. 4.01-4.03; and RASCM, Lawrence, 'The South Vietnam Campaign', p. 2.3.

36 AWM 95, CD - 709 Sig Troop War Diary, June 1965.

37 Lieutenant Colonel R.L.C. Twiss, Interview, August 1989.

38 AAFV 'Quarterly Report to 30 September 1965', p. 4, AHQ. file 723/R5/18, DD cited in McNeil, *To Long Tan,* p. 134.

Colonel O.D. Jackson, the Commander Australian Army Force Vietnam (AAFV) from May 1965 to May 1966, to send a message to Army Headquarters in June concerning the unsatisfactory state of communications, focusing on the long delays that were occurring and the frequent outages of the transmitter.[39] Problems associated with the AN/TRC-75s and Bucknell generators continued to recur whilst Signals units were on operations in Vietnam.[40] These problems arose because much of the equipment in use was different to that used by the Americans, so it could not be repaired by them.[41]

709 Signal Troop also included a Heavy Radio detachment, consisting of a powerful commercial E514 transmitter mounted on a truck painted green. It also started falling apart very rapidly and was not maintainable within Vietnam until more support arrived from Australia.[42]

Moreover, the telegraphic equipment in the communications centre was drawn from unit stocks at 2 Signal Regiment in Melbourne. These were old and worn out and required to be virtually rebuilt. The problems were compounded by the tight manning of the Troop. Twiss recalled

> when we got to South Vietnam, this number [2 officers and 46 other ranks] was required to provide [on a 24 hour basis]: a communications centre at HQ AAFV; a despatch rider facility in Saigon and to Bien Hoa; manning for a transmitter site; manning for a receiver site; [later on they also had to provide] a detachment at Bien Hoa to provide communications centre facilities and a radio link using the AN/TRC-75

---

39 Message 19621 and 19626 of 10 June 1965 in AHQ file 723/R5/17 DD cited in McNeil, *To Long Tan,* p. 134.

40 See for instance, AWM 95, 145 Sig Sqn, 'Report on Reliability of Medium Radio Station - AN/TRC-75', September 1966.

41 RASCM, Lawrence, 'The South Vietnam Campaign', p. 2.6.

42 Twiss, interview, August 1989; RASCM, Lawrence, 'The South Vietnam Campaign', p. 2.4; and AWM 95, CD - 709 Sig Troop War Diary, October 1965.

> [, as well as a VHF manpack detachment for the battalion rear link when away from Bien Hoa]; an alternate rear link to Australia using an AN/ TRC-75 located at Bien Hoa; and a switchboard facility at HQ AAFV and at Bien Hoa![43]

The difficulties experienced were perhaps felt most acutely by 1 RAR in the field when, at times, poor communications prevented them from relaying the details of casualties for the immediate notification of next of kin. On other occasions, HQ AAFV was unable to obtain information of operations for its purposes and for forwarding to Canberra.[44]

To make matters worse, the Troop had only two line men for all the Troop's lines in Saigon and Bien Hoa and the Troop had to provide its own workshop facility. Furthermore, out of the 48 members of the Troop, six of them were administrative staff (comprising the Troop Commander, his Second in Command, Lieutenant N.G. (Neville) Grieve, a sergeant technical storeman, a technical storeman, a clerk and a driver).[45]

The technical storemen also had a major problem on their hands as they were responsible for the maintenance of three months' worth of spare parts for every piece of equipment in use by the Troop. The technical storemen had to control approximately 6,000 line items, and this also had to be done according to peacetime accounting procedures![46]

Colonel O.D. Jackson recalled what he described as

---

43 Twiss, interview, August 1989; AWM 95, CD - 709 Sig Troop Monthly Report, August 1965; and RASCM, Lawrence, 'The South Vietnam Campaign', p. 2.6.

44 Messages Ops 561 of 9 August 1965 and Ops 1160 of 14 September 1965, AHQ file 723/R5/17, DD cited in McNeill, *To Long Tan*, p. 135.

45 Twiss, interview, August 1989; and AWM 95, CD - 709 Sig Troop War Diary, June 1965.

46 Twiss, interview, August 1989; and AWM 98, R579/1/2, 'Repair Parts Support - 709 Sig Tp', September 1965.

> The initial inefficiency of our strategic communications. We relied too heavily on United States resources without appreciating just how overloaded they would be by American requirements generated by the rapid buildup of American Forces in Vietnam. This aspect of Australian planning fell well short of the mark. Certainly, one should not rely on friendly force resources, no matter how friendly those friendly forces may be.[47]

The problems of the theatre itself and of the technical situation did not seem to be adequately known in Canberra, and 709 Signal Troop was virtually operating in a tactical role, with all its equipment still mounted in vehicles, whilst its primary task was to provide a link for the AUSTCAN strategic communications network.[48]

Colonel (Later Major General) R.P. (Bob) Woollard was the Director of Signals at the time.

> As seen by me, as DSigs, the basic cause of the early problems arose from the initial criteria set by the Government in early 1965 for Australia's first combat force involvement. Because of the desire to achieve the 'maximum political kudos for minimum expenditure', a numerical ceiling was set for the total Australian commitment.....[and as] the US agreed to provide 'all logistic backup support that was not peculiar to Australia'...it was claimed that [except the rear link to Australia and perhaps an administrative emergency link to the battalion from the Force Headquarters]

47 RASCM, Letter from Brigadier O.D. Jackson to Lieutenant Colonel G.J. Lawrence, July 1977.

48 RASCM, Lawrence, 'The South Vietnam Campaign', p. 2.6.

> all other communications support would be provided by the Americans. Because of this belief, I believe, no Signal Corps representative was permitted on the initial reconnaissance to Saigon, nor was a Signals Staff Officer included in the HQ AAFV.[49]

In response to observations such as those made by Colonel Jackson, and the repeated requests for better support from Australia by Twiss, the Directorate of Signals at Army Headquarters in Canberra commissioned Major W.R.T. (Bill) Bodger, then a staff officer with the Directorate of Signals, to go for a week to assess the signals situation in Vietnam. He was also tasked to seek better sites and obtain improved in-theatre technical support. He arrived in Vietnam on 4 July 1965 and ended up staying in Vietnam for six weeks in order to complete the tasks assigned to him.[50]

One of the first sights witnessed by Bodger was melting solder on telegraph machines in the un-airconditioned communication centre vehicle, which had to be parked in the sun alongside the HQ AAFV building.[51] Twiss recalled later that within 24 hours, Bodger was sending back to Australia the same type of messages that he had already been sending. He recalls: 'we then started getting people who would listen to us at HQ MACV, as it was field grade officers talking to field grade officers.[52]' Later reports acknowledged the limitations that someone in Twiss' position must face. The Signals Directorate acknowledged that

> A unit [commander] can not be expected to plan and provide the necessary communications quickly, solve his troops immediate accommo-

49 Woollard, letter, August 1990.
50 RASCM, Lawrence, 'The South Vietnam Campaign', p. 2.6; and AWM 95, 709 Sig Tp War Diary, July 1965.
51 RASCM, Lawrence, 'The South Vietnam Campaign', p. 2.7.
52 Twiss, interview, August 1989; and RASCM, Lawrence, 'The South Vietnam Campaign', p. 2.7.

> dation, messing and transport problems in a relatively chaotic environment, if he has to be absent for long periods on liaison tasks with US and other forces trying to obtain telephone lines, real estate for antennas, frequency allocations and the host of other vital things required. These should be obtained by the staff, and a [signals staff officer] should be provided to do it.[53]

Lieutenant General Sir John Wilton, the Australian Army's Chief of the General Staff from January 1963 to May 1966, commented, saying

> I well remember visiting Captain Twiss on the fringes of the airfield. He was very hot and worried and looked harassed. I do not suggest that Twiss was in any way to blame - the circumstances were beyond his control - rather I feel he was to be commended because he eventually... got the link working.'[54]

According to Woollard,

> Bodger's report, which was supported by HQ AAFV, changed the attitude of Operations Branch [in AHQ] to that of listening to advice from the Directorate. Thus, authority was received from 527 Signal Troop to move to Saigon in September 1965 and for a Signals Staff Officer to join HQAAFV.[55]

Another outcome of Major Bodger's visit was that a communication centre was planned and later built in the newly allo-

---

53 D Sigs Report in *Signals Bulletin*, March 1968, p. 3.02.

54 RASCM, Letter from General J.G. Wilton to Lieutenant Colonel G.J. Lawrence, April 1977.

55 Woollard, letter, August 1990.

cated HQ AAFV building, adjacent to the old one. Another more significant outcome of the visit was the increase in manning and equipment.[56]

The Army's Directorate of Maintenance also sent out an officer, Lieutenant Colonel Marsh, to consider the problems of stock holdings and resupply of repair parts for equipment operated by 709 Signal Troop. The solution offered was not for the Troop to be relieved of the responsibility of looking after its own repair parts, but to be given better storage facilities. This was because 709 Signal Troop was the only unit in the theatre at the time that used the particular items in question.[57]

Shortly after these visits, on 14 September 1965, 527 Signal Troop arrived as part of an extra force increment codenamed Tanton.[58] The Troop included Second Lieutenant R.M.A. (Jock) Lonie and 29 other ranks tasked to relieve 709 Signal Troop of its communications centre roles.[59] According to Twiss they arrived virtually as bodies to support the equipment already there, but they did bring with them a number of equipment items such as brand new Perkins diesel generators, and a second transmitter, raising the number of equipment to two sets of each type. Ionospherics notwithstanding, the Troop was then able to maintain communications for up to 20 hours per day despite the lack of reserve capacity in case of breakdown or a protracted maintenance requirement.[60]

The AN/PRC-47 HF radios, 'which none of us had used before', also arrived with 527 Signal Troop. This was a 'godsend', according to Twiss, and they were deployed within 24 hours of arrival, as a Signals Corps element was now required to be

56 RASCM, Lawrence, 'The South Vietnam Campaign', p. 2.7.

57 AWM 98, R579/1/2, 'Repair Parts Support - 709 Sig Tp', September 1965.

58 McNeill, *To Long Tan*, pp. 113-114 & p. 134.

59 RASCM, Lawrence, 'The South Vietnam Campaign', p. 2.7; and AWM 95, 'Monthly Report - AUSTCAN Vietnam, Sep 65'. McNeill, in *To Long Tan*, p. 134, describes the Troop's role as operating 'signals and cryptography'.

60 Twiss, interview, August 1989; and Minute Sigs 2037/1, Colonel R.P. Woollard to Director of Staff Duties, 15 October 1965, DMO&P file 4/13, AWM 121 cited in McNeill, *To Long Tan*, p. 135.

provided to the infantry battalion. 1 RAR had been tasked with three general roles: to defend a part of the northern perimeter of the Bien Hoa air base; to 'strike out against National Liberation Front (Viet Cong)[61] strongholds to the north'; and to 'reinforce South Vietnamese units in combat'.[62] The battalion was under the command of the 173rd Airborne Brigade, but there was a requirement for its commander to have direct contact with the Headquarters AFV, and, as the battalion had no rear link communications capability of its own, the Signal Corps had to provide the link using the AN/PRC-47 HF radios (see Diagram 13). Consequently, two operators were deployed with the battalion when it deployed on operations further away from Bien Hoa than the 40-kilometre limit which the battalion's own VHF manpack radios had imposed.[63] On many such operations, the Viet Cong were forewarned of the locations and timings of the operations because of the Brigade's lax security. Viet Cong agents had infiltrated the South Vietnamese armed services and US bases where they were employed on domestic duties. This was to be a recurring problem throughout the war.[64]

The purchase of the new transistorised man-pack AN/PRC 25 VHF radio set in October 1965 helped to improve the radio communications operated by the infantry signallers within the battalion. The new radio had a higher output and greater range, and far more reliable batteries. Significantly, the replacement of valves by transistors represented a dramatic advancement in man-pack radio technology, which resulted in greater durability and dependability, which in turn generated an increase in usage of radio communications.[65] Moreover, there was a marked

61 Throughout the text the term 'Viet Cong' has been used for members of the National Liberation Front of South Vietnam, communist cadre and the North Vietnamese Army Regular Forces.

62 R. Breen, *First to Fight: Australian Diggers, NZ Kiwis and US Paratroopers in Vietnam, 1965-1966,* Allen & Unwin, Sydney, 1988, pp. 73-74.

63 Twiss, interview, August 1989; RASCM, Lawrence, 'The South Vietnam Campaign', p. 27; and Frost, *Australia's War in Vietnam,* p. 73.

64 Breen, *First to Fight,* p. 254.

65 McNeill, *To Long Tan,* p. 136.

increase in the number of frequencies available for use with the AN/PRC 25, having a frequency coverage of 30-75.95 MHz as compared to only 27-38.9 MHz for the AN/PRC 9 and 36-54.9 MHz for the AN/PRC 10 radio.[66] The increase in reliance on radio communications made possible through the increase in durability and availability of frequencies was to be a marked feature of the Vietnam War, giving commanders at all levels greater and easier access to information concerning higher, subordinate and adjoining units with the minimum of delay.

While operations continued, at the end of October, Headquarters AAFV moved into the Free World Military Assistance Organisation (FWMAO) building in Saigon. This was a large, relatively new building, showing signs of disrepair. It was not well-maintained, with filthy latrines and floors. The Australians were tenants there along with New Zealand, Filipino, Korean, Thai, Vietnamese and US Servicemen. There, a new communications centre was planned and installed by December 1965, by a detachment of 127 Signal Squadron under the command of Lieutenant Owen Richards. An AN/TRC-24 radio relay set and antenna were located on the roof of the building. The rear link facilities now had online cipher (automatic encryption), and the rear link to Watsonia, Melbourne, now had one classified circuit plus one unclassified and one engineering circuit.[67]

Construction work also took place for the receivers and transmitters. A prefabricated building was constructed for the receivers at Tan Son Nhut in November 1965, enabling the Troop to demount the equipment from the vehicles.[68] A transmitter site was harder to find, but eventually one was found in a military compound at Phu Tho (otherwise known as the 'Banana Plantation') in Saigon. This site became available in

66 The Royal Australian Corps of Signals Reference Manual, 1968, p.3-16.

67 RASCM, Lawrence, 'The South Vietnam Campaign', p. 2.7-2.8; D Sigs 723/3/4, R.P. Woollard, 'D Sigs Notes: South Vietnam-Malaysia visit Dec 66', p. 2; and AWM 95, CD, 'Monthly Report - AUSTCAN Vietnam Nov 65'.

68 AWM 95, CD 'Monthly Report, AUSTCAN Vietnam', October and November 1965.

February 1966. Australian engineers loaned a bulldozer to the signals troops who promptly cleared the area, constructed a 'rhombic' antenna to fixed-station standards, built an air-conditioned 'Kingstrand' hut for two new E513 transmitters and drive equipment, with a shed to house the three new generators. Construction was completed and the station commissioned by April 1966. Circuit availability subsequently improved, generally to about 22 hours per day.[69]

Ian McNeill, in his volume of the Official History of Australia's involvement in South East Asian conflicts, noted that

> Budget restrictions, the haste with which elements of the force were assembled, and political imperatives to minimise manpower had a severe impact on the signals network. Communications were militarily and politically crucial, yet it took several months before the system was able to meet the demand.[70]

Despite the complications and difficulties experienced by 709 Signal Troop and 527 Signal Troop when they arrived, they still managed to provide a national rear link communications facility for Australia to the battalion in the field. Several lessons were learned during the initial deployment, particularly concerning the significance of: incorporating a detailed communications plan for any operational deployment, including Signals Corps representation on any reconnaissance for future force deployments, and including at least a field grade Signals Corps staff officer on the headquarters to provide effective liaison with the allied forces' headquarters.

The employment of new technology represented a significant step and also helped to teach several lessons. It enabled Australia to maximise the benefit of contributing only a relatively

---

69 AWM 198, R193/1/1, HQ AFV, 'AUSTCAN Transmitter Installation', February 1966; and RASCM, Lawrence, 'The South Vietnam Campaign', p. 2.8-2.9

70 McNeill, *To Long Tan,* p. 136.

small force while still maintaining effective national control over it through a direct link back to Australia. By being able to exercise national sovereignty in this way, Australia effectively ensured control over the employment of its forces in Vietnam. The new technology was to alter the Australian armed forces' operating procedures even more significantly within Vietnam itself. There, the deployment in 1966 of a Task Force on operations and the establishment of what was to become for Australia an unprecedentedly complex and dispersed command and control and logistic infrastructure in Saigon, Vung Tau, Nui Dat and beyond presented a significant challenge to the Army's communicators.

CHAPTER FIVE

VIETNAM OPERATIONS 1966-67

Strategic and In-Theatre Signals for an Increased Force 1966 - 1967

THE UNITED STATES' plan of operations for 1965 to 1966 did not achieve the desired results and so plans were laid for the rapid expansion of forces in Vietnam in 1966. From the Australian viewpoint, it was clear that the Americans 'took risks one could not afford to take' and 1 RAR found their command, communication and logistics difficult to work with while they formed an integral part of the US 173rd Airborne Brigade.[1] Because of this, the Australians welcomed the opportunity to operate within a sector which they could control under an Australian field commander. Australia's response, therefore, was to offer a second battalion in March 1966 (a third was added in late 1967 along with a tank squadron) and further logistic and communications support.[2] As a result, the newly-designated First Australian Task Force (1 ATF) was assigned the province of Phuoc Tuy and the logistic support company was expanded to become the First Australian Logistic Support Group (1 ALSG), based at Vung Tau, where

1 F. Frost, 'Australia's War in Vietnam 1962-1972', in King, *Australia's Vietnam*, p. 59. See also Breen, *First To Fight* for an account of the experience of working as part of the US 173rd Airborne Brigade.

2 GOC Eastern Command: 'GOC's Exercise Hindsight 1971, Exercise Papers', Chapter 2- Australian Force Vietnam Chronology; and *Current Notes*, Vol 38, 1967, pp. 413 & 416.

the Australian Force gained sea port access to Vietnam. At the same time, Headquarters AAFV was upgraded to Headquarters Australian Force Vietnam (AFV), a joint Australian headquarters with RAAF and later, RAN representation. A squadron of RAAF Iroquois helicopters, a flight of Caribou transport aircraft and a squadron of Canberra bombers were added to the Australian forces in theatre. The Royal Australian Navy maintained a destroyer on station off Vietnam.[3] This was, according to the Prime Minister Harold Holt, the 'upper limit of Army capacity' that could be mustered to deal with the 'expansionist activities of Communist China'.[4]

Australia's military contribution to the war in Vietnam was becoming more substantial and increasingly complex. In doing so, the expanded force posed challenges for the Royal Australian Signal Corps far greater than those experienced on operations in Malaya or Borneo. This chapter shows how the Corps responded to the increased demands, providing an expanded strategic and in-theatre Signals Squadron as well as a Task Force Signal Squadron with attached specialist communications elements. This increased support, in turn, affected the way the Australian Task Force and supporting elements were able to conduct military operations in 1966-1967; enabling commanders to deploy further away and in a more dispersed manner, relying on improved communications as a lifeline for support when it was needed.

The responsibilities of the Royal Australian Corps of Signals were to expand in line with the Australian Force increase to the point where, by October 1968, approximately 500 Signals tradesmen would be employed in providing these communications,

---

3 GOC Eastern Command: 'GOC's Exercise Hindsight 1971, Exercise Papers', Chapter 2- Australian Force Vietnam Chronology. See D. Fairfax, *Navy in Vietnam: A Record of the Royal Australian Navy in the Vietnam War 1965 - 1972,* Department of Defence, AGPS, Canberra, 1980 for a more detailed coverage of the involvement of the RAN in the war.

4 Barclay, *A Very Small Insurance Policy,* pp. 106-133. Pemberton has argued that the failure to have made a greater effort suggests that once the US was deeply committed to Vietnam, the rationale for greater Australian support was weakened. See Pemberton, *All The Way,* p. 320.

spread over two major Signals units and one Signals troop. The main concentrations of Australian troops, from 1966 until 1972, were to be as follows: Headquarters AFV, located in the 'Free World' building, Cholon, Saigon; 1 ATF, based at Nui Dat, near Baria in the Phuoc Tuy Province, 65 kilometres from Saigon to the east; and 1 ALSG, based at Vung Tau on Cape St Jacques on the north eastern edge of the Mekong river delta, also 65 kilometres from Saigon but to the south east. Vung Tau to Nui Dat was about 25 kilometres. The RAAF bombers were to occupy a base at Phan Rang, about 275 kilometres north east of Saigon but other RAAF units were to be based at Vung Tau (see Maps 7, 8 & 9 and Diagrams 14, 15 & 16).[5]

The Australian contribution to the war effort in Vietnam fitted into the overall structure organised by the USMACV. Vietnam was divided into four regions or tactical zones, numbered I to IV from the northern end of South Vietnam to the southern edge of the Mekong Delta region. Australia's primary contribution, from 1965 onwards, was in the III Corps Tactical Zone (or Military Region) where its combat units came under the command of the American II Field Force, Vietnam, headquartered at Long Binh, north west of Nui Dat. The AATTV worked directly to Headquarters MACV in Saigon. Only the logistic units to be based at Vung Tau were to remain directly under Australian operational control.[6]

In the meantime, technical problems and escalating communications needs led the US Army Signal Corps to install the Integrated Wideband Communication System, provid-

5 RASCM, Lawrence, 'The South Vietnam Campaign', p. 3.1; AWM 98, R193/1/67, HQ AFV, 'Signals in Malaysia and Vietnam', August 1967; and *Defence Report,* 1968, p. 27.

6 Rienzi, *Communications-Electronics,* p. 25; D.M. Horner, 'A complex command: the role of the Commander, Australian Force, Vietnam', in *Journal of the Australian War Memorial,* No 13, October 1988, p. 22; and AWM 95, CD, 145 Sig Sqn: 'Communications Supplement to Military Working Arrangement Between Chief of the General Staff Australian Army and Commander, United States Military Assistance Command, Vietnam, dated 17 March 1966'.

ing multi-channel telephone cable, tropospheric-scatter and multi-channel line-of-sight radio systems such as microwave, covering the length and breadth of South Vietnam. Satellite communications were also installed and this provided the US Army with the critical link between Vietnam and Washington D.C. for three years.[7]

These new communications technologies were to prove critical in providing the flexibility required by commanders to respond to varied and dispersed operational requirements. The technologies facilitated changes in the methods of conducting military operations but they themselves were also adapted to meet the needs peculiar to the situation experienced in South Vietnam.

'Electronics has never been so vital in a war as it is here in Vietnam' argued Brigadier General W.E. Lotz, the United States' assistant chief of staff for communications and electronics in South Vietnam, in early 1966. His three main reasons were: the need to 'blanket' the whole country with communications because of the dispersed nature of operations; because of the way the war was run, in that decisions involving smaller units were being made at higher levels; and because of the need for accurate and rapid electronic surveillance and identification processes to know who the enemy was and where he was.[8]

The Australian response to these needs was to provide communications in four categories: national rear link communications, connected to the AUSTCAN system in Australia; in-theatre command and logistic communications; tactical communications; and 'special wireless' interception.[9]

The national rear link and in-theatre command and logistic communications was to be provided by 145 Signal Squadron (and, subsequently, from 1967 onwards, by 110 Signal Squadron). The

---

7 Coker & Rios, *A Concise History of the US Army Signal Corps*, p. 30.

8 W.E. Lotz (Brigadier General), 'The war that needs electronics' in *Electronics*, 16 May 1966, p. 96.

9 K.P. Carey (Major), 'Swift and Sure: The Royal Australian Corps of Signals in Vietnam' in *Command Communications*, HQ US Army Vietnam, October 1968, USARV Pam No 105-4, pp. 6-7.

headquarters of 145 Signal Squadron was raised in January, 1966, by 2 Signal Regiment, the Army's Line-of-Communications Signal Regiment in Melbourne, and despatched to Saigon in March 1966. 709 and 527 Signal Troops became part of 145 Signal Squadron upon its arrival.[10] There the Squadron Commander, Major J.H. (John) Bird and his Second in Command, Captain R.S.N. (Bob) Denning set about preparing for 'Operation Hardihood' - the deployment of the Task Force and Logistic Support Group to Vung Tau, set to take place in May 1966.[11] The advance party detachment of 145 Signal Squadron deployed at Vung Tau, on 9 April 1966, to provide the communications link for the ALSG and the Force reception. The telegraph link from Vung Tau to Saigon was opened on 12 April 1966. The main body arrived as this circuit was converted to a secure link (capable of operating with classified information automatically encrypted) on 20 April. Telegraph communications were later also opened, in June 1966, over US and Australian bearer circuits, from Headquarters AFV (Saigon) to Headquarters 1 ATF (Nui Dat), from 1 ATF (Nui Dat) to 1 ALSG (Vung Tau) and from 1 ALSG (Vung Tau) to the RAAF communications centre (also at Vung Tau).[12]

These teleprinter circuits were made secure and operated on a 24 hours-a-day basis. This meant that all messages could be transmitted immediately over the circuits with no delays for encoding or deciphering. This was essentially because of the triangulated links within Vietnam that were established during this time. Basically, there was a triangle joining together Saigon, Vung Tau and the Nui Dat areas. These links terminated in the Saigon Major Relay Station, a Minor Relay Station at Vung Tau and in the Headquarters 1 ATF signal centre at Nui Dat. From the Major Relay

---

10 AWM 95, 145 Signal Squadron CD, Cable 'Raising and Movement HQ 145 Sig Sqn, AAFV', 27 January 1966.

11 Twiss, interview, August 1989; and AWM 95, 145 Sig Sqn CD, 'Operation Hardihood' April 1966.

12 AWM 95, CD, 145 Sig Sqn War Diary April 1966; and 145 Sig Sqn Comms Report- Jun 66'. The RAAF had their own radio system but Army provided the joint AFV communications links.

Station in Saigon, circuits existed to the Headquarters AFV signal centre, to RAAF at Phan Rang (over hired US lines) and to Australia (the AUSTCAN relay station at Watsonia). From the Minor Relay Station at Vung Tau circuits existed to the 1 ALSG signal centre and to the RAAF at Vung Tau airfield. All stations on the network were able to originate messages to any other station and the triangular structure provided the facility for alternate routing of messages if one link was faulty. Similarly, telephone communications were provided over the radio relay system connecting the switchboards at the Free World Building in Saigon, the Vung Tau base and Nui Dat.[13] HF back-up radio circuits, initially using AN/PRC-47 HF radio sets and later the more modern and powerful AN/GRC-106 HF sets, were also installed and regularly tested. These were required to provide emergency communications links between the three main Australian areas.[14] These facilities were a marked improvement on the communications links that the British Commonwealth forces had been able to obtain during the Korean War. On-line encryption, telephony and the triangulated communications network using radio relay and tropospheric scatter represented a marked increase in the communications capabilities of the force, giving commanders greater access to information across, up and down the chain of command.

By 14 June 1966, the Australian communication centre and switchboard at Bien Hoa had closed and the despatch rider service from Saigon to Bien Hoa ceased, as the Australian infantry battalion (1 RAR) was no longer working as part of the 173rd Airborne Brigade there. US bearers also provided two voice channels from Headquarters AFV to Headquarters 1 ATF. A voice channel was established to 1 ALSG switchboard, code-named 'Emu' on 8th May 1966, and the 1 ATF switchboard was code-named 'Ebony'. These

13 AWM 98, R193/1/67, HQ AFV 'Signals in Malaysia and Vietnam, August 1967'.

14 AWM 95, CD, 110 Sig Sqn, Monthly Report Jun 1967.

names remained unchanged when the switchboards closed more than six years later.[15]

145 Signal Squadron was formed on the 'brick' system and, by mid 1966, had a headquarters and five signal troops with a total strength of 162 all ranks (increased to 221 in 1967).[16] The Squadron Commander, Major John Bird, also had the task of being the Staff Officer Signals with Headquarters AFV at the Free World building. In November 1966, he moved the Squadron Headquarters and co-located it with the transmitter site at Phu Tho. He continued acting as the Signals Staff Officer until March 1967, when Major D.J. Cummerford was posted into that appointment to relieve Major Bird's successor, Major E.H. (Ted) Hynes, of that particular responsibility and enable him to concentrate on commanding his squadron.[17]

Once 1 ATF and 1 ALSG arrived in Vietnam, there was no time to consider even transferring the rear link terminals to Vung Tau and so the pattern of communications, which was to continue for a number of years, was set by the initial commitment made in 1965. The communication centre and HF radio terminals to Australia already existed in Saigon in support of 1 RAR at Bien Hoa and, hence, it seemed logical that on the build up of a Task Force in the Vung Tau peninsula these facilities be expanded and improved, rather than built from scratch in a new area. This configuration was to provide the Signals units with significant problems of command and control for their remaining years in Vietnam.[18]

With the increase in the force being effected, it was not surprising that the number of messages handled as part of both the

---

15 RASCM, Lawrence, 'The South Vietnam Campaign', p. 3.2; and AWM 95, 145 Sig Sqn, CD, War Diary May 1966; and 145 Sig Sqn War Diary Jun 66.

16 RASCM, Lawrence, 'The South Vietnam Campaign', p. 3.1; and AWM 200, R579/1/1, 'Raising/Reorganisation Instruction 1/67, Australian Force Vietnam (Army Component)'.

17 RASCM, Lawrence, 'The South Vietnam Campaign', p. 3.1; AWM 95, 145 Sig Sqn CD, War Diary March 1967; and AWM 95, 145 Sig Sqn, CD, War Diary, November 1966.

18 *Signals Bulletin,* March 1968, pp. 3.04 & 3.10.

rear link and in-theatre message traffic, rose from approximately 116 messages daily in January 1966 to nearly 430 per day by June.[19] Following the Long Tan battle in August 1966, so much traffic was generated that a 'minimise' [instruction to reduce low priority message traffic] had to be instituted to allow the overworked staff in Saigon to clear a huge backload.[20] Concern was expressed soon after, in February 1967, about the steadily increasing volume of electrically transmitted message traffic and the large proportion of high precedence messages which was reducing the speed of electrical transmissions by overloading the communication centres and systems. This included what some considered to be 'blatant abuse of precedence and sometimes classification as well as the use of the Signals system to pass very long messages on mundane matters that should have been sent by postal channels.[21] The matter was viewed so seriously that the US regional military commander requested a substantial reduction in the number of messages transmitted by electrical means and a concerted effort to eliminate the abuses which were degrading the speed and efficiency of the common user teletypewriter system.[22]

At that stage, the traffic level seemed high but in a few years traffic levels were to rise significantly, so much so, that by September 1967, 2140 messages would be passed daily.[23] By comparison, in August 1943, the 17th Australian Infantry Brigade Signal Section, which provided communications support for more than two infantry brigades in combat against the Japanese in New Guinea, averaged 450 messages per day and peaked at only 600 messages in one day - using mostly line links rather than radio.[24]

---

19 AWM 102, 192/4/1, 'Monthly Traffic Statistics, AUSTCAN Relay Stations'; and RASCM, Lawrence, 'The South Vietnam Campaign', pp. 3.2-3.3.

20 Weir, letter, July 1990.

21 Weir, letter, December 1993.

22 AWM 103, 193/1/1, Pt 1, Memo: 'Communications Improvements' 25 February 1967.

23 AWM 102, 192/4/1, 'Monthly Traffic Statistics, AUSTCAN Relay Stations'; and RASCM, Lawrence, 'The South Vietnam Campaign', pp. 3.2-3.3.

24 Barker, *Signals*, p. 263.

In general, line had been the backbone of divisional and brigade communications in New Guinea with wireless being the main carrier for links above divisional level.[25] The dispersed nature of operations in Vietnam accentuated the reliance on radio communications links to a greater extent than had been experienced before. The improved and reliable communications facilities made possible by advances in communications technology also generated a greater reliance on the system which in turn fed demands for further improvements and increases in capacity.

By July 1966, 145 Signal Squadron had installed a communications centre at Vung Tau and VHF radio relay trunks, using the AN/MRC-69 system. These trunks connected Headquarters AFV at Cholon, Saigon, to 1 ALSG at Vung Tau via a repeater on 'VC Hill' in the Vung Tau area. Another link from Nui Dat to Vung Tau provided channels from Nui Dat to Vung Tau and Saigon.[26]

A daily Aerial Despatch Service was set up, using a Sioux helicopter from 161 Reconnaissance Flight based at Nui Dat. No establishment existed for the provision of this service so it was organised on an 'ad-hoc' basis because of the urgent need for it. This run included:

- Headquarters AFV in Saigon;
- Headquarters 1 ATF at Nui Dat;
- 1 ALSG and RAAF Base Operations Centre (both at Vung Tau): and
- the operational controller of 1 ATF, the US Headquarters of II Field Force, Vietnam, located at Long Binh.

The Sioux was later replaced by a Caribou aircraft which conducted regular flights from Vung Tau to Nui Dat, to Saigon and back to Vung Tau. There was never any lack of volunteers

25 WO 244/144 Notes on the Australian Army Signals in the North east Sector of the South West Pacific Area, 1944.

26 AWM 95, CD, 145 Sig Sqn Comms Report - Jun 66'; and RASCM, Lawrence, 'The South Vietnam Campaign', p. 3.3.

for a courier trip, although some couriers arrived as green as the bags they carried after a rough, wet Caribou trip in the bumpy tropical air.[27] By January 1970 the Aerial Despatch Service would be handling about 190 bags and 130 safe-hand bags per month.[28]

The Vung Tau detachment of 145 Signal Squadron was tasked with providing the following facilities: a tape relay station at Vung Tau; a signal centre for 1 ALSG; and HF Communications to all Australian Naval and commercial shipping using Vung Tau facilities, Australian Army Small Ships operating in coastal waters and inland waterways and the RAAF units at Phan Rang in northern South Vietnam.[29]

The US Defence Communications Agency system provided a 4-wire 3 KHz duplex channel between Headquarters AFV in Saigon and the RAAF Canberra bomber squadron (2 Squadron) in Phan Rang on the central eastern coast of South Vietnam.[30] However, whenever the Saigon - Phan Rang circuit failed, a Vung Tau - Phan Rang HF circuit was activated. This link was later established operationally in April 1967 because of the US link's unreliability and it was to receive particularly frequent use in 1970 and 1971. A secure radio teletype mode would be used when conditions permitted and, alternatively, voice or morse code would be used for passing traffic.[31]

Quite frequent contact also was made with the visiting RAN ships including the troop carrier HMAS *Sydney*, the supply ship

27 AWM 98, R193/1/67, 'Signals in Malaysia and Vietnam', August 1967, p. 7; AWM 95, CD, 145 Sig Sqn Comms Report - Jun 66'; AWM 95, CD, 103 Sig Sqn Monthly Report, July 1966; and RASCM, Lawrence, 'The South Vietnam Campaign', pp. 3.3 & 7.4.

28 AWM 95, CD, 110 Sig Sqn, Monthly Report, Jan 1970.

29 "'Swift and Sure': The Royal Australian Signal Corps in Vietnam", in *Command Communications*, October 1968, p. 12.

30 D Sigs 5-LC-2, 'Fixed Installation Equipment Assessment', 110 Sig Sqn, Oct 1967; and RASCM, Lawrence, 'The South Vietnam Campaign', p. 7.4.

31 RASCM, Lawrence, 'The South Vietnam Campaign', p. 7.4; AWM 95, CD, 110 Sig Sqn Monthly Reports Jan, Mar, May, Jun, Sep, Nov, Dec 1970, & Jan, Feb, Mar, Apr, May, Jun 1971; and AWM 95, CD, 145 Sig Sqn, Summary of Operations Apr 1967.

HMAS *Jeparit* and occasionally other RAN ships and Army vessels including the AV *Clive Steele* and AV *Harry Chauvel*. HF CW (morse) radio contact was usually made with these ships when they were two days out from Vung Tau and the links were usually maintained until two days after their departure from Vung Tau harbour on their return trip to Australia. Occasionally the link was maintained until the vessel arrived safely back in an Australian port.[32]

A communications link was also established with the Royal Australian Navy's Helicopter flight (aircrew and groundstaff) which operated with the US Army 135 Aviation Company, from October 1967 to June 1971.[33]

The Vung Tau detachment of 145 and then 110 Signal Squadron was located with the 1 ALSG. 1 ALSG had established itself in the sand hills on the edge of the beach, about 5 miles from the Vung Tau airstrip and about 6 miles from Vung Tau town and port.[34]

In 1966, the Signal installations left room for improvement. Equipment was set up in very hot and airless tents and was continually covered with fine dust.[35] Attempts were made to reduce this by laying topsoil and grass seeds recommended by the Australian CSIRO. In addition, mango and eucalyptus saplings were received and planted in an effort to stabilise the sand.[36]

Two well-constructed 'sloping V' HF antennas were also set up in the Vung Tau area to provide an alternative rear link to Melbourne, Australia. Later, in October 1967, a vertically polar-

32 RASCM, Lawrence, 'The South Vietnam Campaign', p. 7.4; AWM 95, CD, 110 Sig Sqn Monthly Reports Dec 1967, May 1968, Jun, Sep & Oct 1970, Dec 1971 & Jan 1972.

33 McNeill, 'An Outline of the Australian Military Involvement in Vietnam July 1962 - December 1972'; and Cowley, letter, 1990.

34 D Sigs 723/3/4, R.P. Woollard (Colonel), 'D Sigs Notes: South Vietnam - Malaysia Visit Dec 66.'

35 D Sigs 723/3/4, R.P. Woollard (Colonel), 'D Sigs Notes: South Vietnam - Malaysia Visit Dec 66'; AWM 95, CD, 110 Sig Sqn - Monthly Report, October 1967; and AWM 95, CD, 110 Sig Sqn - Monthly Report, August 1967.

36 AWM 95, CD, 110 Sig Sqn - Monthly Report, July 1967; and AWM 95, CD, 110 Sig Sqn - Monthly Report, Aug 1967.

ised 'log periodic' antenna was to be constructed because of the high noise level encountered on one of the sloping V antennas.[37] This was part of the solution to the problem of severe overcrowding of the HF frequency spectrum and lost signal power, captured by more powerful antennas. Other measures required to overcome the difficulties included dual transmissions and frequency diversity.[38]

The noise level problems were not surprising, considering that the Vietnam-Melbourne link was a 3800 mile two-hop path, crossing several time zones and the geomagnetic equator, and terminating in the most naturally noisy HF radio area in the world at the time and one which was also the most frequency congested.[39] The problems experienced were compounded by the effect of the fine beach sand at Vung Tau on the equipment.[40] A small hut was constructed to house a dismounted AN/TRC-75 and generator sets which were controlled from the Saigon end of the link. It was clear, in late 1966, that the Force was going to remain where it was for quite some time and so plans were made to relocate the key equipment items into air-conditioned pre-fabricated huts, but these were not immediately forthcoming.[41]

To help cope with the difficulties that were being encountered in managing the Australian communications network in Vietnam and to increase operational efficiency, 145 Signal Squadron was reorganised on a functional basis in March 1967. From this point on, the detachment at Vung Tau was to have an officer commanding, a communications officer, a technical officer and an administrative officer. Located with them at Vung Tau

---

37 D Sigs 723/3/4, R.P. Woollard (Colonel), 'D Sigs Notes: South Vietnam - Malaysia Visit Dec 66'; AWM 95, CD, 110 Sig Sqn - Monthly Report, October 1967; and AWM 95, CD, 110 Sig Sqn - Monthly Report, August 1967.

38 D Sigs 5-LC-2, Major E.J. Hynes, 'Fixed Installation Equipment Assessment, 110 Signal Squadron - South Vietnam', 4 October 1967.

39 Ibid.

40 Weir, letter, July 1990.

41 D Sigs 723/3/4, R.P. Woollard (Colonel), 'D Sigs Notes: South Vietnam - Malaysia Visit Dec 66.'

was a Minor Relay Station, with a tributary station located at the ALSG headquarters, also at Vung Tau.[42]

By early April 1967, the communications system had developed to the point where a comprehensive system of telegraph and voice channels, with a number of alternatives, existed over radio relay, line and radio circuits supported by the despatch rider service. However, according to reports, the system, including the fixed facilities at Nui Dat, hovered near collapse with marginal radio relay circuits, switchboards being used for tasks beyond their designed capacity, keying lines subject to damage, unreliable generators and overworked battery chargers, insufficient men for all the tasks required, equipment defects and other difficulties. The Signals Staff Officer and his squadron commanders from 145 (in-theatre and national rear link) and 103 (Task Force) Signal Squadrons wondered if they could hold on long enough until the problems were overcome.[43]

Shortly after this, on 18 April 1967, the advance party of 145 Signal Squadron's replacement - 110 Signal Squadron, arrived at Vung Tau and Saigon. The main body arrived on 28 and 29 April and, on 1 May 1967, the official handover was completed.[44]

Personnel shortages continued, however, with 110 Signal Squadron arriving 50 men understrength. The situation became critical when 145 Signal Squadron returned to Australia and this factor, combined with major transmitter failure on the Melbourne link, was instrumental in causing a 'minimise' order to be issued on 12 May concerning the use of the Melbourne link.[45]

In an assessment made in September 1967 by Major E.H. (Ted) Hynes, the Squadron Commander of 110 Signal Squadron, it was made clear that the static nature of the Australian Forces commitment in the theatre, with large headquarters permanently based in Saigon, Nui Dat and Vung Tau, had demanded

---

42 AWM 95, 145 Sig Sqn, Summary of Operations- Mar 67.
43 RASCM, Lawrence, 'The South Vietnam Campaign', p. 3.4.
44 AWM 95, 145 Sig Sqn, Summary of Operations - April 1967; and *Signals Bulletin,* Vol XVI, No 3, December 1968, p. 3.01.
45 AWM 95, CD, 110 Sig Sqn, Summary of Operations - May 1967.

a fixed communication system of a capacity and standard not attainable using the squadron's field equipment available at the time.[46] Improvements were needed but they took time to plan and, when the equipment was available, time to install. 133 Signal Squadron (the Base Signal Park in Penrith, NSW) contributed here, and throughout the campaign rapidly met demands for all items that it was responsible for supplying.[47]

One example of this was the AN/MRC-69 radio relay repeater that had been located on VC Hill near Vung Tau in late August 1967.[48] VC Hill was the only hill in the area that was higher than 50 metres and practicable to use as a radio site to provide for longer radio relay shots from Vung Tau. The link established from this hill provided for a triangulated communications system between Vung Tau, Nui Dat and Saigon by December 1967.[49] However, because of the unsatisfactory radio relay facilities that were later provided for the 1 ATF tactical headquarters in January 1968, 110 Signal Squadron was required to 'heli-lift' a radio relay shelter into the area of operations from Vung Tau. Prior to this, the radio relay resources had already been stretched thinly and this step forced the issue. It was becoming clear that more equipment was needed, so the decision was made to introduce the Siemens Halske 24 channel, 400 MHz UHF (fixed station) commercial microwave radio relay equipment that had initially been bought for the Army's requirements in Queensland and Papua New Guinea. It was envisaged that this would release the tactical radio relay equipment for possible future tactical operations out of Nui Dat.[50]

The Siemens equipment was subsequently successfully trialed in April 1968 on the Saigon-Vung Tau link (which had never

46 D Sigs, 'Fixed Installation Equipment Assessment 110 Signal Squadron - South Vietnam', dated 4 October 1967.

47 RASCM, Lawrence, 'The South Vietnam Campaign', p. 6.6.

48 AWM 95, CD, 110 Sig Sqn, Monthly Report, Aug 1967.

49 AWM 95, CD, 110 Sig Sqn, Monthly Report, Dec 1967.

50 AWM 95, CD, 110 Sig Sqn, Monthly Report, Jan 1968; D Sigs 5-LC-2, 'Communications SVN', D Sigs (Col J.D. Honeysett) Jan 1968; and D Sigs 723-3-4, Vietnam D Sigs Notes - South Vietnam - Malaysia Visit Dec 66.

worked properly using the tactical radio relay equipment) and the equipment was then permanently installed on VC Hill.[51] By January 1969 the installation of these triangulated commercial radio relay links was completed at all three sites and their performance rated as excellent.[52]

Another example of the problems encountered with providing fixed communications-type links was the Australian major tape relay centre and Headquarters AFV communication centre at the Free World building in Saigon, which was becoming increasingly overloaded. The volume of traffic handled by the installations increased by approximately 500 per cent during 1966 and this traffic growth continued into 1967.[53]

This increased traffic volume meant that the main link to Melbourne from Saigon had to be upgraded from 50 to 75 Baud operation, thus resulting in improved traffic handling because of the greater speed of message delivery.[54] This upgrade was to be short lived as problems were later experienced in maintaining the Kleinschmidt telegraph equipment that had been procured in 1961 simply because of the constant use and resultant wear and tear.[55] The cryptographic protection on this circuit

---

51 AWM 95, CD, 110 Sig Sqn, Monthly Report, Apr 1968; and AWM 98, R193/1/67, HQ AFV, 2 Apr 68.

52 AWM 95, CD, 110 Sig Sqn, Monthly Report, Jan 1969; and AWM 95, CD, 110 Sig Sqn, Monthly Report, Aug 1969. The Saigon-Nui Dat link did not change to Siemens equipment in 1969, although Siemens equipment was installed at both places for their links to Vung Tau. Tactical equipment continued to be used for the Saigon-Nui Dat link. See AWM 95, CD, 110 Sig Sqn, Monthly Report, Jan 1970; and AWM 95, CD, 110 Sig Sqn, Monthly Report, Dec 1969.

53 AWM 98, R193/5/2,. 'New Tape Relay Centre and Sigcen-H.Q. AFV', Memo dated 14 April 1967.

54 AWM 95, CD, 110 Sig Sqn, Monthly Report, July 1967; and D Sigs, 'Fixed Installation Equipment Assessment 110 Signal Squadron - South Vietnam', dated 4 October 1967..

55 Hockney, letter, 1989; and AWM 102, VN, Misc, 5800/1/1, 'Kleinschmidt Telegraph Equipment Comparison of 50 and 75 Baud Working', 7 Nov 69.

was also upgraded, reducing the time for maintenance and 'key' changes.[56]

While deliberations concerning the transition of the network from a field to a fixed type were taking place, arrangements had to be made to ensure that reliable means of communications were maintained. In order to provide reliable alternate routing to Australia, in the event of a loss of the Saigon-Melbourne link, the Major Relay Station in Saigon had been connected to the US STRATCOM Primary Relay Station at Phu Lam, in the Long Binh military complex, on 21 June 1967. This link allowed for alternate routing to Australia via Okinawa, Japan, through RAAF channels, or Honolulu via RAN channels.[57] This link had become necessary because of the difficulties experienced on the Vung Tau back-up to the Melbourne-Saigon link. The transmitter located at Vung Tau never quite managed to get good communications with Melbourne because of the basic incompatibility of the AN/TRC-75 field equipment at Vung Tau with the fixed station equipment at 6 Signal Regiment in Melbourne. The matter could not have been left unresolved because of the increased demands placed upon it due to the ever increasing volume of message traffic.[58]

Also in October 1967, a proposal was put forward by Captain P.G. (Phil) Skelton (later Lieutenant Colonel and Director of Communications - Army), the then Saigon detachment commander, outlining the advantages that would accrue from moving the AUSTCAN Saigon receiver station and co-locating it with the Phu Lam US Strategic Communications (US STRATCOM) Signal Battalion at Long Binh, 15 miles away from the Free World building. He argued that this move would alleviate the problems of local man-made noise levels, low received signal strength, high man-made radio interference, the shortage of space for antenna

56 AWM 95, CD, 110 Sig Sqn, Monthly Report, July 1967; and D Sigs, 'Fixed Installation Equipment Assessment 110 Signal Squadron - South Vietnam', dated 4 October 1967..

57 AWM 95, CD, 110 Sig Sqn, Monthly Report, June 1967; and D Sigs 723-3-4, Australian Force Vietnam, Monthly Report, June 1967.

58 AWM 95, CD, 110 Sig Sqn, Monthly Report, June 1967.

construction and technical maintenance, and limited accommodation, encountered at the Ba Queo site. Captain Skelton noted that the remoteness of Long Binh from Saigon was not a major difficulty as a new highway connected the two, and besides, the Phu Lam Signal Battalion's commander had indicated his support for the proposal.[59] Approval was given for this move and the planned completion date, given in January 1968, was for early March 1968.[60] The move was to be overtaken by events shortly after this announcement, as the Viet Cong and North Vietnamese offensives of early 1968 showed up the difficulties experienced in ensuring the physical security of the Ba Queo station.[61]

Thus, as the Tet offensive of 1968 was about to commence, 110 Signal Squadron was a substantially sized unit (See Diagram 18); and plans were well underway to establish the intra-theatre and overseas communication links into more secure, permanent and modern facilities.

The level of communications support to the Australian forces on operations in South Vietnam had already surpassed that experienced in earlier conflicts and yet further improvements were still to come. The introduction of modern semi-permanent equipment would improve the reliability and capacity of the network. It would also free up high-technology equipment previously used for the in-theatre and national rear links for use by the Task Force on operations remote from the Task Force base at Nui Dat. How this equipment would be used and how it would affect the operating procedures of the Task Force is described below. A particularly difficult challenge for the Task Force Signal Squadron was how it would cope with the unexpected additional pressures and strains of operating as an independent Task Force, while carrying out many of the responsibilities that would normally have

59 D Sigs 5-LC-2, 'AUSTCAN Saigon Receiving Station Proposed Relocation', 31 October 1967.

60 D Sigs E8-2-33A, 'AUSTCAN Installation Projects - 1968', 110 Signal Squadron, Phu Tho, 4 January 1968.

61 AWM 102, 192/6/1, 'HQ AFV Signal Instruction 1/68', 28 Feb 68.

been carried out by a supporting divisional signals unit. The challenges were enormous.

Task Force Signals - 1966 to 1967

As plans were being made for increasing Australia's military contribution to Vietnam from a battalion force to a brigade-sized Task Force, 103 (Task Force) Signal Squadron was formed, under the command of Major K.J. (Ken) Taylor who handed over to Major P.D. Mudd just before the unit departed overseas as he was promoted to command 1 Signal Regiment. This squadron was formed at Ingleburn in 1965, where it commenced training for operations. On 10 March 1966, the unit received a warning order for deployment to South Vietnam, while it was conducting a training exercise called 'Exercise Shakedown'. At that time, the squadron was nearly fully manned but its equipment was inappropriate. It held the British C11 and C42 radios, discussed in the Appendix, but needed AN/PRC-25 VHF manpack radios and the AN/PRC-47 and AN/GRC-106 HF radios, so as to be able to operate alongside its American counterparts, before it could leave. These items arrived only one month prior to the unit's departure from Australia.[62]

It was envisaged that the Squadron would provide communications support for the Task Force while other communications support would come from the higher headquarters' supporting Signals unit. In Australian terms, this support would normally have come from the Divisional Signals Regiment. The divisional signals regiment provides the necessary communications support to the divisional headquarters, the divisional administrative area, and to supported units and formations (including task forces) as well as to flanking formations. So the Task Force Signal Squadron, in theory, would be responsible only for the Task Force's own headquarters and subordinate units

62 RASCM, Lawrence, 'The South Vietnam Campaign', p. 4.1; and AWM 95, CD 103 Sig Sqn War Diary 10 Mar - 31 May 1966.

and for maintaining links with flanking formations. The situation in Vietnam was much more complex than this and the difficulties arising from the unique arrangements in Vietnam were to create numerous strains and challenges for the members of the Royal Australian Corps of Signals.

The instructions for 'Operation Hardihood' - the deployment of the Force to Vietnam in May 1966, were received as formation training was being conducted with Headquarters 1 Australian Task Force, 5 RAR, 6 RAR and supporting units in the Gospers area in New South Wales for ten days at the end of March. This exercise exposed several problems in the Squadron such as the inadequate capacity of line and switchboards. In light of this, extra capacity was taken to Vietnam.[63]

The Squadron moved to South Vietnam in groups from mid April through May 1966. Major Mudd, along with Sergeant Evans and Corporal Humble went by air, arriving in Saigon on 20 April, to effect liaison with the Australian and American commanders already there in the 173rd Airborne Brigade, based at Bien Hoa, and the Task Force's future higher command, II Field Force Vietnam, also at Bien Hoa. Men and equipment began arriving at Vung Tau on 4 May, 1966. There the vehicles and equipment were unloaded from HMAS *Sydney* into the squadron's staging area.[64] Of the 22 Fitted-For-Radio (FFR) Landrovers on board, eight had been towed aboard in Sydney because they were un-serviceable, thus indicating a potential source of problems that would be encountered once the Task Force deployed to Nui Dat.[65]

While at Vung Tau, 103 Signal Squadron also had its communication centre fully operational as, over the following few weeks, the Task Force prepared for its deployment to Nui Dat. The squadron deployed radio detachments with 5 RAR, which was operating under 173rd Airborne Brigade in clearing operations in the area 1 ATF was to occupy. Other detachments were deployed with 1

63 Ibid.
64 AWM 95, CD 103 Sig Sqn War Diary 10 Mar - 31 May 1966.
65 RASCM, Lawrence, 'The South Vietnam Campaign', p. 4.1.

Field Squadron, Royal Australian Engineers, 1 Field Regiment, Royal Australian Artillery, and 161 Reconnaissance Flight, of the Australian Army Aviation Corps. Major Mudd expected that manpack operations might become quite common, so he was anxious that his men become familiar with manpacking both PRC-25 and PRC-47, together with full personal equipment, rations and ammunition.[66]

On 5 June 1966, the Squadron made the move from Vung Tau to Nui Dat where preparation of the Squadron area then commenced. A rear detail of eight men remained at Vung Tau to control the move forward of stores held there. Some of the Squadron travelled in their unit vehicles but most of the move was by Chinook helicopter landing on what was soon known as 'Kangaroo Pad'. This was located immediately in front of the squadron's lines and was guarded and patrolled by the Squadron and its successors for the remaining six years.[67]

In selecting Nui Dat as the base camp for the Australian Task Force, communications requirements played a significant role. Major Mudd wrote, in his June 1966 report, that the choice of the base camp site was based on a wide variety of factors. One of these was the requirement for a position which would allow communications coverage of the area of operations... so we were forced towards [Nui Dat hill].[68]

Brigadier (later Major General) S.C. Graham, Commander of 1 ATF from January to October 1967 stressed that almost total reliance on VHF and radio relay usually was the largest single factor in deciding where a headquarters could be located. Headquarters have traditionally been located where control can best be assured but the VHF requirements made this a particular problem in Vietnam.[69]

66 AWM 95, CD 103 Sig Sqn War Diary 10 Mar - 31 May 1966.
67 AWM 95, CD 103 Sig Sqn War Diary June 1966; and RASCM, Lawrence, 'The South Vietnam Campaign', p. 42.
68 AWM 95, CD 103 Sig Sqn War Diary June 1966.
69 RASCM, Letter from Major General S.C. Graham to Lieutenant Colonel G.J. Lawrence, June 1977.

There were other criteria used to determine the site. Nui Dat best answered these requirements.[70] Two later Commanders of the Australian Force Vietnam (COMAFV) were to criticize this decision, claiming that the operations could have been conducted from the security of Vung Tau, thus relieving 1 ATF of the immense burden of securing and administering the Nui Dat base.[71]

Nui Dat was described as a 'pimple in a large flat area.' Unfortunately, the hill was a steep and rough piece of ground - unattractive as a headquarters site. Consequently, 'we finished up with the situation where the VHF radio elements of the squadron [were] operating on the top of Nui Dat, remotely controlled from the Task Force HQ, over about ¾ mile of cable.'[72] In the long run, it was argued, the advantages of having a good comfortable site in a rubber plantation would outweigh the disadvantages of having to operate some radio detachments a little remote from the rest of the squadron.[73]

The Signals area, including that of 547 Signal Troop (the signals intelligence troop discussed in Chapter Eight), formed a sizeable portion of the Task Force Headquarters perimeter (see Diagram 17). The Signals installations were all dismounted from the vehicles and installed in dug-in tents, providing uncomfortable operating conditions. A new Signal Centre was planned and

---

70 The first requirement was that it be located between the enemy's main force bases and the areas of dense population. It had to provide an airfield, for a maintenance area, and provide sufficient room to fight should the base come under heavy attack. It had to be on ground which would be above water level during the wet season. It needed to be on relatively open ground and not too near densely settled areas so that maximum use of firepower, rather than manpower, could be used for its defence. It needed to be located close enough to Vung Tau/Baria area to avoid diversion of forces for line of communication security duties or, alternatively, maintenance by air. McNeill, *To Long Tan,* p. 197. See also Frost, *Australia's War in Vietnam,* pp. 77-78; and T. Burstall, The Soldier's Story, University of Queensland press, St Lucia, 1986, p.37.

71 D.M. Horner, 'A Complex Command: The role of the Commander, Australian Force, Vietnam', in *Journal of the Australian War Memorial,* No 13, October 1988, p. 21.

72 AWM 95, CD 103 Sig Sqn War Diary June 1966.

73 AWM 95, CD 103 Sig Sqn War Diary June 1966.

constructed, in late 1966, using a 60 x 24 feet pre-fabricated hut and a concrete floor, poured with some assistance from the local engineers. It became operational on 21 January 1967. A fixed signals installation was also being established by late 1966 for the base camp, in order that unit equipment might be retained for mobile operations if and when they were to occur.[74] The communication centre was classified as a Minor Relay Station and its circuits operated to Headquarters AFV in Saigon, 1 ALSG at Vung Tau, and 1 ATF (Forward) - that is, the Task Force headquarters when deployed away from the base at Nui Dat.[75]

In addition, plans were made, and subsequently implemented, for using the telephone central office (MTC-7), and teletypewriter central office (MGC-17) shelters, which had been denuded of their equipment components, as an emergency signal centre in the event of the main signal centre being damaged in an attack.[76]

Problems soon surfaced with equipment becoming unserviceable because of insufficient spare parts. The Squadron was particularly hampered by the high incidence of vehicle failure, which had been predicted. Another matter which required rapid attention was the provision of air conditioners for all the equipment shelters. They had gone to Vietnam without any and high capacity communications simply could not run efficiently in a tropical environment without reliable air conditioning.[77]

The Squadron was also poorly equipped to provide an efficient Task Force switchboard. The telephone switchboard service was operated by two signalmen during the day and one at night, and the operators usually bore the brunt of the frustra-

---

74 D Sigs 723/3/4, R.P. Woollard (Colonel), 'D Sigs Notes: South Vietnam-Malaysia Visit Dec 66'; and AWM 95, HQ AFV-Communications Report, January 1967..

75 '104 Signal Squadron - Unit History' (unpublished manuscript).

76 AWM 95, CD, 103 Sig Sqn, Monthly Report, December 1966.

77 AWM 95, CD 103 Sig Sqn War Diary August 1966; AWM 95, CD, 110 Sig Sqn, Monthly Report Mar 70; and D Sigs 723/3/4, R.P. Woollard (Colonel), 'D Sigs Notes: South Vietnam-Malaysia Visit Dec 66'.

tions of those trying to use the service. Initially the SB-86 tactical switchboard, with a maximum of 60 subscribers, was used for the Task Force headquarters and the operators had to contend with the unreliable 'eyeball' indicators and cords that came with it. But this switchboard was not big enough. 200-line TC-10 switchboards, of 1941 vintage, were to be eventually despatched in 1967 from the Base Signal Park in Penrith, Australia for use at Vung Tau and Nui Dat (known as Emu and Ebony respectively).[78] Due to their 'healthy appetite for spares', difficulties were soon encountered in obtaining spare parts for the frequently un-serviceable and out-of-production switchboards.[79]

This equipment was to be eventually replaced, in late 1970, by a more advanced and reliable AN/MTC-3 switchboard, eventually servicing 143 local subscribers, including 29 unit switchboards.[80] The need for a field operable automatic switchboard was clear, but this need would have to wait until after the Vietnam War to be satisfied.

The Director of Signals, Colonel R.P. Woollard, visited the Task Force Signal Squadron in Vietnam in December, 1966. In his report, written later 'to assist and guide [the Directorate of Signals] staff in taking appropriate action', he clearly indicated the inadequacy of the equipment they had been provided with.[81] The then Lieutenant Colonel Barry Hockney, working at the Directorate at the time, later commented that the independent nature of the operations conducted by the Task Force and the massive communications support that was required in that sort of warfare had not been fully realised...

78 RASCM, Lawrence, 'The South Vietnam Campaign', p. 7.8.

79 AWM 98, R193/1/67, HQ AFV, 'Signals in Malaysia and Vietnam', August 1967; AWM 95, CD, 104 Sig Sqn Monthly Report, May 1967; AWM 95, CD, 104 Sig Sqn, Monthly Report, May 1971; and AWM 103, R981/1/10, HQ 1ATF, 'Switchboard-Ebony Central'.

80 AWM 95, CD, 104 Sig Sqn Monthly Report, November 1970; and AWM 95, CD, 104 Sig Sqn Monthly Report, July 1971.

81 D Sigs 723/3/4, R.P. Woollard (Colonel), 'D Sigs Notes: South Vietnam-Malaysia Visit Dec 66'.

> It was just so much bigger than we had appreciated. We had seen the Task Force as part of a division and it was equipped for that role - but it was operating independently and it took some time [for us] to get a true perspective of the real needs.[82]

The response of the technical maintenance staff to all these problems was admirable. Their performance, working on equipment that was unreliable in very difficult and trying conditions, spoke very highly of the capability of the Australian technicians and operators.[83]

In the technical maintenance area, the staff were also running up against a problem which seemed to be almost universal there at the time. The squadron's technical maintenance section was equipped for mobile operations and not for establishing a base camp workshop, and as far as could be seen, the technical maintenance workshop would probably not have to move from its location at Nui Dat. Consequently, a dismounted workshop was set up, thus making available the use of an extra, much-needed, vehicle.[84]

With the Task Force having taken over operational control in the Phuoc Tuy area, it commenced operating a number of nets, including

- the Task Force command net on VHF and HF (as a back-up),
- a telegraph net,
- a command net rear (activated when HQ 1 ATF was deployed forward of Nui Dat),
- a tactical air net and

82 Hockney, interview, August 1989.
83 Ibid.
84 AWM 95, CD, 103 Sig Sqn, Monthly Report, July 1966.

- a base defence net.[85]

The VHF voice command radio net, using the ubiquitous AN/PRC-25 (replaced in 1969 by the AN/PRC-77 with secure speech mode) and the HF telegraph nets (using the AN/PRC-47) began operating as had been envisaged by the unit establishment. Substations on both nets were provided from every major unit in the Task Force and minor units were connected only to the voice command net. In addition, the squadron provided two non-standard radio links. One was to the US Army advisers based at the provincial capital of Baria to the south of Nui Dat. The other was to alert the casualty evacuation helicopters, relocated each night from Nui Dat to Vung Tau, because of the scarcity and value of the machines.[86]

As there were a number of radio nets to operate and maintain, radio operators were required in many places. During a tour of service in Vietnam radio operators were employed in a number of areas. They were attached to units such as infantry battalions, the field (artillery) regiment or the reconnaissance flight detachment. Personnel working in battalion detachments required a good working knowledge of battalion internal communications, fire-support-base procedures, defence and patrolling activities.[87] An alternative was for an operator to be sent, along with two others, as part of a liaison officer's detachment, to provide the communications. Liaison officers were dispersed throughout the Phuoc Tuy Province and attached to US or other allied units working alongside the Australians on operations, so as to minimise misunderstandings and mistakes.[88] Another place

85 1st Australian Task Force Vietnam, Standing Operating Procedures for operations in Vietnam, Nov '69 edition; and '104 Signal Squadron Unit History'.

86 AWM 95, CD, 103 Sig Sqn, Monthly Report, June 1966; RASCM, Lawrence, 'The South Vietnam Campaign', p. 4.3; and AWM 98, 5820/130/8, 'Radio Equipment - AN/PRC-77' 23 December 1968.

87 RASCM, '104 Signal Squadron Unit History'.

88 RASCM, Lawrence, 'The South Vietnam Campaign, p. 7.7; and '104 Signal Squadron Unit History'.

of employment for operators was in the radio detachment on Nui Dat Hill, where some 18 radio sets were operated from the squadron's radio bunker on the hill. An operator was required to supervise a large number of radio nets, including antenna systems and remote facilities. A usually less hectic task was as a retransmission detachment operator. For this work, operators were deployed to remote areas to set up and operate an automatic radio retransmission facility. The final alternative was to serve as a command post operator in the 1 ATF Command Post. There the operators had to be highly skilled in radio procedures, security codes and map marking. As the 1 ATF command net was a large operational net, with widely distributed sub-stations, the command post operator had to be capable of handling a large volume of traffic and maintaining strict radio discipline on the nets. They were thus able to prove to commanders and staff officers that, apart from performing their communications jobs well, Signalmen performed vital tasks other than those for which they had specifically trained.[89]

Headquarters 1 ATF operated two principal types of command posts: namely the main command post in the Nui Dat base and a forward command post for forward deployments of the Task Force Headquarters. An emergency command post was also maintained, duplicating the main command post functions for emergency use. There were three different means of communication employed in the command posts: radio, telephone and intercom set. Most HF and VHF radios were operated by remote lines carried by cable.[90]

Despatch rider services also began upon arrival at Nui Dat with four runs a day, as well as daily meeting the RAAF flight from Vung Tau and the US flight from Headquarters II Field Force,

89 '104 Signal Squadron Unit History'; and RASCM, Lawrence, 'The South Vietnam Campaign', p. 7.8.

90 AWM 103, 193/1/1 Pt 1, Command Post Communications - 1 ATF', Report by Major G.J. Lawrence, November 1967.

Vietnam.[91] As the war progressed, 104 Signal Squadron's despatching responsibilities expanded to included daily aerial runs to all Task Force units deployed outside the Nui Dat base, using aircraft from 161 Reconnaissance Flight, as well as a daily air letter service to other bases such as at Bien Hoa.[92]

The sheltered telegraph terminal, AN/MGC-17, described in the Appendix and used by the squadron for its circuits to Saigon and Vung Tau, was of limited use due to high fault rates and the confined space which did not allow for a technician and an operator to be working simultaneously. The telegraph terminal also could not take extra connections. Consequently, the equipment was dismounted and set up in a tent.[93]

US procedures and field codes were introduced into service with the Australian Army in Vietnam for the sake of uniformity. Monthly Signal Operating Instructions (SOI) were issued to give the information necessary to operate the various signals facilities in the Task Force. The US operations code and their numeral & authentication codes were also adopted. It took some time for the radio operators to adjust to these changes. The adoption of the US codes had wide ramifications for Signals Corps and other Corps, particularly in terms of training.[94]

Major Mudd confronted the Task Force Commander, Brigadier O.D. Jackson, over the issue of the frequency of changing call signs, address groups and codes. In his June 1966 report, he noted:

> Last month we changed every second day. Now, on request from units and after my arguments to

---

91 AWM 95, CD, 103 Sig Sqn, Monthly Report, June 1966; AWM 95, CD, 104 Sig Sqn, Monthly Report, November 1967; RASCM, Lawrence, 'The South Vietnam Campaign', p. 4.3; and AWM 98, R193/1/67, HQ AFV - 'Signals in Malaysia and Vietnam, August 1967'.

92 AWM 95, CD, 104 Sig Sqn, Monthly Report, July 1971.

93 AWM 95, CD, 103 Sig Sqn, Monthly Report, June 1966; and RASCM, Lawrence, 'The South Vietnam Campaign', p. 4.3.

94 RASCM, Lawrence, 'The South Vietnam Campaign', p. 4.4; and AWM 95, CD, 103 Sig Sqn, Monthly Report, June 1966.

> the contrary, the Commander has directed that we will change only every seven days. This is based on the argument that [Viet Cong] reaction time is of the order of a month to produce anything of the order of battalion size reaction. I am sure this subject is one which will receive further discussion. It is the old one of balancing speed and simplicity against security.[95]

Later revelations, discussed in Chapter Seven, were to show how misplaced this confidence was. Within a year the frequency had been reduced to monthly changes for the sake of convenience, especially for infantry battalions which increasingly conducted month long (or longer) operations.[96] The security benefits were too strongly countermanded, in the minds of many, by the benefits of having a clear and reliable frequency knowing that the enemy could instantly identify their nets rather than have unreliable communications each time the frequency was changed.[97]

The initial operations conducted from Nui Dat were designed to secure the base area, to extend the controlled area and destroy Viet Cong installations in the Tactical Area of Responsibility (TAOR). These operations required radio communications over ranges of only five or six kilometres, presenting no difficulties to 103 Signal Squadron. Securing the base area involved clearing a zone, around the base, of any civilians. Consequently, the village of Long Phuoc, about four kilometres to the south east of the base, was relocated and the bunker, trench and tunnel system found there was destroyed by 6 RAR in June 1966.[98]

---

95 AWM 95, CD, 103 Sig Sqn, Monthly Report, June 1966.

96 RASCM, Lawrence, 'The South Vietnam Campaign', p. 4.4

97 Ibid., p. 7.10.

98 AWM 95, CD, 103 Sig Sqn, Monthly Report, June 1966; and AWM 95, CD, 103 Sig Sqn, Monthly Report, August 1966. See also McNeill, 'An Outline of the Australian Military Involvement in Vietnam July 1962-December 1972', p. 53.

From mid-July to mid-August, battalion group operations were conducted to expand the controlled area. On 17 August, Nui Dat came under a mortar attack, receiving about 60 rounds of 82mm mortars. None landed in the Signals area and no damage was suffered except for one line being cut by a piece of shrapnel.[99]

The following day, D Company of 6 RAR fought its famous battle in the Long Tan rubber plantation three kilometres to the east of Nui Dat, where they encountered a significantly larger force than had been anticipated, thought to be on its way to attack the base camp. The battle was closely fought and by the following day the enemy dead totalled 245 whilst D Company 6 RAR suffered 18 dead and 25 wounded.[100] The signals intelligence aspects of the battle are discussed in Chapter Eight. It is sufficient to say at this stage that the Battle of Long Tan was a turning point for the acceptance of the value of signals intelligence by Australian commanders. There remains, however, a significant aspect of the battle which was to have wide ramifications for the conduct of operations for the remainder of the war.

The Task Force Signal Squadron Commander, Major Mudd, reported that throughout the battle, communications on the Task Force, artillery and 6 RAR nets were good, though there were some troubles within D Company itself. Because the communications were good, the artillery was able to give very close fire support and there was good control over the despatching and movement of the reinforcements for D Company. Major Smith, the infantry company commander stated that 'we in D Company headquarters virtually just sat there calling in the artillery and watching events as if we were impartial observers.'[101] Mudd concluded by writing 'I believe this had a fairly important role in the

99 Tonkin, letter, October 1989.

100 The battle of Long Tan is covered in detail by McNeill in *To Long Tan,* Part Four. See also L. McAulay, *The Battle of Long Tan,* Arrow Books, London, 1987; and T. Burstall, *The Soldier's Story,* St Lucia, University of Queensland Press, 1986 for further accounts of this battle.

101 Cited in Burstall, *The Soldier's Story,* p. 107.

successful outcome of the battle.'[102] Major H.B. Honner, the New Zealand Commander of 161 Field (Artillery) Battery supported this contention by stating that 'without these [AN/PRC-25 VHF radios] and the far more reliable communications than those provided by the replaced HF radios, there might well have been a different outcome to this whole battle.'[103]

The impact of effective radio communications and electronic intercept experienced during this time set the tone not only for the subsequent operations but for the remainder of the Australian Task Force's deployment to Vietnam. The critical role played by the Signals Corps in establishing that there was a presence followed by the effectiveness of the radio equipment in ensuring the provision of effective and accurate fire support and reinforcements was not lost on commanders throughout the Force. Not surprisingly, it served to engender greater confidence in the ability of the Force to respond quickly and effectively to any threat that could emerge. It also served to confirm the combat power multiplying effect of ready access to artillery and other arms support due to the effective communications provided. Even for elements as small as company sized and platoon sized groups deployed on unprecedentedly dispersed operations against substantial enemy forces. As Horner wrote

> Better communications meant the more flexible use of artillery. In turn, this enabled the infantry commanders to take more risks with their patrol tasks, secure in the knowledge that artillery support would be available should the infantry patrol meet an unexpectedly superior force.

This then had a flow on effect on the tactics employed by infantry battalion commanders who began predicating their plans

102 AWM 95, CD, 103 Sig Sqn, Monthly Report, August 1966.
103 Cited in McAulay, *The Battle of Long Tan,* p. 71. See also McNeill, *To Long Tan,* p. 323; and Horner, *The Gunners,* p. 479.

on the rapid and effective employment of artillery to block or channel the enemy into ambush sites.[104]

Subsequent operations began taking place a little further out and in some more difficult country. For instance, during 'Operation Vaucluse' in September, 6 RAR was deployed in the Nui Dinh hills area, to the north west of Baria, and south west of Nui Dat. During this operation it was necessary to use a rebroadcast station on the hills above Vung Tau, when the infantry battalion moved down to the south-western slopes of the hill area. This worked well and communications were maintained throughout the operation. This procedure was to be frequently used over the remaining five and a half years in Vietnam by the Task Force Signal Squadron.[105]

Prior to the operation, in case difficulties should arise, the Squadron had arranged with 161 Reconnaissance Flight for tests of aerial rebroadcast using both the Sioux helicopter and Cessna light fixed-wing aircraft. Mudd reported that the Squadron provided the AN/PRC-25s, suitable mounting arrangements were developed for both aircraft, the trials were carried out successfully and the facility was subsequently used on numerous occasions. However, some difficulty was experienced with faulty retransmission cables. This problem was not fully overcome until the more powerful AN/VRC-12 radios were received in 1967 and used for this purpose.[106]

In October 1966, it was decided that Headquarters 1 ATF would deploy a tactical headquarters, for the first time, during a cordon and search operation of the Hoa Long village between Nui Dat and Baria a few kilometres to the south. The deployment provided no communications problems because of its simplicity and proximity to Nui Dat. Only one radio net with operators was required. Nevertheless, this operation set a prec-

104 Horner, *The Gunners,* p. 490.
105 AWM 95, CD, 103 Sig Sqn, Monthly Report, September 1966.
106 AWM 95, CD, 103 Sig Sqn, Monthly Report, September 1966; AWM 95, CD, 103 Sig Sqn, Monthly Report, October 1966; and RASCM, Lawrence, 'The South Vietnam Campaign', p. 4.6.

edent, marking the start of a new phase of communications for the squadron. This new phase concerned Major Mudd, as his Squadron was already fully committed and the deployment of the Task Force even further away than Hoa Long would prove to be a strain on resources beyond that with which they had been intended to cope. [107]The manpower provided for the Task Force Signal Squadron was not adequate for the commitments, given that it was required to provide communications both for a large semi-permanent base camp and tactical communications on operations away from Nui Dat.[108]

Early in November 1966, the Main Headquarters of 1 ATF once again deployed away from the base area on Operation 'Hayman'. This was a search and destroy operation on Long Son Island, lasting five days, just off the coast to the south west of Nui Dat. Voice and telegraph circuits were again provided by the Task Force Signal Squadron.[109]

By March 1967, during Operation 'Portsea', increasing problems were being experienced with VHF radio communications. In this exercise, a radio link was operated over about 20 kilometres using the AN/PRC-25. Although the radio path was good, VHF frequencies provided a problem, as the limited number of frequencies allocated to the Australians were widely duplicated in other formations working alongside them. The Americans also used the AN/VRC-12 radio (a more powerful set than the AN/PRC-25 which used a similar frequency range), and this added to the frequency problems.[110] On the other hand, the AN/PRC-25 radios were considered to be more reliable than the HF AN/PRC-47 sets used to provide the telegraph net for the Task Force. The AN/PRC-47 sets were subsequently withdrawn from operations in April 1967. Unit headquarters did not appreciate being slowed

107 AWM 95, CD, 103 Sig Sqn, Monthly Report, October 1966; and RASCM, Lawrence, 'The South Vietnam Campaign', p. 4.6.

108 AWM 95, CD, 103 Sig Sqn, Monthly Report, August 1966; and AWM 95, CD, 103 Sig Sqn, Monthly Report, January 1967.

109 AWM 95, CD, 103 Sig Sqn, Monthly Report, November 1966.

110 AWM 95, CD, 103 Sig Sqn, Monthly Report, March 1967.

down by the telegraph detachments and the little administrative traffic to Task Force Headquarters, generated on operations, could easily be handled on the Task Force command net, using the AN/PRC-25s.[111]

The closure of the telegraph net to the Task Force's subordinate units was one of the first changes introduced at Nui Dat with the arrival of the new Task Force Signal Squadron, 104 Signal Squadron, under Major G.J. (Gerry) Lawrence. There, the newly arrived squadron took the place of 103 Signal Squadron, which left for Australia in the same month. The squadron's advance party of 14 all ranks had arrived at Nui Dat on 17 April 1967, with the main group of 45 all ranks arriving on 26 April 1967.[112]

Shortly after their arrival, in August 1967, Lawrence wrote back to Australia proposing that the Task Force Signal Squadron adopt the practice of individual relief entirely, rather than complete unit replacement. It was considered that this would ensure no disruption to communications, as had been experienced in April 1967, as the newly arrived squadron had taken some time to adjust to the routine at Nui Dat and to the operations away from the base.[113] Subsequently, 139 Signal Squadron, the Brisbane based Regular Army Task Force Signal Squadron, became responsible to train and prepare reinforcement personnel sent as individual replacements to 104 Signal Squadron at Nui Dat.[114]

The Task Force Signal Squadron, although not operating radio relay links themselves, had a number of radio relay links terminating in their telegraph terminal and switchboard. [115]Operations in Vietnam, unlike in previous wars, had resulted in the need to supplant the traditional wire links with mobile multichannel radio relay systems across the miles that separated the

111 AWM 95, 104 Sig Sqn Monthly Report, April 1967.

112 Ibid. For a witty, detailed and anecdotal account of 104 Signal Squadron in Vietnam in 1967 see Weller, *The Cream of Her Youth*.

113 AWM 95, CD, 104 Sig Sqn, Monthly Report, August 1967.

114 '139 Signal Squadron Unit History'; and interview with Brigadier K.P. Morel, August 1989.

115 AWM 95, CD, 103 Sig Sqn, Monthly Report, July 1966.

base camps of subordinate units and, as a result, multichannel radio systems were extended to lower levels than ever before. Moreover, in Vietnam, a US division usually operated in an area of responsibility covering 3,000 to 5,000 square miles whereas in 'conventional' warfare it was envisaged that a US division would operate in an area of 200 to 300 square miles. In addition, battalion-sized and at times company-sized bases were sometimes established and these all had to be tied together with reliable, responsive communications to ensure quick reaction in an emergency.[116] These dispersed operations were made possible only because of the improvements in range, durability and flexibility of the increasingly sophisticated communications technology being introduced.

The three links that were established early at Nui Dat to provide these lower level responsive communications links were:

- a four-man detachment from 145 (later 110) Signal Squadron, providing switchboard-to-switchboard, 'sole user' and telegraph circuits to Saigon and Vung Tau.
- 39th (US) Signal Battalion, providing a back-up for the 145 Signal Squadron system which was available for deployment when the Task Force moved forward on operations. They also provided telephone trunk access direct from Task Force Headquarters into the US communications base system.
- A 10-man detachment from 53rd (US) Signal Battalion, providing switchboard-to-switchboard, sole user and telegraph circuits to the Task Force's higher operational command, II Field Force, Vietnam. These circuits to II Field Force, Vietnam, were essentially tactical circuits for control by and reporting to the superior headquarters.[117]

116 Rienzi, *Communications Electronics*, 1962-1970, pp. 60 & 62.
117 AWM 95, CD, 103 Sig Sqn, Monthly Report, July 1966.

By late 1967, there were increasing occasions when 1 ATF had to operate outside its normal area of operations in Phuoc Tuy Province; and the Squadron was forced to grow to meet the demands (see Diagram 19). To ensure satisfactory communications over longer distances, it became necessary to establish a forward headquarters close to where the combat units were deployed.[118] Radio relay detachments also accompanied the Task Force Headquarters when it deployed away from Nui Dat. Major K. P. (Keith) Morel, the squadron commander of 104 Signal Squadron from January 1969 to January 1970, commented that 'we had what was termed the first XI and second XI'. The first XI deployed with the Task Force's Forward Headquarters and the second XI remained at the Nui Dat base.[119] In order to function effectively at both locations at one time, an increase in the size of the signal squadron was asked for by Major Mudd and eventually granted.[120] When operating away from Nui Dat, alongside other allied formations, the advanced headquarters normally required 28 speech and telegraph radio relay or radio circuits operating to the flanks and rear of 1 ATF (excluding standby and signal-engineering channels). This figure stood in marked contrast to the requirement for seven rear and flank links projected, in 1965, for a Task Force operating under the command of a division.[121] Improvements in communications technology had not only facilitated more flexible and dispersed operations but they had served to generate an even greater demand for communications where previously there had been no expectation of such means being available.

Operation 'Paddington' was the first of such large operations conducted some distance from Nui Dat. This operation was a search and destroy mission, instigated at the behest of the

118 AWM 103, 193/1/1, 1 ATF 'Communication Requirements of a Task Force Headquarters.'
119 Morel, interview, August 1989.
120 AWM 103, R310/1/7, 'Amendment to TF Sig Sqn Establishment', Jul 1967.
121 AWM 103, 193/1/1, 1 ATF 'Communication Requirements of a Task Force Headquarters.'

US Commander in Vietnam, General Westmoreland, into the May Tao mountain area, over thirty kilometres to the north east of Nui Dat, in July 1967.[122] For this operation, an ad-hoc Signal Troop of 36 all ranks, including US signalmen attached, was formed. It moved with Headquarters 1 ATF Forward and stayed with the forward headquarters throughout the operation. The signals arrangements proceeded as a normal 'step-up' operation (whereby a detachment went ahead of the main headquarters to establish the links for the remainder of the headquarters while it was on the move) and no difficulties were encountered. However, a number of 'lessons' came from this operational deployment. The amount of D10 cable was underestimated and over 10 miles was finally laid, after having taken only 8 miles initially. Users also underestimated their requirements for telephones and, 'everyone re-learnt that tracked vehicles cut cable!' But perhaps the most significant lesson from this deployment was the advantages of using radio relay. On this operation, the MRC-69 radio relay shelters were provided by the 9th US Division's Signal Battalion (the Task Force's superior headquarters for this operation).[123]

As a result of this experience and at the request of the Task Force's 104 Signal Squadron, a radio relay deployment detachment was constituted, in August 1967, from technical maintenance personnel, by the in-theatre and national rear link unit (110 Signal Squadron) for use when the Task Force Headquarters split for operations. The detachment used a reconstituted MRC-69 radio relay shelter that had been used on VC Hill near Vung Tau.[124] The radio relay detachment from 110 Signal Squadron at Vung Tau, which operated the Nui Dat terminal of the in-theatre radio relay system, was brought down from Nui Dat Hill at the same time and installed in the signal centre adjoining the Task

122 Frost, *Australia's War in Vietnam*, p. 93.
123 AWM 95, CD, 104 Sig Sqn, Monthly Report, July 1967.
124 AWM 95, CD, 110 Sig Sqn, Monthly Report, August 1967.

Force Headquarters at Nui Dat.[125] This had been made possible by the installation of a '102 Hills' mast with 'top mounted Yagis' outside the signal centre. At the same time, the radio relay attendants became the System Control Officers (SCOs) at 1 ATF, thus providing the Australian in-theatre links with centralised control and training by 110 Signal Squadron. This move proved successful and the circuits improved considerably. Eventually, the system control room was re-modelled, allowing for three AN/TRC-24 radio relay systems and associated equipment to be installed.[126]

The first deployment for the newly constituted deployment detachment occurred when a nine-channel link was established between Headquarters 1 ATF at Nui Dat and the Task Force Forward Headquarters during Operation 'Kenmore'. This took place just outside the Task Force's normal area of operations, in late September 1967.[127] This operation involved the two infantry battalions, the field regiment and the cavalry squadron deploying in thick country some 20 kilometres east-south-east of Nui Dat.[128] Two MRC-69 shelters and crews, each of three technicians, were detached to 104 Signal Squadron to man both ends of the radio relay link. These technicians were detached from 110 Signal Squadron's Vung Tau detachment's technical maintenance workshop. Captain Joe (later Lieutenant Colonel) Viksne, the detachment's commander, reported that the shelter was 'heli-lifted' into position. No problems were encountered in bringing the system into operation. However, two generators failed on the first day. They were immediately replaced, and a high quality circuit, carrying eight channels, was in operation at the end of [September].[129]

---

125 K.P. Carey (Major), 'Swift and Sure', in *Command Communications*, USARV Pam No 105-4, October 1968, p. 12; and AWM 5, CD, 110 Sig Sqn, Monthly Report, August 1967.

126 AWM 95, CD, 110 Sig Sqn, Monthly Report, February 1968.

127 AWM 95, CD, 110 Sig Sqn, Monthly Report, September 1967.

128 AWM 95, CD, 104 Sig Sqn Monthly Report, September 1967.

129 AWM 95, CD, 110 Sig Sqn, Monthly Report, September 1967.

Three CH-47 Chinook helicopters and three armoured personnel carriers (APCs) were required to move all the men and stores that were deployed with the Task Force Signal Squadron.[130] The artwork on the cover of the book captives much of this. One significant effect of the introduction of radio relay to the Task Force's deployed headquarters was that the secure telegraph circuit, using the AN/PRC-47 (and later the more reliable AN/GRC-106) HF radio provided by the Task Force Signal Squadron, received little use (four to six messages a day). This was because the staff used the radio relay telephones for most communications to the rear, whereas, for the topics discussed, a secure circuit should have been used. This lack of security was pointed out on several occasions but it did not have the desired effect - people continued doing so. Besides, many of them believed that the Viet Cong were well on the way to being defeated.[131]

Lessons and Reflections

From the time that the Australian Task Force arrived in Vietnam in May 1966 until the end of 1967, a number of lessons had been learnt. It was clear from early on that the deployment of a Task Force to Phuoc Tuy Province would present the Signals Corps with a number of challenges. Firstly, a substantial communications network had to be established between Australian forces deployed in Saigon, Vung Tau, Nui Dat and further afield. Secondly, the network had to be tied in with the massive US military communications network then being expanded across South Vietnam. To do this 145 Signal Squadron and 103 Signal Squadron deployed to Vietnam along with 547 Signal Troop (signals intelligence) and elements of 152 Signal Squadron (in support of the SAS). In 1967, 145 Signal Squadron was replaced by 110 Signal Squadron as the in-theatre and national rear-link

130 AWM 95, CD, 104 Sig Sqn Monthly Report, September 1967.

131 AWM 95, CD, 104 Sig Sqn Monthly Report, Oct 1967; AWM 95, CD, 110 Sig Sqn Monthly Report, Aug 1967; and AWM 95, CD, 104 Sig Sqn Monthly Report, Aug 1967.

signal squadron and 103 Signal Squadron was replaced by 104 Signal Squadron as the Task Force Signal Squadron. From then on there were to be no further whole-squadron rotations because of the complexity of operations that could not be handed over complete without an unacceptable degradation in the service provided.

The quantity of message traffic transmitted caught everyone by surprise. It had not been fully appreciated just how great the staff requirement for greater capacity would be - particularly given the need to be responsive to 'real time' news reporting and television coverage of events. It had also not fully been realised how much traffic the new communications technology would allow to be passed, particularly in such a complex and dispersed operational setting. Consequently, the system had to be repeatedly modified and improved to meet the demand. The problems associated with this were compounded by the decision to retain the communications infrastructure in Saigon once the decision was made to establish a separate Australian Task Force Base, removed from Saigon, in Phuoc Tuy Province. These problems, in turn, were exacerbated by the need to replace overloaded field communications equipment with permanent communications facilities in Saigon and Vung Tau.

In the meantime, other lessons were being learnt at Task Force Headquarters in Nui Dat. Firstly, the reliability of VHF radio communications had played an important part in the process of deciding to place the base at Nui Dat where the hill provided a good site for radio antenna reception and transmission. Secondly, the increase in circuits to, from and within the Task Force presented the Signals personnel with a hefty workload requiring periodic increases in unit manning to cope with the increase in responsibilities.

Perhaps the most significant lessons learnt arose as a consequence of the battle of Long Tan. There the availability of reliable and durable radio communications provided a company sized group with the link to the necessary forces to defeat a much larger enemy force. Also, signals intercept played a critical role in detecting the enemy force.

The period also saw the commencement of tactical deployments of Task Force Headquarters on operations away from the base at Nui Dat. This led to a further reorganisation of the Task Force Signal Squadron given that it was responsible for manning the semi-permanent facilities at Nui Dat while also providing the necessary communications support to the remotely located Task Force Headquarters. Its ability to fulfil effectively its increased responsibilities was to be put to the test in the months that followed.

The command arrangements and the semi-permanent nature of the base at Nui Dat, combined with the forward deployment of the Task Force Headquarters and the lack of a supporting divisional headquarters and supporting divisional signals unit had severely challenged the resources and ingenuity of the Signals Corps. Operating as an independent Task Force with a main base and a forward operating base was only feasible because of the access to sophisticated communications equipment above the normal entitlement of a Task Force Signal Squadron. Operating this equipment, with limited manpower and trying circumstances stretched the capability of the Corps, but even greater challenges lay ahead in the years following.

By the end of 1967, the communications infrastructure in support of Australia's deployed forces was unprecedentedly sophisticated. At platoon level, widely-dispersed infantry soldiers in the field could use VHF radios to link up with armour, artillery or air force supporting elements to coordinate the battle and maximise their combat power. They could also call for resupply, reinforcements or aero-medical evacuation assistance and be assured of reliable communications - frequently at distances of six to ten kilometres from their fire-support base where, typically, the battalions or Task Force headquarters was deployed. Even greater distances were achievable if necessary using helicopters for aerial rebroadcast. In terms of medical support alone, these improvements were to dramatically improve the survival rate of casualties who were able to be removed from the battlefield to a hospital within phenomenally short time-frames. Rear links from the Task Force or battalions' forward headquarters back to Nui Dat over 20 kilometres and more were feasible and were to become commonplace thanks to the availability of such items as the new and powerful radio relay equipment. To support these forces, a triangulated network between Saigon, Vung Tau and Nui Dat was established with links into the US integrated wideband communications network. This served to guarantee the link from Australia to the highest level commanders as well as between those commanders and their smallest and most remote sub-units on operations. 1968 would confirm many of these lessons and reflections, but it would present a range of additional challenges. In particular, they would call into question the quiet confidence of progress towards ultimate victory in Vietnam up to this point.

CHAPTER SIX

VIETNAM OPERATIONS IN 1968

DURING LATE DECEMBER 1967 and January 1968, advantage was taken of the dry season as VC and NVA troops were infiltrated southwards into forward positions and into the population centres of South Vietnam. The initial assaults, employing about 84,000 troops, were mounted against 42 cities and provincial capitals, 64 district capitals and 50 hamlets. These attacks were quickly repulsed in most places, but in Saigon and the old imperial city of Hue, near the North Vietnamese border, the fighting was protracted.[1]

The 1968 offensives removed any sense of complacency that may have been emerging in Saigon and stretched the resources and capabilities of the two Signal Squadrons supporting the Australian forces deployed on operations. The strategic and in-theatre Signal Squadron was forced to relocate while maintaining the vital rear link to Australia and forward link to the Task Force Headquarters. In the meantime, the Task Force Signal Squadron developed new techniques to overcome the difficulties of maintaining the Nui Dat base while also deploying forward on operations.

Saigon

For the Australian Signalmen in Saigon, the VC Tet offensive brought about increased security precautions which demonstrated

1 Frost, *Australia's War in Vietnam*, pp. 102-3; and Rienzi, *Communications-Electronics*, pp. 103-4.

to many, for the first time, the fact that they were fighting an active enemy. The offensive also made people aware of the vulnerability of the communications facilities in Saigon. So what could be done about it? To begin with, all personnel in the Saigon detachment were moved into the Free World Building Area, as vehicle restrictions permitted, to provide extra hands for guard duty, digging of bunkers and command posts and normal station operation.[2]

During the night of 30-31 January, heavy attacks were carried out by the Viet Cong on installations at the nearby Tan Son Nhut airbase. At 3am on 31 January, the HF radio receiver site technician on duty at Ba Queo (on the perimeter of the Tan Son Nhut base) was forced to evacuate the station and assist local US Signal Corps personnel in defence of the area from a large Viet Cong force which had broken through the air base perimeter defences less than 200 metres from the site.[3]

By 7am the following morning, all hired US lines between the Free World Building and the transmitter station had been cut, as well as the lines between the transmitter station at Phu Tho and the receivers at Ba Queo. No road movement at all was possible between the three localities. One old cable connecting the Free World Building with the Phu Tho station was found to be still through and this was used to restore the Melbourne circuit, using AN/GRC-106 radios for reception and the AN/TRC 75 field HF radio for transmission at the Free World Building. Antennas for these consisted simply of lengths of D10 cable strung to the top of a 250 feet tower. The effectiveness of this improvisation was indicated by the traffic figures to and from Australia for the 24 hour period from 8am 31 January to 1 February. These were at an all time high. By the end of 31 January, the Squadron was also providing a three-station AN/PRC-25 net between the Australian ambassador's residence, the residence of the Commander AFV, and Headquarters AFV in the Free World Building.[4]

2 AWM 95, CD, 110 Sig Sqn, Feb 68.
3 AWM 95, CD, 110 Sig Sqn, Jan 68.
4 AWM 95, CD, 110 Sig Sqn, Jan 68; and AWM 95, CD, 110 Sig Sqn, Feb 68.

The decision was taken, after the offensive had eased, to evacuate the Ba Queo receiver station completely. The equipment was then transported to Long Binh, 12 miles north east of Saigon and three miles east of Bien Hoa, and co-located with the US STRATCOM Signal Station.[5]

After the Tet offensive of February 1968, the immediate action taken was to complete the installation of a reserve HF E513 (fixed station) transmitter, in place of the AN/TRC-75 in the 110 Signal Squadron area at Vung Tau. In late March, Intelligence reports indicated a threat of another offensive on Saigon. The squadron had initiated the development of a defensive area at Phu Tho by constructing a smaller compound around the unit's area to include the hardening of the watch towers, defensive wire fences and the construction of a number of bunker positions.[6]

On the night of 5 May 1968, the Viet Cong commenced their second offensive on Saigon, and on 6 May, the Phu Tho area was close to this attack.[7] At about 6am on 6 May, a Viet Cong force reached the north west corner of the compound and then moved through the houses to the north of it. Helicopter gunships engaged the force and it was eventually halted by a South Vietnamese Army force. No direct action was taken by the Viet Cong against the compound during this move, but during their withdrawal weapon fire was directed at the compound without inflicting any casualties. When the first contact occurred at 6am, the total duty staff of the US transport compound in Phu Tho appeared in the 110 Signal Squadron inner compound, including one man in a state of shock.[8]

---

5 AWM 95, CD, 110 Sig Sqn, Jan 68; and AWM 98, R193/5/5, 'HQ AFV Signal Instruction 1/68, AUSTCAN Receiver Station- Long Binh'.

6 RASCM, Lawrence, 'The South Vietnam Campaign', p. 5.2; AWM 98, R193/1/67, 'Signals in Malaysia and Vietnam', August 1967; and AWM 95, CD, 110 Sig Sqn, Mar 68.

7 AWM 95, CD, 110 Sig Sqn, May 68.

8 RASCM, Lawrence, 'The South Vietnam Campaign', p. 5.3 (based on notes provided by Captain I Willoughby).

As the battle proceeded between the Viet Cong and South Vietnamese Army force in the village to the north of the compound, a number of explosions occurred just outside the compound and some form of explosive projectile struck the unit's shower block, beside the watch tower. The explosions in the village started a fire which spread quickly along the huts beside the compound. After the mast supporting the emergency V antenna collapsed, because its rear guy wire had burnt through, it became obvious that the end mast of the rhombic antenna would also suffer the same fate. The staff in the Free World Building were informed of this situation and a signal was sent to Melbourne. As the fire reached the antenna guy wires, the transmitter was closed down and the unit's personnel watched as the mast fell. A small party from the unit moved out into the antenna field to cut the broken wires of the fallen Type 1 antenna away from the still standing Type 2 of the nested rhombic antenna. This resulted in the circuit being re-established to Melbourne on the one remaining antenna after an outage of only two hours. This was, according to Chief Signals Officer (CSO) in Vietnam at the time, Major Hook, an excellent effort bearing in mind that there was still considerable shooting in the vicinity at the time.[9]

The Phu Tho area was cut off from the rest of Saigon for two days, during which time the Viet Cong were inside some sections of the perimeter fence. Small arms fire, mortars, grenades and rockets were all received in the compound during the action. After this time, the unit riggers reconstructed the outer rhombic antenna within seven days. However, by then, planning had commenced for the re-location of the rear-link transmitters to a more secure location.[10]

Shortly after the attack on the transmitter station at Phu Tho, the newly re-located receiver station at Long Binh came in

9 Willoughby, letter, August 1990; RASCM, Lawrence, 'The South Vietnam Campaign', p. 5.4; and AWM 95, CD, 110 Sig Sqn, May 68.

10 RASCM, Lawrence, 'The South Vietnam Campaign', pp. 5.4-5.53 (based on notes provided by Captain I. Willoughby).

for a 120mm rocket attack on the evening of 11 May 1968. A single 120mm rocket impacted in the receiving station grounds close to the barracks. Most members on the station were in the club, but Signalman J.W. Cole was lying in bed in the barracks. He was blown out of bed but luckily suffered only minor shrapnel wounds. Three US Signal Corps soldiers were also wounded. In his monthly report, Major Hook indicated that as the receiving station was in close proximity to the US 3rd Ordnance Depot (a huge ammunition dump), the area was a favourite target of the Viet Cong and several rounds of rockets were received in the antenna field or depot area at least once a week. The warped roof of the receiver station bore testimony to the effect when a rocket actually hit an ammunition dump.[11]

Because of the events in February and May 1968 in Saigon, Major General A.L. MacDonald, Commander AFV, directed, upon the advice of Major Hockney (who had been sent to Vietnam in May 1968 to recommend the best alternatives for the communications links), that the main AUSTCAN transmitter site was to be relocated from Phu Tho to Nui Dat.[12] Hockney later recalled that the best and most secure site for the transmitter was within the 1 ATF base at Nui Dat. No one enjoyed the thought of 102 feet masts for the rhombic antennas at the Task Force base particularly when we had to put lamps on the top of each because of their proximity to Luscombe [air]field.[13]

The move was accomplished in August 1968 with no loss of circuit time on the rear link to Australia. The Nui Dat transmitters eventually took over the primary function on the rear link while the building they occupied was still under construction by a detachment from 127 Signal Squadron based in Melbourne. Major P.L. (Phil) Perman, the new commander of 110 Signal

---

11 AWM 95, CD, 110 Sig Sqn, May 68; and AWM 98, R193/5/5, 'AUSTCAN Receiver station Long Binh 120mm Rocket Attack - 11 May 1968'.

12 AWM 93, R193/5/15, 'Provision of Additional Buildings - 110 Sig Sqn', 14 June 1968; and AWM 98, R193/5/4, HQ AFV, 'User requirements For Engineer Works New Transmitter Station - Nui Dat', 21 May 1968.

13 Hockney, letter, September 1989.

Squadron as from August 1968, stated that it was a credit to 127 Signal Squadron that they were able to provide a reliable, well installed communications facility in such haste and under trying conditions.[14]

Headquarters 110 Signal Squadron, which had been collocated at Phu Tho, was also required to move during this period. This move was to be to Vung Tau to be collocated with the Vung Tau detachment of the Squadron.[15] Whilst the Squadron headquarters was in transit, a temporary headquarters was set up in the Free World Building in Saigon and, although there were minor difficulties, the system functioned quite well.[16] In the meantime, the Australian Task Force, based at Nui Dat, was to experience fighting of a greater intensity.

Nui Dat

Between January 1968 and May 1969, there were three major periods of Task Force operations outside Phuoc Tuy Province: during and after the Tet offensive, between January and March 1968; the Viet Cong's (National Liberation Front) May 1968 offensive; and Tet 1969. On each of these occasions the Task Force contributed to the protection of the major bases outside Phuoc Tuy, at Bien Hoa and Long Binh.[17]

The first of these deployments was on Operation 'Coburg', conducted in the Bien Hoa Province about 50 kilometres from Nui Dat. It commenced shortly before the onset of the Tet offensive, on 28 January 1968, and continued until 2 March 1968.[18] This operation was the most extensive independently undertaken by 1 ATF since its arrival in South Vietnam and it involved the main

14 AWM 95, CD, 110 Sig Sqn Monthly Report - Aug 68.

15 AWM 93, R193/5/15, 'Provision of Additional Buildings - 110 Sig Sqn', 14 June 1968; and AWM 98, R193/5/4, HQ AFV, 'User requirements For Engineer Works New Transmitter Station - Nui Dat', 21 May 1968.

16 AWM 95, CD, 110 Sig Sqn Monthly Report - Aug 68.

17 The phases are as outlined in Frost, *Australia's War in Vietnam*, p. 107.

18 AWM 95, CD, 104 Sig Sqn Monthly Report - Mar 68.

headquarters of 1 ATF being away from its base longer than on any previous occasion.[19]

Throughout this operation the usual Task Force communications nets were established including the command VHF net and various links to the Task Force base at Nui Dat and the higher operational command, II Field Force, Vietnam, as well as adjoining units.[20] 110 Signal Squadron also deployed a MRC-69 radio relay terminal shelter in support of Operation Coburg after the Task Force had experienced inordinate delays over four days with the US supplied radio relay detachments. Major N.C. (Norm) Munro, the Task Force Signal Squadron commander at the time, reported that these US detachments were meant to provide the links to II Field Force, Vietnam and the base at Nui Dat, but had suffered from numerous equipment breakdowns which were compounded by inexperienced technicians and operators. In his monthly report for April 1968, he concluded by stressing that reliance on the US Army network for 1 ATF internal communications 'should be avoided wherever possible'. The deployment of 110 Signal Squadron radio relay resources stretched equipment reserves to breaking point as, at the time, more than one fault in any location would have disrupted the whole system and replacements could only have been delivered by helicopter. A medium radio detachment was also deployed from 110 Signal Squadron with the Task Force and their AN/GRC-106 HF radios were used to maintain this telegraphic link.[21] The radio relay detachments from 110 Signal Squadron continued deploying with the Task Force during subsequent operations and the service they provided was simplified with the arrival of three more MRC-69 radio relay shelters in April 1968.[22]

19 RASCM, Lawrence, 'The South Vietnam Campaign', p, 5.5.

20 AWM 95, CD, 104 Sig Sqn, Monthly Report, Mar 1968.

21 AWM 95, CD, 110 Sig Sqn, Monthly Report - Jan 68; AWM 95, CD, 104 Sig Sqn, Monthly Report - Apr 68; and AWM 95, CD, 104 Sig Sqn, Monthly Report - Mar 68.

22 AWM 95, CD, 110 Sig Sqn, Monthly Report, Apr 1968; and AWM 95, CD, 110 Sig Sqn, Monthly Report, Oct 1968.

The increasing complexity of the communications requirement of the Australian Task Force, given its deployment remote from its base at Nui Dat and the varied and complex control and coordination requirements meant that new resources were required to increase efficiency. Thus, an innovation suggested by Captain G.R. Arnold was introduced by 104 Signal Squadron at Nui Dat. This was the establishment, on 1 March 1968, of a signals operation centre, designated as the Communications Control, located in the squadron headquarters building and manned on a 24 hour basis. This idea had first been tried by 139 (Task Force) Signal Squadron in Australia and was successfully introduced on the same basis at Nui Dat.[23] It became the focal point of the unit's operations and thus of the Task Force's communications system. The office effectively co-ordinated and centralised the operational, technical and fault control of the Task Force's communications systems, including frequency management and the issuing of signal operating instructions (SOIs) in accordance with the Squadron commander's policy. It also detailed current and future communications planning for the Task Force, as directed by the Squadron commander.[24]

In 1968 an effort was also made to streamline the telephone system at Nui Dat, in view of the then deteriorating condition of the line system. As a result, the D10 cables that had been laid by linesmen when the Task Force first occupied the site at Nui Dat were replaced in April and May 1968 by a complete underground cable system. On completion, all circuits had alternate routings in the event of the destruction of any cable, and 100 per cent spares were available.[25] The squadron commander, from January 1969 to January 1970, Major K.P. (Keith) Morel, com-

23 AWM 95, CD, 104 Sig Sqn, Monthly Report, Feb 1968; AWM 103, 193/1/1, Pt 1, 'Communications Control' dated 27 February 1968; and N.C. Munro, letter, August 1990.

24 AWM 95, 104 Sig Sqn, Monthly Report, March 1968; and AWM 95, 104 Sig Sqn, Monthly Report, April 1968.

25 AWM 98, R193/5/9, '1 ATF Area-Underground Cable Project, 16 May 1968'; and AWM 98, R193/1/67, HQ AFV.

mented subsequently that 'the engineers were adept at finding signal cables. If Signals laid cable underground, you could bet your life that they would find it and dig it up.'[26] One instance was to occur in 1969 when the Engineers severed a cable. To the dismay of the squadron's linesmen, it was found that the cable was not colour coded. Moreover, the break occurred in an area which was to be the site of the Nui Dat dam, so a complete new section of cable was laid out ready to be put in the trench when a tank neatly severed it again![27] The problem was added to by the extremes of wet and dry seasons which made it difficult to maintain the buried cables.[28] Later on, in October 1970, a multi-pair aerial cable was to be completed, thus eliminating many of the problems that were being experienced with underground lines, which had been laid only a few inches below the surface and which in places had been poorly repaired, allowing water to get into them.[29] This line work occurred after the Task Force's return to Nui Dat from Operation Coburg.

Shortly after the Task Force's return, the Viet Cong's May 1968 offensive commenced affecting both the Australian Signals units in Saigon and the Australian Task Force which soon featured prominently in battle. In fact, the Task Force had deployed on Operation 'Toan Thang' (or 'Complete Victory'), on 21 April 1968, in anticipation of the May offensive. The Task Force had been tasked to delay and disrupt enemy infiltrations and preparations to attack vital targets around Saigon and to its north.[30]

By 12 May 1968 the Task Force Headquarters was moving to the Fire Support Base 'Coral' north of Bien Hoa. From the first night the Base came under extremely heavy enemy attacks.

---

26 Morel, interview, August 1989.

27 RASCM, Lawrence, 'The South Vietnam Campaign', p. 7.8; and AWM 95, CD, 110 Sig Sqn, Monthly Report, April 1969.

28 AWM 95, CD, 110 Sig Sqn, Monthly Report, November 1967.

29 AWM 95, CD, 104 Sig Sqn, Monthly Report, March 1970.

30 The Royal Australian Regiment Standing Orders, Chapter 2, Annex B. See L. MacAulay, *The Battle of Coral,* Hutchinson, Melbourne, 1988 for a detailed account of this battle.

Radio nets were becoming clogged at this time and II Field Force Vietnam, the Task Force's superior operational headquarters, introduced two new message precedences to overcome the overuse of the 'flash' precedence (ordinarily the fastest type available). These new precedences were 'red rocket flash' and flash override'. One officer recalled that it was 'probably the only time a daily 'situation report', authorised by the Duty Officer, was sent 'flash'.[31] In contrast, the Vietnamese Communist forces relied on runners in daylight and lights and flares at night time.[32] This predicament was indicative of the problems encountered, particularly at the higher levels of command, whereby the communications system became overloaded with information, leaving commanders unable to cope adequately with, and respond in a timely manner, to the mass of information received.

Fire Support Base Coral was developed and new bunkers constructed, radio relay vehicles lowered and bunkered and strong defences were prepared.[33] As part of the attack experienced on 16 May, the Signals sector of Fire Support Base Coral was mortared and rocketed, killing Signalman A.H. Young and wounding another.[34] Further action continued throughout the rest of May, with another Fire Support Base called 'Balmoral' being established approximately 6,000 metres away from Coral. Further battles, involving both these bases, ensued in late May 1968.[35]

These operational deployments away from Nui Dat tested the Task Force Signal Squadron under most contingencies likely to arise in the theatre. Whilst maintaining all communications facilities at Nui Dat base, the Task Force had undertaken a deployment from Nui Dat to one field base and from there to Coral, in an enemy controlled area. It had also required the pro-

---

31 MacAulay, *The Battle of Coral,* p. 192.

32 Ibid, p. 145.

33 RASCM, Lawrence, 'The South Vietnam Campaign', p. 5.8.

34 AWM 95, CD, 104 Sig Sqn Commanders Diary, May 1968; and MacAulay, *The Battle of Coral,* p.338.

35 'The Royal Australian Regiment Standing Orders', pp. 2B-16 - 18; and MacAulay, *The Battle of Coral,* chapters 4 -13.

vision of the full range of communications facilities available to the squadron, and it involved the maintenance of communications whilst under enemy fire. It was a credit to those concerned that at no time was there a loss of command communications.[36]

This performance was all the more remarkable, considering the fact that the telegraph equipment had been exposed to so much heat and dust that it became difficult to keep operating. Early in the morning on 22 May, accurate enemy mortar bombardment hit the Task Force Headquarters area resulting in radio antennas being damaged. The antennas were quickly replaced and as a result of problems of command experienced in such circumstances (that is, being well dug in, but with no communications) it was decided that Armoured Command Vehicles (ACV) should be employed in future for Task Force Headquarters and Artillery nets.[37] An interim solution had been to use a 'conex' metal shipping container. This was subsequently insulated and fitted for use as a signal centre for deployments, using the AN/GRC-106 HF radio, which, when combined with cryptographic equipment, provided the Task Force with its secure radio-teletype link to Nui Dat. It also provided a forward headquarters switchboard facility. The teletype link further served as the back up to the radio relay links. It was hoped, however, that an armoured command vehicle (a modified M113 armoured personnel carrier) would be issued to the squadron to be fitted out as the forward signal centre vehicle, along similar lines to the configuration developed by the Americans, and this occurred in October 1968.[38] To do this the key components of the telephone switchboard and teletypewriter shelters were also incorporated in the armoured command vehicle.[39] Thereafter, during Operation 'Capitol II', conducted from 3 November to 27 November 1968, and Operation 'Goodwood' which commenced on 3 December 1968, the armoured command

36 RASCM, Lawrence, 'The South Vietnam Campaign', p. 5.8.
37 MacAulay, *The Battle of Coral*, p. 201; and Munro, letter, August 1990.
38 AWM 95, CD, 104 Sig Sqn, Monthly Report, Oct 1968; and AWM 95, CD, 104 Sig Sqn, Quarterly Report, Jul 1968.
39 AWM 103, R981/1/10, 'Request for Modification of Equipment Shelters'.

vehicle proved 'eminently successful with the cryptographic and telegraphic equipment providing a completely reliable service in the air conditioned environment'.[40]

Reflections on 1968

1968 had proved to be a significant year, not least because of the press coverage of events. 1968 also proved to be the turning point in Australian political rhetoric about the war when Prime Minister Gorton asserted that no more Australian troops would be sent to Vietnam.[41]

The deployments of the Australian Task Force in 1968 presented a challenge to commanders and communicators alike. For the Task Force Signal Squadron, the changed circumstances, combined with newly-available equipment combinations, enabled them to make the necessary changes and thus to continue to 'get through'. This was despite the almost exponential growth in demands placed upon them. From makeshift shelters to secure radio teletype and from forward line switchboards to the establishment of a Signals Operations Centre, the Task Force Signal Squadron adapted to the circumstances and provided the Task Force Commander with the flexibility and clear control of his subordinate units - despite the dispersed and difficult nature of the operations. It is fair to say that these changes represented a quantum leap in terms of communications capability at brigade (task force) level from that experienced by equivalent force commanders at the end of World War II.

The events of 1968 also had a profound effect on the functioning of 110 Signal Squadron. The squadron faced circumstances which stretched their resources and capabilities. Yet despite the attacks and the difficulties in maintaining communi-

40 AWM 95, CD, 104 Sig Sqn, Monthly Report, Nov 1968; AWM 95, CD, 104 Sig Sqn, Monthly Report, Nov 1968; and 'The Royal Australian Regiment Standing Orders', p. 2B-20.

41 R. Tiffen, 'News Coverage of Vietnam', in King, *Australia's Vietnam*, pp. 176-179.

cations with Australia and the Task Force, 110 Signal Squadron managed to provide the Force Commander, and through him the Australian Government, the means for directly and closely monitoring developments in Vietnam to a level not experienced in earlier conflicts, even though some had involved far larger Australian force deployments. On reflection, 1968 came to be seen as an inflection point in the war. And yet combat operations would continue apace for several years more. It is to that experience, from 1969 to 1971, to which we now turn.

CHAPTER SEVEN

VIETNAM OPERATIONS 1969 to 1971

THE PERIOD FROM 1969 to 1971 saw Australia's contribution to the Vietnam War focus on supporting 'pacification' both within Phuoc Tuy Province and beyond. Task Force deployments beyond Phuoc Tuy Province were to continue throughout 1969, 1970 and into 1971 when, with the changing political mood in Australia and the US, the Task Force was reduced in size from a three-battalion to a two-battalion Task Force in late 1970, and the focus shifted towards supporting the 'Vietnamization' of the war. Vietnamization allowed the war-weary US (and Australian) government to back out of the conflict. This was followed by a withdrawal of the force in 1972. Pacification, Vietnamization and then the withdrawal of the Australian Task Force altogether presented further challenges for the Signal Corps units supporting it. This chapter looks at how the Royal Australian Corps of Signals managed to cope with these wide-ranging responsibilities in addition to those it was already dealing with. The chapter also examines the impact this had on the conduct of Australia's military operations.

Theatre-Wide Communications 1969 - 1971

By 1969, the Australian military communications infrastructure was working effectively, having been refined and developed progressively. Technical oversight over the maintenance and development of this infrastructure was exercised by the Chief

Signals Officer (CSO) - a Lieutenant Colonel - on Headquarters AFV. He was not only the adviser to the Commander AFV on Signal matters, but was also the staff officer representing the Army's Aviation Corps elements in Vietnam - essentially 161 Reconnaissance Flight.[1]

The CSO was also responsible for liaising with the US Signals elements of the US 1st Signal Brigade commanded, between 1968 and 1970, by Brigadier General Thomas Rienzi. This Brigade was responsible for the two main types of communications systems, namely the Corps Area Communications System and the Combat Communications System.[2]

Observers noted that the relationship with the Americans was very close. They were genuinely friendly towards the Australians and repeatedly did favours for their Australian counterparts. It was not a matter of the wealthy big brother giving things, but 'a very genuine respect between professional military people'.[3] Morel recounted that at one stage they had a tow truck to offer, but this 'would have been a bit difficult to hide'.[4]

On one occasion, Rienzi visited the Task Force at Nui Dat and showed a 'deep interest' in 1 ATF communications and met many members of 104 Signal Squadron and Detachment of 110 Signal Squadron who were 'most impressed by his effervescent personality'.[5]

The main direct contact that Australian Task Force Signalmen had with the US Signal Corps was with the members of the 39th US Signal Battalion detachment of about 20 men, mentioned earlier, that was virtually permanently located with the Task Force Signal Squadron. They provided radio teletype and radio relay communications to II Field Force, Vietnam, at Long Binh, the Task Force's superior operational headquarters. In addition, channels on the radio relay system were established

1 Cowley, interview, August 1989; and Hockney, interview, August 1989.
2 Rienzi, *Communications Electronics, 1962-1970*, p. 57.
3 Hockney, interview, August 1989.
4 Morel, interview, August 1989.
5 AWM 95, CD, 104 Sig Sqn Monthly Report, Apr 1970.

to the 9th US Infantry Division and US advisers at the Baria sector headquarters, to the south of Nui Dat in Phuoc Tuy Province.[6]

The US Integrated Communications System-South East Asia (ICS-SEA) consisted mainly of fixed stations, high-capacity (60 - 300 channels) wideband line-of-sight microwave and tropospheric scatter radio and auxiliary equipment. This system supplemented the combat communications and provided the 'backbone' communications links up and down South Vietnam and across to Thailand. It was by far the largest communications complex the US Army had ever undertaken, creating an equivalent of a national telephone system for South Vietnam and Thailand.[7]

The wide dispersion of both logistic and combat forces in Vietnam had increased the need for these interconnecting communications facilities.[8] Within the combat communications system, there were matching facilities for the various echelons of command. Each of the four Field Force Headquarters in South Vietnam had a US Corps Signal Battalion; each Division had a US Division Signal Battalion, and each separate US Brigade had a Signal Platoon or Detachment.[9]

Brigadier General Rienzi was responsible for this massive organisation, which, in 1969, consisted of almost 20,000 people (down from a high of 23,000 in late 1967) operating nearly US $500 million worth of communications hardware at over 200 sites in two countries. His Brigade was the most sophisticated signal unit that had ever gone to war, and the communication system it had established was the most complex ever placed

---

6 AWM 98, R193/1/67, 'Signals in Malaysia and Vietnam', August 1967.

7 Rienzi, *Communications Electronics, 1962-1970,* pp. 76 & 132; and 'AUTODIN', in *Command Communications,* October 1968, USARV Pam No 105-4, pp. 49-50.

8 AWM 98, R197/8/1 'Corps Area Communications Systems' HQUSARV, Dec 1968, Regulation No 105-15. See also Rienzi, *Communications Electronics,* 1962-1970, pp. 57-59.

9 J.E. Trotter (Master Sergeant), 'The Integrated Communications system in Vietnam', in *Signal, Journal of the Armed Forces Communications and Electronics Association,* April 1967, p. 15.

in a combat theatre. The resources under his command made those under his World War II or Korean War counterparts pale by comparison due to the tremendous growth in communications technology that had taken place in the intervening years.[10] An estimated one-third of all the major items of US equipment used in Vietnam were communications-electronics items, and over 50,000 different types of replacement and repair parts had to be stocked by the US supply system in the theatre. As Commander, Rienzi commanded six Signal Groups and 23 Signal Battalions throughout South Vietnam and Thailand.[11]

The size and scope of the US communications network in Vietnam at the time were indicative of the rise in significance of communications in the conduct of military operations in Vietnam. The massive nationwide communications network made possible by dramatic improvements in communications technology enabled US or allied forces to be in contact with each other, no matter where they were. It would appear that the almost mesmerising marvel of high-powered and virtually ubiquitous communications equipment contributed to an inability to appreciate the fundamental flaws in the overall campaign strategy.[12]

110 Signal Squadron 1969-1971

Following the offensives of 1968 and before the Force was to be withdrawn in late 1971 and 1972, several developments took place within the Australian Army's theatre-wide and rear link communications network in Vietnam. By 1970, 110 Signal Squadron had gained an establishment of nine officers and 243 other ranks (as compared to 162 all ranks in mid 1966). Elements of the squadron were deployed in Vung Tau, Saigon, Nui Dat and

10 T.M. Rienzi, 'Management Tools for dynamic communications-electronics', in *Signal, Journal of the Armed Forces Communications and Electronics Association,* October 1969, p. 24; and Rienzi, *Communications Electronics, 1962-1970,* p. 95.

11 Rienzi, *Communications Electronics, 1962-1970.,* pp. 114 & 138.

12 For a more detailed discussion on this aspect of the Vietnam War see Van Creveld's *Command In War.*

temporarily in other places. The unit averaged 90,000 message transactions per month at the Major Relay Station, with average daily telephone calls in Vung Tau of 6,800 while still operating on a continuous basis in all locations.[13]

One of the developments referred to above occurred with the AUSTCAN HF receivers. The receiver station, which had been moved to Long Binh during 1968, provided excellent results for the Melbourne-Saigon HF circuit. However, its location contributed to the complexity of the Army's communication system, which had expanded and changed in direct response to immediate needs rather than long-term planning.[14] The relocation took place in November 1969 and, as a result, the Siemens 12/400 UHF radio relay bearer system that had been installed between Saigon and Long Binh was reinstalled between VC Hill, near Vung Tau, and the Signals Sector in the 1 ALSG base at Vung Tau instead of the AN/TRC-24 field radio relay system. The Siemens equipment was so reliable that it was placed, unmanned, within the US tropospheric scatter radio facility compound on VC Hill, further reducing the complexity of squadron deployment.[15]

Telegraph traffic and the maintenance of relay stations in Vietnam continued to be a significant problem for the Army's communicators between 1969 and 1971. Traffic through the Saigon Major Relay Station reached 104,900 messages in September 1969, and telephone calls through Vung Tau reached a high in April 1970 at 7,000 per day, or one every six seconds in the busy daylight hours.[16] By way of comparison, the BCOF Signal Regiment in Japan in 1946, supporting a force of over 40,000 troops in the British-Indian Division, New Zealand Brigade and Australian Brigade as well as naval and air force units,

---

13 AWM 102, VN, Misc, 110 Sig Sqn, Annex C to 1 ALSG R128/1/4, 31 Jul 1970.

14 AWM 95, CD, 110 Sig Sqn Monthly Report Aug 1969.

15 RASCM, Lawrence, 'The South Vietnam Campaign', p. 7.2.

16 RASCM, Lawrence, 'The South Vietnam Campaign', p. 7.1; AWM 95, CD, 110 Sig Sqn Monthly Report Apr 1970; and AWM 95, CD, 110 Sig Sqn Monthly Report Sep 1969.

peaked at only about 600 messages in one day.[17] The contrast demonstrates that by 1970, demand had more than kept up with the ability to supply communications facilities for an Australian Force deployed on operations.

It was about this time that a telegraphic teleconference facility was established, linking the Chiefs of Staff in Canberra with Headquarters AFV in Saigon. From that point on, staff officers at Army Headquarters in Canberra could talk to the staff at Headquarters AFV via the teleprinter.

It was a useful facility, but it was not easy to convince the staff to use it because it was new, and sometimes the questions were a bit tough, and people didn't warm very easily to having written questions on the spot.[18]

Meanwhile, in November 1970, the Major Relay Station and Senior System Control facility were transferred from Saigon to Vung Tau.[19] By that stage, the transmitters and receivers were at Vung Tau, and it seemed sensible to have the Tape Relay Station there as well. The Saigon installation was then reduced to become a communication centre off the Tape Relay network from Vung Tau.[20]

The installation of an AUSTCAN transmitter at Vung Tau, completed in January 1971, initially provided alternative transmitting facilities to those provided from Nui Dat. In due course, the Vung Tau station was upgraded, using equipment from Nui Dat, and the Nui Dat station was eventually closed down.[21] The

---

17 Murphy, 'History of the Post War' Army, p. 238; Grey, *Australian Brass*, p. 155; and Nalder, *British Army Signals in the Second World War*, Annex 3. Murphy cites '56,000 groups' as being the highest figure in one day in Japan in 1946. Given that a 'group' is: 'any word or combination of symbols signalled consecutively to form one entity'; and that Nalder's Annex 3 indicates, on average, that about 95 groups made up an average message; then the figure of 56,000 groups equates to about 600 messages.

18 Hockney, interview, August 1989.

19 AWM 95, CD, 110 Sig Sqn Monthly Report, Nov 1970.

20 Hockney, letter, September 1989.

21 AWM 103, R571/1/32, 'CSO AFV Signal Instruction No 1/70: AUSTCAN Transmitting Station Vung Tau'; and AWM 95, CD, HQ AFV CSO Monthly Report December 1970 (18 Jan 1971).

transmitters had been moved to Nui Dat from Phu Tho to Nui Dat in August 1968 because Nui Dat appeared to offer the best security. With hindsight, though, it appears that had the reduction and eventual withdrawal of the force been anticipated, then Vung Tau would have been chosen for the move from Phu Tho in the first place.

The move of the Major Relay Station was made with no break in communications from Australia except for an inability to pass highly classified messages for a period while on-line cipher equipment was transferred.[22] Various other communications links were maintained by elements of 110 Signal Squadron during the Vietnam campaign, including the 1 ALSG Defence net.[23]

110 Signal Squadron also maintained a back-up communications link between Vung Tau, Nui Dat and Saigon, using AN/GRC-106 HF radios. An AFV emergency frequency, available for any callsign to use in a real emergency, had a permanent listening watch manned by 110 Signal Squadron.[24]

In 1971, a VHF link was also installed at the 1 Australian Field Hospital at Vung Tau to warn of the arrival of 'dust off' helicopters carrying the wounded. The hospital used the facility regularly, and the effectiveness of the link rose considerably.[25]

Change in various forms was taking place within the base at Vung Tau during this period. An operations room was also established to provide central coordination and supervision of Squadron operations on a continuous basis. This move, according to reports, 'proved its worth' as rapid reaction to any situation became practicable at any hour of the day or night and problems could often be predicted and even, at times, prevented. Moreover,

22 Cowley, interview, August 1989.
23 AWM 95, CD, 110 Sig Sqn Monthly Reports Sep & Oct 1970.
24 AWM 95, CD, 110 Sig Sqn Monthly Report Oct 1970.
25 AWM 95, CD, 110 Sig Sqn Monthly Report Jan 1971.

it allowed for better, longer-term analysis of performance and for centralised implementation of local defence measures.[26]

Since 1967, approval had been sought to establish radio telephone facilities so that members could speak to friends and relatives in Australia.[27] The matter was taken up by the Australian Interdepartmental Telecommunications Advisory Committee - AIDTAC but it was not until September 1970 that a system was finally approved (the US equivalent had been operating in Vietnam as early as late 1965).[28] Telephones were subsequently located at Nui Dat, Saigon, 1 ALSG Vung Tau and 1 Australian Field Hospital, Vung Tau.[29] This was the first time Australian soldiers serving overseas could talk to their homes in Australia via an Army-provided service.[30] One month later, 140 calls had been made from Nui Dat to Australia and New Zealand.[31] This service continued until 19 January 1972, when the force withdrew from Vietnam.[32]

The in-theatre and national rear link communications unit - 110 Signal Squadron had managed to restructure and introduce new facilities with limited resources under considerable pressure to maintain the operational communications links without disruption. They achieved this successfully. The move of the receivers from Long Binh, and the streamlining of the major relay stations and system control centres was carried out without interruptions to the vital link which the Royal Australian Corps of Signals provided. The introduction of the telegraph teleconference facility, as well as the hospital radio link and the radio

26 AWM 95, CD, 110 Sig Sqn Monthly Report Jul 1969; AWM 95, CD, 110 Sig Sqn Monthly Report Feb 1970; and AWM 95, CD, 110 Sig Sqn Monthly Report Aug 1969.

27 AWM 103, R198/1/5, 'Radio Telephone Facilities, AFV - Australia', Nov 1967.

28 Rienzi, *Communications Electronics 1962-1970,* p. 128.

29 RASCM, D Sigs 5-LC-2, 'Military Affiliate Radio System (MARS): SVN', December 1967; and AWM 103, R198/1/8, 'Overseas Telephone Service For AFV', 10 September 1970.

30 RASCM, Lawrence, 'The South Vietnam Campaign', p. 7.11.

31 AWM 95, CD, 104 Sig Sqn Monthly Report Nov 1970.

32 AWM 103, R198/2/1, Cable: 'Call Home Phone Service'. Jan 1972.

telephone link to Australia for personal use, were added features that were greatly valued and added to the overall effectiveness of the force in Vietnam. Medical support was further streamlined, morale was boosted with the personal phone facility, and commanders had greater access to the resources of the Army's headquarters in Australia. In turn, commanders in the field knew that they were more easily reached by, and thus answerable to, their superiors than had previously been possible.

Task Force Signals 1969 - 1971

As Task Force operations continued in late 1968, 104 Signal Squadron began a major rebuilding programme, in view of the way radio facilities for the force had grown without any master plan. This plan was to provide for the Task Force's radio requirements, including those of 110 Signal Squadron and other units in the Task Force.[33] Earthworks construction progressed through September and October for the Task Force Signal Squadron's radio bunker on Nui Dat hill. The bunker was operational by December 1968, and an 'other units' bunker was built alongside it to house the other radios that would not fit in the signal squadron's own bunker.[34] Major Keith Morel, 104 Signal Squadron commander from January 1969 to January 1970, recalled that

> We had our radio relay rigs up there as well as a lot of VHF radios. We devised a system whereby we had the local net antennas remoted, and one of my operators maintained the radios for the other units' operators via a patch panel connecting them to their compounds. That way, I had control over all those defence-type communications. Even when they were on

33 AWM 95, CD, 104 Sig Sqn, Monthly Report, Oct 1968; and RASCM, Lawrence, 'The South Vietnam Campaign', p. 6.4.

34 AWM 95, CD, 104 Sig Sqn, Monthly Report, Oct 1968; AWM 95, CD, 104 Sig Sqn, Monthly Report, Nov 1968; and AWM 95, CD, 104 Sig Sqn, Monthly Report, Dec 1968.

> operations, their [radio] links were maintained on the hill via the bunker, by my operators and that provided excellent communications.[35]

Because of the risks involved in having all the radio links concentrated in one place on the hill, Munro and his successor, Morel, had a hole dug, nearby the squadron headquarters, with heavy overhead cover. This was designed to provide a duplicate radio bunker for all the antennas and radio duplicates on all the command nets with a switch in the bunker. Morel recalled that

> this was tested every day, and once we had these alternate communications, I was happy. We didn't have the equipment to do this, but we had a 'pass' from the Americans under my control to get into the Long Binh dump, and we would get extra radio sets, being thrown out, and which I never declared on my equipment tables, which we repaired.[36]

Clearly, there were limits to the capacity of the communications technology then available as well as to the generosity of the Australian Army's supply system. Nevertheless, with some initiative and ingenuity, these limitations were minimised. That initiative and ingenuity would come in handy as the tempo of Task Force operations proceeded apace.

Prolonged deployments of the Task Force Headquarters away from the Nui Dat base continued in 1969, with the Task Force deploying to a number of fire support bases such as 'Julia' and 'Kerry', between January and April 1969.[37] From then on, most operations were closer to the Task Force base, and as a result, the

---

35 Morel, interview, August 1989.

36 Ibid.

37 RASCM, Lawrence, 'The South Vietnam Campaign', p. 7.6; AWM 95, CD, 110 Sig Sqn, Monthly Report - Feb 69; and AWM 95, CD, 110 Sig Sqn, Monthly Report - Feb 69.

need for the Task Force Headquarters and the Signal Squadron to deploy away from Nui Dat was reduced. Nevertheless, occasional deployments of Task Force Headquarters continued. Full deployment communications (including tactical radio, radio relay, communication centre, switchboard and line) were provided and worked successfully throughout these deployments.[38] Radio relay deployments occasionally also occurred to the 'Horseshoe' feature, immediately to the north of the township of Dat Do, east of Baria and southeast of Nui Dat. According to Brigadier S.P. Weir, M.C., the Australian Task Force Commander from September 1969 to May 1970, 'The secure, ever reliable communication links made my task easy... and, during my tour of command of 1 ATF, the Signal Corps component could not have been more effective.'[39]

In April 1970, the style of operations conducted by the Task Force changed once again. This time, their efforts were focused on ambushing the Local Force Viet Cong in Phuoc Tuy Province.[40] The reduction of the Task Force from three down to two infantry battalions, which was to occur in November 1970, would further reduce the level of operations that the Task Force could embark.

Throughout 1970 and 1971, the need remained for liaison officer detachments around the countryside. In January 1970, for instance, ten liaison officer detachments had been deployed with neighbouring US, Thai, and Vietnamese units in neighbouring provinces, as well as the various district headquarters within Phuoc Tuy Province. There was also a retransmission detachment providing automatic and manual retransmission facilities at Nui Chua Chan for the Special Air Service (SAS) Squadron's long-range reconnaissance patrols as well as other infantry battalions operating, on occasions, in the vicinity.[41]

---

38 AWM 95, CD, 110 Sig Sqn Monthly Report, Dec 1969; and AWM 95, CD, 104 Sig Sqn Monthly Report - Apr 1970.

39 RASCM, Letter from Brigadier S.P. Weir to Lieutenant Colonel G.J. Lawrence, 1977.

40 Frost, *Australia's War in Vietnam*, p. 134; and AWM 95, CD, 104 Sig Sqn, Monthly Report - Aug 1970.

41 AWM 95, CD, 104 Sig Sqn, Monthly Report - Jan 1970; and AWM 95, CD, 104 Sig Sqn, Monthly Report - Jun 1970.

Daily life at Nui Dat, according to Morel, became 'pretty well set'.

> We had the normal administration, and we were always operational, so we had to keep up the maintenance of equipment and workshops. The signal centre was a very busy [office] - phenomenally busy. We had operators at Task Force Headquarters, linesmen, operators on the hill and radio relay operators. We worked shift work all the time, and to us, life became quite routine for the 2nd XI (those remaining at Nui Dat). We were stretched fairly thin when split, but when all together, on weekends, we had swim runs down to Vung Tau. It was a very happy unit.[42]

From 1969 to 1971, the Task Force Signal Squadron's operating procedures had become quite refined. The communications links had been streamlined and duplicated. Also, the prolonged deployments of the Task Force Headquarters were efficiently accommodated for with radio relay detachments and liaison officers in support wherever they were required. Despite these increased procedural and technical efficiencies, there remained a significant shortfall in the area of communications security. The enemy repeatedly was able to elude Australian forces despite carefully planned and conducted operations launched on the basis of normally reliable information from intelligence sources.

Communications Security (COMSEC)

Control of the airwaves was of vital significance for Australian, US and other allied forces during the War in Vietnam, as radio communications were relied upon to a far greater extent than in previous conflicts. It was also clear, throughout the campaign, that the enemy had the capability to conduct both active and passive

42 Morel, interview, August 1989.

electronic warfare - that is, to jam radio nets and use imitative deception on English-speaking nets as well as passively to intercept radio traffic. Telephone lines were also known to have been tapped, and it was possible for information to have been gathered from radio relay circuits and civilian switchboard operators.[43] All of these types of measures were taken against the Australian Forces in Vietnam at varying stages throughout the campaign.[44]

According to one American paper, written in early 1965, the North Vietnamese had established a Central Research Agency, which maintained 'a large and expanding radio communications network'.[45] In an Australian Army information bulletin, it was estimated that upwards of 4,000 personnel were employed in the Central Research Agency's radio intercept network covering South Vietnam and its immediate environs. The bulletin correctly asserted that

> Radio intercept is then, a cumulative function; the mistakes of yesterday will therefore manifest themselves not only today, but for a long tomorrow... One cannot pay lip service to security training and expect good security in war.[46]

It was obvious that, if the enemy could know the Task Force's plans in advance, surprise would never be achieved, and without surprise, it was acknowledged, the prospect of carrying out successful 'counter-insurgency' operations could only be

---

43 1st Australian Task Force Vietnam, Standing Operating Procedures for Operations in Vietnam, Revised Nov 1969; and G.W. Arbogast, (Captain) 'Radio Communications on Vietnam', in *Signal, Journal of the Armed Forces Communications and Electronics Association, April 1967,* p. 22.

44 HQ AFV, Communications Report - Feb 1967; AWM 103, R193/1/5, 'VC/NVA Electronic Warfare Capability', September 1967; and AWM 103 193/1/1 pt 2, cable 'Imitative Communication Deception' 20 Mar 1970.

45 Department of State, *Aggression from the North: The Record of North Viet-Nam's Campaign To Conquer South Viet-Nam,* Washington DC, February, 1965, p. 25.

46 Australian Army, Training Information Bulletin, No 19, pp. 1.4 - 1.7.

slight.[47] In fact some commanders felt that that was why Task Force units rarely saw enemy 'main force' units on operations.[48] Rienzi, reflecting on the Vietnam campaign in 1972, implicitly acknowledged the significance of the Allied Force's ineptitude in this matter when he wrote, 'We could do a great deal, I am sure, to increase the security of our voice traffic and to improve the quality of transmission.'[49]

Lieutenant Colonel Ken Whyte, an Australian Signals Corps electronic warfare expert in Vietnam, supported this view, thus:

> I always had a fairly low opinion of the [communications security] of our own forces. One can only draw the conclusion that a great deal of useful information would have been gained by the North Vietnamese Army's intercept service.[50]

In his report on the enemy's electronic warfare capability, written in September 1967, the Task Force Signal Squadron commander, Major Lawrence, observed

> They are thoroughly familiar with our communication procedures and equipment. Some of their intercept units have English-speaking personnel. They realise that [electronic warfare], especially intercept, is a subtle yet highly effective means of gathering intelligence and causing confusion.

It is not known how much [electronic warfare] has aided the enemy, but it must be concluded that if the [Viet Cong] can mon-

47 Signal Training (All Arms) Pamphlet No 6, *Communications Security,* 1971, Australian Army pamphlet.
48 Cowley, interview, August 1989.
49 Rienzi, *Communications-Electronics,* p. 160.
50 Interview with Lieutenant Colonel K. Whyte, August 1989.

itor our communications, they have the potential to exploit any compromised information.[51]

Concrete evidence for this conclusion was to be found repeatedly in the Vietnam War. One instance of this occurred when the Task Force deployed on operations in 1969. One of the battalions captured a Viet Cong soldier who had a captured American AN/PRC-25 VHF radio set tuned to the 1 ATF Artillery Command Net on which all the fire orders were transmitted. Thus, when fire orders were heard, they had enough warning time to take shelter in their tunnels and escape the onslaught.

Despite the effectiveness of the enemy's radio monitoring, it was acknowledged that this could have been greatly reduced by the correct use of voice procedure, adherence to Signal Operating Instructions and voice codes and by constantly remembering the principles of communication security (that is, a denial of information valuable to the enemy which could be derived from the interception and study of electronic communications transmissions). To help ensure that these steps were being taken and to improve communications security, a communications security (COMSEC) monitoring team was established from 110 Signal Squadron resources and used predominantly alongside the Task Force Signal Squadron from Nui Dat.[52]

Their job was immense as they faced a wall of antipathy from most quarters. This was compounded by the fact that most members of the three-man detachment were not trained in communication security duties.[53] Their reports were duly forwarded

---

51 AWM 95, CD, 104 Sig Sqn Monthly Report, Sep 1967.

52 1st Australian Task Force Vietnam, Standing Operating Procedures for Operations in Vietnam, Revised Nov 1969; and Signal Training (All Arms) Pamphlet No 6, Communications Security, 1971, Australian Army pamphlet.

53 AWM 95, CD, 104 Sig Sqn Monthly Report, Jun 1970; and AWM 95, CD, 110 Sig Sqn Monthly Report, Aug 1970.

to the respective unit commanders, who were often embarrassed by the results.[54]

Unfortunately for the Americans and their allies, efforts aimed at stemming the flow of valuable information to the enemy over the airwaves remained all but completely ineffective as late as mid-1970. Allied communications provided the enemy with information regarding unit movements, B-52 [bomber aircraft] strikes, artillery employment, ambushes and other tactical operations.[55]

Some improvements were noted in the Australian Task Force in July and August 1970, as the issue of Signal Operating Instructions was reduced in quantity and simplified in format, and a radio telephone procedure training programme was implemented for Signals and other unit radio operators.[56] Nevertheless, many security breaches continued to occur.[57]

One measure taken to reduce the security violations via radio relay systems was introduced in August 1970: the switchboard operators cautioned all trunk telephone users with the phrase - 'Radio Relay, no security, Sir' before the trunk connection. This measure had the effect of making a surprising number of users aware that they were speaking over a radio bearer and not a cable, as they had supposed.[58]

Not only did these old problems continue but others came up as well. In 1971, complete compromises occurred when, on 31 March, copies of codes and Signal Operating Instructions were captured by the enemy in a 'contact' with 3 RAR. Then, again, on 18 April 1971, a further compromise occurred when a

---

54 See for instance AWM 103, 193/1/3, 'COMSEC Report: 6 RAR Bn Comd Net', Jan 1967; AWM 95, CD, 104 Sig Sqn Monthly Report, Aug 1967; AWM 95, CD, 110 Sig Sqn Monthly Report, Mar, Sep Dec 1968; AWM 95, CD, 104 Sig Sqn Monthly Report, Aug 1971.

55 AWM 103, 193/1/3, cable, II Field Force Vietnam to subordinate commands, May 1970.

56 AWM 95, CD, 104 Sig Sqn Monthly Report, Jul 1970; and AWM 95, CD, 104 Sig Sqn Monthly Report, Aug 1970.

57 AWM 95, CD, 104 Sig Sqn Monthly Report, Sep 1970.

58 AWM 95, CD, 110 Sig Sqn Monthly Report, Aug 1970.

helicopter of 9 Squadron RAAF was shot down in the Long Hai hills whilst supporting a Vietnamese 'Regional Force' battalion. The complete re-issue of all 1 ATF codes and [Signal Operating Instructions] was necessary on both occasions.[59]

The number of recorded security breaches on the Task Force Command radio net was to drop off dramatically in July 1971, when the speech security devices were fully introduced and used in conjunction with the new AN/PRC-77 radios, which replaced the AN/PRC-25 sets. These were identical radios except that the 77 set had the additional capability of accepting the American KY-38 on-line digital voice encryption device, which allowed users to speak freely on the radio, without fear of being monitored.[60] These reportedly made an enormous difference, saving hand written codes. The Task Force Secure Command Net could then be used 'like a public telephone, as commanders didn't have to use veiled speech or codes - it was the best thing since sliced bread!'[61] Nevertheless, on radio nets that did not yet have the voice encryption facility, security breaches continued. In 104 Signal Squadron's final report, written in November 1971 before departing for Australia, it was reported that

> A number of breaches have arisen as a result of the widespread use of so-called veiled speech. Transmissions on one or more radio nets may, each taken in isolation, be sufficiently veiled as not to constitute breaches of security. When a number of nets are monitored simultaneously, veiled speech on one is often rapidly unveiled by a transmission on another. Such has been the case in 1 ATF, leading on one occasion

59 AWM 95, CD, 104 Sig Sqn Monthly Report, Apr 1971.
60 AWM 95, CD, 104 Sig Sqn Monthly Report, Jul 1971; and Cowley, interview, August 1989.
61 Morel, interview, August 1989.

> to the night location of a battalion being prejudiced.[62]

One member of 547 Signal Troop aptly summarised what was perhaps the underlying misconception that served to hinder good communications security:

> [The Australians and Americans] just regarded the Vietnamese as what they used to call them - slope heads or nogs or gooks. I just don't think they gave the Vietnamese credit for having any intelligence or courage or military skills, which was surprising given the history of the Viet Minh and the experience of the French and Chinese.[63]

Clearly, the issue of communications security was important. The dramatic improvements in communications technology that had facilitated the operating procedures employed in Vietnam had left commanders overawed as to the equipments' and their own capabilities. It appeared that with technology on their side, a degree of invincibility had been given to them. This bred a harmful complacency, the effect of which only began to be contained with the introduction of encryption equipment. By then, most of the damage had already been done, and the decision to withdraw the force had already been taken. Nevertheless, the introduction of voice encryption equipment marked a new stage in the combat power multiplying effect of modern communications technology. Until then, the communications technology had allowed for dispersion and the rapid concentration of forces as and when required - but with no way of preventing the enemy from listening in. From then on, radio transmission security would

62 AWM 95, CD, 104 Sig Sqn Final Report - Sep/Oct 1971.
63 Cited in Rintoul, *Ashes of Vietnam,* p. 80.

deny the enemy this privilege. The full effect of this benefit would not be seen until after the end of the Vietnam War.

Reflections

The period from 1969 to 1971 provided many challenges to the members of the Royal Australian Corps of Signals serving throughout South Vietnam.

Much of the work of the Corps in Vietnam was overseen by the CSO, who played an important role in advising on communications matters and liaising between the disparate Australian and US elements involved. His role served to ensure the provision of effective and efficient communications to the Australian forces in Vietnam.

It is also significant to note that effective communications for 1 ATF would not have been so readily obtained had there not been access to the vast US military communications network established across South Vietnam. Numerous occasions arose where Australian deficiencies were met by the generosity of their American counterparts and the surplus provisions of the US infrastructure.

For 110 Signal Squadron, the period saw Squadron resources streamlined and concentrated at Vung Tau while its communications tasks diversified. The provision of a telegraphic teleconference facility enabled prompt responses to requests but also gave the Chiefs of Staff in Canberra unparalleled access to the Commander AFV and to the day-to-day conduct of operations in theatre. For the first time, Australian commanders overseas potentially could be monitored and probed on a minute-by-minute basis. Thus, although the Chiefs of Staff and their political masters now had far greater access, this access served to add greater complexity of command for the commanders concerned. 110 Signal Squadron also provided support to the medical and welfare facilities, which contributed to the morale and well-being of the Force.

For 104 Signal Squadron, it involved further Squadron deployments in support of 1 ATF operations while adapting to the introduction of voice encryption equipment and coping with the problem of insecure voice communications links. It also involved providing liaison detachments to neighbouring provinces, providing retransmission facilities for the more remote deployments of the Task Force and at the same time assisting in the provision of manpower for the local security patrols. In essence, they continued to provide critical support in ensuring the Task Force was able to conduct dispersed and diverse operations by providing reliable communications wherever the need arose, even though the Force was being reduced and the general war effort was being scaled down.

The most significant communications capability shortfall during the Vietnam War concerned the poor standard of communications security evident throughout the war. Repeated efforts to draw commanders' and their subordinates' attention to the problem did not have the dramatic impact required to overcome the problem. The introduction of effective communications security equipment certainly helped to lessen the problem at the Task Force Headquarters level, but its introduction occurred too late to counter the effect of the damage already done by repeated security violations in the face of a vigilant enemy. The evidence indicates that poor communications security was critical in undermining the positive effect of the introduction of much of the modern communications technology, which had facilitated the dramatic changes in the methods of conducting operations during the Vietnam War in the first place. Even with the introduction of encryption technology, the problems of communications security persisted at lower levels through to the end of the war.

This chapter has illustrated the vulnerabilities of Task Force operations to poor communications security. What follows is a review of Australia's own special and eavesdropping capabilities deployed in support of 1ATF, coupled with some broader reflections on Australia's experience in Vietnam.

CHAPTER EIGHT

VIETNAM SPECIALIST COMMUNICATIONS AND WITHDRAWAL

WHILE THE ENEMY was gathering valuable information from our electronic communications, the Australian Army had its own electronic warfare and special action force elements deployed against the Viet Cong and North Vietnamese Army (VC and NVA). 547 Signal Troop was the electronic warfare troop detached from 7 Signal Regiment in Australia. Many people argued that the VC and NVA had little electronic communications equipment of any note; however, what they did have served as key indicators of their capability and intent in a given tactical situation. This chapter examines the role played by 547 Signal Troop and the 152 Signal Squadron Detachment in exploiting these indicators and concludes with an overview of the withdrawal of the Force. The first section of the chapter considers why an electronic warfare troop was sent to Vietnam, and the contribution that they were able to make. It also seeks to make an assessment of the value of their contribution to the overall commitment of Australian forces to Vietnam.

547 Signal Troop

The need to deploy 547 Signal Troop to Vietnam was largely due to the recommendations of two officers, Lieutenant Colonel Ken Whyte and Major Colin Cattanach. Before the commitment of the

Task Force in 1966, Cattanach had visited Vietnam, and he did much to ensure that the troop would be committed to a firm role within the Task Force.[1]

547 Signal Troop consisted of two officers and 35 other ranks and was operationally maintained alongside the Task Force Signal Squadron at Nui Dat in South Vietnam, from 1966 until December 1971. 547 Signal Troop left Borneo Barracks, Cabarlah, on 21 May 1966 and landed on Vung Tau beach, South Vietnam, two weeks later.[2] Shortly thereafter, the Troop moved up to Nui Dat, where the Task Force Signal Squadron Commander, Major Peter Mudd, allocated a position that was 'quietly tucked away in the Signals sector so that others in the Task Force didn't realise who [they] were or what [they] were doing'.[3]

As part of the First Australian Task Force based at Nui Dat, the Troop provided special communication facilities, often under hazardous conditions, over the next five and a half years. It was committed to a role of intermediary between allied radio research units and the Task Force Headquarters. The troop was to operate under the strategic control of the US signals intelligence organisation in Vietnam and work through an Australian Defence Signals Directorate liaison officer, but it was also to provide very valuable tactical information enabling effective counteraction to be taken against the Viet Cong and North Vietnamese Army.[4] Within a couple of weeks, it had commenced independent 'communications research', using antennas made of 'star pickets' welded together to a height of about 35 feet. The up-to-date information gathered clearly indicated, by contrast, the extent to which information provided from other sources was out of date. Their information was to prove of immense value to the

1 RASCM, 'A History of 547 Signal Troop in Vietnam', unpublished manuscript, p. 1.

2 RASCM, 7th Signal Regiment Unit History', prepared by Captain S.W. Foley, June 1986; and Davies, RASCM, 'A History of 547 Signal Troop in Vietnam', p. 2.

3 Murray, interview, August 1989.

4 Whyte, interview, August 1989.

Australian and allied commanders, and this reputation was first established at the Battle of Long Tan in August 1966.[5]

Signals intelligence officers from 547 Signal Troop considered that the battle was a 'stuff-up', despite the apparent Australian success at Long Tan.[6] The enemy 275 Regiment had been found by 547 Signal Troop operators monitoring enemy radio traffic, but a mistake was made in interpreting the information gathered. It was thought that the enemy force was much smaller than it actually turned out to be. According to Murray, the problem was that the enemy regiment's headquarters travelled with a battalion as protection and therefore did not need radio communications with them, and the second battalion was close enough to have the use of a runner. The third battalion in the Regiment was the only radio link discovered, and as a result, no network could be pieced together from the radio intercept. [Brigadier O.D.] Jackson[, the Task Force Commander,] thought that it was only a part of the third battalion, but [his signals intelligence officer, Captain Trevor Richards] informed him that it was in fact a regiment.[7]

The official historian, Ian McNeill has written that signals intelligence had begun revealing unusual activity to the east of Nui Dat as early as 29 July but the information was poorly used partly because of the habitual secrecy, which resulted in many officers, including Brigadier Jackson, having no experience in interpreting and evaluating or even understanding the strengths and limitations of this kind of information. Consequently, Jackson appeared to take little notice of Richards while Major John Rowe, the Intelligence Officer, was sceptical of the reliability of radio direction finding. In contrast to Rowe, his deputy, Captain Bob Keep, believed Richards' assessments. Then on 14 August, Richards informed Jackson that the enemy 275 Regiment had

5 RASCM, 7th Signal Regiment Unit History'; and RASCM, 'A History of 547 Signal Troop in Vietnam', pp. 2-4.

6 Murray, interview, August 1989.

7 Ibid.

reached 5000 metres from the Task Force base. In response to this, and in the absence of Rowe, who had been evacuated to hospital with hepatitis, Jackson ordered company patrols out specifically into the area where 547 Signal Troop had reported 275 Regiment as being located. The patrolling did not reveal any signs of enemy activity until the engagement at Long Tan on 16 August.[8] Consequently, as the official historian asserts,

> Jackson can hardly be criticised for ignoring supposed evidence regarding the presence of the 275 Regiment. Priorities were nevertheless convoluted, and in this instance could have led to catastrophe, when the need to protect the nature of signals intelligence weighed more heavily than the need to inform[the Commanding Officer of 6 RAR, Lieutenant Colonel] Townsend of the suspicions it had engendered.[9]

Whichever way the events are interpreted, it is clear that from that time on, Captain Richards was to find Australian and allied staff officers 'increasingly attentive to his advice'.[10]

The Troop manned a 'set room' with complex equipment, including computers for processing information gathered. The Troop's other elements included technical maintenance, air operations and a single station locator 'cell' described below.[11] In July 1967, a proposal was put forward for the unit's establishment to be increased from fifteen operators to twenty. This was due to the increased effort required to maintain the unit's support to the Task Force. As the demand for their 'products' increased so did their workload. It was reported at this time that

8 McNeill, *To Long Tan*, pp. 307-311.
9 Ibid., P. 360.
10 RASCM, 'A History of 547 Signal Troop in Vietnam', unpublished manuscript, p. 4.
11 RASCM, 'A History of 547 Signal Troop in Vietnam', pp. 25-26 & 30.

the Operator Signals personnel in the troop were employed in the set room for a minimum of 63 hours per week, with one rest day every 16 days if operating activity permitted![12] Approval was given for an increase of three corporals and two signalmen operators' signals in September 1967.[13]

In the meantime, back in Australia, a special 'Pre-Vietnam Orientation Course' was set up at Cabarlah in Queensland to prepare replacement operators for the unique and difficult intercept role in that theatre. A South Vietnam training room was established for the course using live tapes sent back by 547 Signal Troop.[14]

Innovative equipment was introduced to enhance the Troop's capabilities. One such development was the Weapons Research Establishment's accurate airborne radio direction finder (ARDF). According to Lieutenant Colonel Peter Murray, the Officer Commanding 547 Signal Troop in 1968, the ARDF was the single most important intelligence gatherer apart from the 'eyeball' of infantry patrols.[15] Early in 1967, aircraft experiments began with this equipment. Operators who seldom ascended more than twenty feet up a mast suddenly found themselves 'hurled around the sky in a Cessna', but these initial tests proved unsuccessful, and it was not until later that the system proved effective.[16]

Once the tests proved the equipment to be effective, the pattern was established, and what came to be called 'shush' flights could be sent out up to six times per day to eavesdrop on enemy radio traffic to pinpoint their headquarters and unit locations. The information so gathered could then be used by Task Force staff in planning operations.[17]

---

12 AWM 103, R310/1/23, 'Request for Amendment to Establishment 547 Sig Tp', Jul 1967.

13 AWM 103, R310/1/23, cable 'Amdt to Estb 547 Sig Tp, 14 Sep 67'.

14 Lieutenant Colonel Steve Hart, letter, June 1993.

15 Murray, interview, August 1989. For a more detailed account on signals intelligence flights see Don Dennis, 'Of Shush Missions, Voice Sorties and Pucker Factors' in *Australian Aviation,* May 1991, pp. 39-41.

16 RASCM, 'A History of 547 Signal Troop in Vietnam', p. 6.

17 Dennis in *Australian Aviation,* p. 39.

The small Cessna aircraft had signals intercept equipment crammed into it without major alterations to the aircraft's exterior, except for the small antenna dome under the fuselage. These flights depended on the aircraft flying steadily at the same speed, height and bearing, for the operator, sitting behind the pilot in the Cessna aircraft, with a manually operated directional antenna pod under the plane, to time it and take a series of 'listening shots' (giving bearings and time along the flight) going on the previous day's general information gathered about the enemy's location. He would then collate the information and return across the bearings given, and this gave very accurate information.[18] Murray recalled that at times,

> If we knew the location and type of unit, we would sometimes say 'leave it alone'. On one occasion, the [Special Air Service] went out and took a radio station that had 14 people. They came out with the radio set and their cipher books. The cipher books opened up a lot of information to us.'[19]

The resources available to the Australians for these missions were small in comparison to their American counterparts, who had at their disposal a 45-aircraft strong aviation arm of the 1st Signal Brigade. The Australians relied on two aircraft from 1 ATF's 161 Reconnaissance Flight, which had to be flown on a level straight course purely on instruments - a dangerous thing for pilots to do at that height over enemy-occupied territory.[20] The Americans had similar aircraft, usually Otters (and later on U-8 and RU-21 aircraft); however, the Americans seemed to neglect one essential factor - the enemy had eyes. The Otters

18 Dennis in *Australian Aviation*, p. 39; and Murray, interview, August 1989

19 Murray, interview, August 1989. See Horner, *SAS: Phantoms of the Jungle*, pp. 255-260 for an account of this action.

20 Rienzi, *Communications-Electronics*, pp. 170-171; and Murray, interview, August 1989.

bristled with antennas, and it was not hard to figure out what was happening when one of these was overhead.[21]

On one occasion, a Viet Cong ambush was discovered in the area over which an Australian reconnaissance aircraft was flying. A US company of infantry was moving toward it. At Troop Headquarters, Captain Hugh Nichols and Corporal Ron Biddle worked frantically to ascertain the identity of the troops in imminent danger. Operational files showed them to be part of the 199th Light Infantry Brigade. Word was immediately passed via the aircraft system to the brigade headquarters, which promptly ordered its company to redeploy and attack the ambush. 30 Viet Cong and one American 'GI' were killed in the ensuing fight. The company commander sent a message of thanks, endorsed by the brigade commander, with the comment 'Instead of a US body count we were able to do a VC count'.[22]

Close liaison with other allied (particularly American) units had been difficult at first, as their US colleagues seemed distant, but as the quality of the Australians' material derived from the sources described became self-evident, American attitudes softened. Some had initially regarded the Troop as enthusiastic amateurs, and as these first impressions were being dismissed, the 'convivial can of beer' proved to be a great unifier.[23]

Relations with their American counterparts were a major feature of the Troop's role. Murray recalled that

> They had masses of people and equipment, but not the standard of training (which was brief, many not even having received Morse training). By contrast, ours were largely long-term servicemen who were well-trained.[24]

21 Dennis in *Australian Aviation*, p. 39.
22 RASCM, 'A History of 547 Signal Troop in Vietnam', p. 24.
23 Ibid., pp. 6-7.
24 Murray, interview, August 1989.

547 Signal Troop's timely results caught the attention of the Commander II Field Force, Vietnam, Lieutenant General Bruce Palmer jnr, and once he realised how accurate the Australians' information was, he insisted on sending a daily flight to collect their information from Nui Dat with a liaison officer to exploit the latest results. He thought his own organisation's reports were too vague, and his feelings were shared by fellow commanders of the US 9th Infantry Division and 11th Armoured Cavalry Division working near the Australians.[25]

Further experimental work was carried out in 1968, when a site to the north east of Nui Dat Hill was cleared for a new installation nicknamed 'cell'. Aerials were erected and air-conditioned equipment shelters installed. The project was of a much more complex nature than previous ones, and these demands were catered for by the use of a computer. Trials showed that there was a lot unknown about the ionosphere.[26]

This equipment was being developed in conjunction with the Weapons Research Establishment and was known as a 'single station locator'. It was based on the World War II German 'Wullenweber' aerial system, which used four acres in a circle and proved to be effective. The single station locator located the source of the radio wave by measuring the phased time and angle difference between two incoming radio waves striking the antennas in the circle and calculating the angle of deflection from the ionosphere, having already established the height of the ionosphere.[27] According to Lieutenant Colonel Whyte, the technique of putting it all together in a container and placing it on a site in Nui Dat was very successful.[28]

---

25 RASCM, 'A History of 547 Signal Troop in Vietnam', p. 13; and Murray, interview, August 1989.

26 RASCM, 'A History of 547 Signal Troop in Vietnam', p. 17.

27 Murray, interview, August 1989.

28 Whyte, interview, August 1989. The artillery's locating battery used a string of sensors to locate enemy guns and these were similar in concept to this. See Horner, *The Gunners,* p. 494.

By late 1968, the Task Force's forward deployments had resulted in the issuing of armoured command vehicles for the Task Force Signal Squadron, and one was issued to 547 Signal Troop as well. This was then fitted and wired for the necessary communications equipment, and by working on 547's own links, the Troop's representatives were able to give rapid replies to questions asked by Task Force Headquarters staff officers. It became a part of standard procedure for the 547 vehicle to go with the Task Force Headquarters and Signal Squadron on their forward deployments. The appellation '547 Airmobile, Cavalry and sometimes Signal Troop' jokingly came into being.[29]

Major General S.C. Graham, AO, DSO, OBE, MC, who had been the Task Force Commander from January 1967 to October 1967, strongly praised the work of 547 Signal Troop when he wrote the following

> The enemy was largely inhibited from widespread use of radio by his knowledge that our intercept capability and technical expertise in the difficult field of communications intelligence were just too good to take risks with. Even when enemy messages could not be decoded, invaluable information was still obtained from them. Task Force signalmen were tireless in their ingenuity in keeping one jump at least ahead of their opposite numbers.[30]

The combined effect of the employment of the airborne radio direction finder, the single station locator and the deployable armoured command vehicle served to make the significance of 547 Signal Troop's contribution to the overall effort of the Australian Task Force out of all proportion to its size. The

29 RASCM, 'A History of 547 Signal Troop in Vietnam', p. 18.

30 RASCM, Letter from Major General S.C. Graham to Lieutenant Colonel G.J. Lawrence, 1977.

high level of skill and dedication of the Troop's members served to ensure that the usefulness of these clever technical innovations was maximised.

152 Signal Squadron Detachment

The specialised intelligence communications of 547 Signal Troop also played a large role in enabling the Special Air Service (SAS) Regiment to operate in an unorthodox and sometimes almost incredible fashion, as the SAS was one of the key recipients of 547 Signal Troop's 'product'. They could be inserted and extracted from positions in the midst of enemy units with a high degree of secrecy and security, and relay their information as they required it.[31] They often operated in places remote from the base at Nui Dat and had unique communications support requirements. This section examines the nature of the support they received and the impact it had on their methods of conducting operations.

The SAS had learned much from their experience in Borneo, where they had worked alongside the British SAS. This experience proved invaluable for the SAS Squadron sent, on rotation, to Vietnam.[32]

152 Signal Squadron was affiliated with this formation, which had as its roles: 'reconnaissance and to seek out, often by deep penetration, report on, harass and disrupt the enemy.' This it did by means of small patrols inserted by foot, by parachuting, including free fall, and rappelling, as well as by sea, using small boats, canoes or swimming. The Signal Squadron detachment was responsible for all aspects of the SAS Squadron's communications, including: the provision of a base station for deployed elements, monitoring and cipher facilities for the SAS Squadron,

31 Ibid.

32 For a detailed account of the role of the *SAS in Vietnam see Horner, SAS: Phantoms of the Jungle,* Chapters 11-21.

advice on communications and liaison with other Signals units, and accompanying SAS personnel on missions.[33]

Horner records that the most important equipment carried on SAS patrols besides the weapons was the radios. The favoured long-range radio was the AN/PRC-64 lightweight HF radio that could be used for voice and Morse. Communications to base were usually by Morse code in cipher using one-time letter pads. The URC-10 ground-to-air UHF radio was the next most important, as it was used in case of emergency and sometimes on extraction. Two were normally carried on each patrol - one on the 'mayday' emergency frequency and the other on the supporting aircraft's frequency. The other radio used was the VHF AN/PRC-25 used by the infantry battalions. These three radios gave the patrols a number of options for establishing communications back to base.[34]

The 152 Signal Squadron personnel had to provide communications for patrols reaching out much further than most infantry patrols did, and this presented them with unique problems in terms of finding the best configuration of equipment to maintain their links. In December 1966, shortly after their arrival in Vietnam, they required antenna masts (to reach above the canopy at the Nui Dat base camp) and reliable and stable HF receivers. They also required crystals for AN/PRC-64 on new frequencies, as the initial issue of frequencies was shared with other units, and this was inadequate for their specialist needs.[35]

These immediate problems were addressed by 1967, but in 1968 and 1969, trials were conducted by Second Lieutenant Peter (later Lieutenant Colonel and CO of 2 Signal Regiment)

---

33 AWM 98, 579/1/70, 'AHQ. General Staff Instruction No 7, Roles and Tasks - Special Air Service Regiment', 30 Oct 1970.

34 Horner, *SAS: Phantoms of the Jungle,* pp. 187, 244 & 280. Horner's book, on page 245, also includes a detailed description of the procedures followed in establishing radio communications and passing messages.

35 RASCM, D Sigs 723-3-4, 'D Sigs Notes South Vietnam-Malaysia visit Dec 66'; AWM 95, CD, HQ AFV, Communications Report, Jan 1967; and AWM 95, CD, HQ AFV, Communications Report, Feb 1967..

Fitzpatrick, the unit's Signals Officer. This involved using tethered balloons carrying AN/PRC-25 radio sets acting as an airborne radio relay station platform, which extended communication ranges by extending line-of-sight distances. The tests showed that VHF radio ranges could be extended, but problems with keeping the balloons inflated meant that the project was not considered to be completely successful.[36] Nevertheless, the facility was appreciated by the patrols operating closer to the Task Force base as they no longer had to use the AN/PRC-64 to send their messages.[37]

Communications between the SAS Squadron at Nui Dat and long-range patrols varying from 5000 to 35,000 metres away from the Nui Dat base were largely by HF skywave using the AN/PRC-47 as the main base transmitter and the AN/PRC-64s carried by the patrols. This allowed patrols to go beyond the range of VHF line-of-sight radios, but it imposed the limitation of having to erect long wire HF antennas while out on patrol, although this problem was partially overcome with the introduction of a versatile, efficient and reliable LPH-17 high-angle HF antenna.[38]

Setting up HF antennas was a risky activity when in close proximity to the enemy, so moves were made to change over to VHF communications. But to do this, there was a need to obtain the maximum height for the VHF antennas at the base station, to overcome the shorter distances of propagation that would otherwise have been imposed by the limited line-of-sight from the base. The balloons had been one way of overcoming this problem, but a more reliable means was the erection of a sturdy, pur-

---

36 AWM 95, CD, 104 Sig Sqn Monthly Report, Dec 1968; and AWM 103, R193/1/8, 'Final Program Report: ARPA Tethered Balloon'; and AWM 103, R193/1/8, 3 SAS Sqn 'Evaluation of the Goodyear Tethered Balloon System AFV', Sep 1969.

37 Horner, *SAS: Phantoms of the Jungle*, p. 269.

38 AWM 103, 193/1/1, part 2, HQ AFV, 'Report on Communications - 3 SAS Sqn', Apr 1969; AWM 103, 193/1/1, part 2, 3 SAS Sqn 'Installation of Antenna Mast by 3 SAS Sqn on Nui Dat Feature', May 1969; and AWM 98, R193/5/7, 2 SAS Sqn 'Report on Hi-Angle Antenna Type LPH-17', 29 Sep 1971to.

pose-built 90-foot-high antenna mast on the top of Nui Dat Hill. By 1970, VHF had become the primary means of communication for the patrol back to base and was used in preference to HF. By using radio relay stations deployed to the north and east of Nui Dat and by gaining maximum height of VHF antennas at the base, quite good VHF communications could be achieved. One of these stations was at Hill 837 (Nui Chua Chan) in a US-defended locality, resupplied by helicopter from Nui Dat. Patrols normally carried the AN/PRC-25 VHF radio set with Morse keyers, and the base station used the more powerful RT-524/VRC radio, which was borrowed from a US source. This equipment proved invaluable as it enabled the base station to reply to a patrol VHF Morse code transmission in voice.[39]

As mentioned earlier, the SAS Squadron also relied upon UHF band communications to contact aircraft in their area by the use of a beacon facility on the international distress frequency and to talk to aircraft during the insertion or extraction phase of a patrol using the American AN/URC-10 radio set. This set had only one channel and, because a patrol required communications on two frequencies (the distress frequency and the SAS frequency), each patrol had to carry two AN/URC-10s. This one-channel capability was considered to be a severe limitation and, as a result, efforts were successfully made to replace the URC-10s with the more advanced AN/URC-64, which had four channels available to the operator.[40] Once these were introduced, they replaced the AN/PRC-25 radios as the primary means of air and ground communications. because they were capable

39 AWM 103, 193/1/1, part 2, 3 SAS Sqn 'Installation of Antenna Mast by 3 SAS Sqn on Nui Dat Feature', May 1969; and AWM 95, CD, 104 Sig Sqn Monthly Report, Jun 1970.

40 AWM 103, 193/1/1, part 2, HQ AFV, 'Report on Communications - 3 SAS Sqn', Apr 1969; and AWM 98, 310/2/141, 'Request for Variation to Establishment - Det 152 Signals Squadron - Vietnam', 19 May 1970.

of doing the jobs that both the URC-10 and PRC-25 had done previously.[41]

The significance of the role played by 152 Signal Squadron Detachment in conjunction with 547 Signal Troop was illustrated by the then Brigadier R.L. Hughes, DSO, Commander of the Australian Task Force from October 1967 to October 1968, who had this to say:

> The intercept unit at Nui Dat was a wonderful source of intelligence, and it achieved its greatest success when, as a result of their information, I was able to send an SAS patrol to destroy a VC logistics element which had been on a rice-buying expedition.[42]

The members of 152 Signal Squadron had effectively learnt and applied the lessons from Borneo, and they went on to adapt again for the changed environment in Vietnam. Innovations with balloons and then tall masts enabled the SAS to switch from reliance on the difficult to conceal and HF communications to VHF, whilst operating at distances far greater than those of their infantry counterparts. The radio was acknowledged to be the SAS's second most important item of equipment, and the men of 152 Signal Squadron ensured that they maximised its capabilities.

Withdrawal of the Force

The Task Force achieved a number of such successes over the years that it was in Vietnam, but most of these were overshadowed by the influence of political factors and flaws in the campaign. This culminated, on 18 August 1971, when the Australian Prime Minister, William McMahon, announced the withdrawal

41 AWM 103, R980/500/107, 'Radio Sets - URC-64', Sep 1970; and 'Report on AFV Trial No 5/70, Radio Set AN/URC-64', 30 Jun 1970.

42 RASCM, Letter from Major General R.L. Hughes to Lieutenant Colonel G.J. Lawrence, 1977.

of the Australian Force from Vietnam. Most of the combat elements were to be home by Christmas 1971, and the shipment to Australia of stores and equipment was to be completed in the early months of 1972.[43] The Prime Minister's announcement was listened to by members of the Force over the Australian Forces Radio Station, which provided the most up-to-date news coverage on Australia. Strangely enough, the announcement was not greeted with wild cheers, despite radio reports to the contrary.[44] One account stated that 'all immediately saw the heavy workload ahead and each person formulated a hundred questions to which firm answers would be required before any movement from Nui Dat.'[45]

In Parliament, the then opposition leader, Gough Whitlam, was sharply critical of the commitment to Vietnam. His statement echoed the feelings of people from many walks of life in Australia as expressed in the first and second moratorium rallies on 8 May and 18 September 1970. Responses such as these in Australia cut deep into the hearts of many men who had served in Vietnam.

President Nixon announced a new US doctrine at Guam in 1969, asserting that 'we shall look to the nation directly threatened to assume the primary responsibility of providing the manpower for its defense'.[46] It had become clear that the US commitment to both Southeast Asia and Australia had been weakened by the Vietnam experience in four short years.[47] This decision was followed up with the 'Vietnamization' programme, whereby the Vietnamese forces were to take responsibility increasingly for their own defence while the US and allied forces pulled out.

43 *Commonwealth Parliamentary Debates,* 20 Eliz II, Vol H of R, 73, 18 Aug 1971, pp. 227-229

44 D.M. Horner, then a platoon commander at Nui Dat recalls: 'I well remember listening to the ABC news on Australian Forces Radio Station in the morning and being told that there was 'wild cheering in Nui Dat'. This was my first information that we were going home.'

45 '547 Signal Troop: June 1971 to December 1971.'

46 Cited in Rienzi, *Communications-Electronics,* p. 142.

47 King, *Australia's Vietnam,* p. 10.

The net effect of a series of problems faced by Vietnamese military forces, including government ineptitude and Viet Cong infiltration, was that the various local Vietnamese forces were in no position to replace the Task Force in 1971 on the eve of its departure.[48]

Nevertheless, the Australian Force had begun its withdrawal with the non-replacement of one of the three battalions of the Task Force in late 1970.[49] By the end of September 1971, the Australia Force withdrawal, code named 'Operation Interfuse', was well underway, with stocktakes of equipment abounding.[50] 104 Signal Squadron was planning to cease operations by mid-October and in the meantime, the Squadron's fixed installations were progressively replaced by mobile equipment and 547 Signal Troop relocated on Logan Hill in 110 Signal Squadron's area at Vung Tau. On 16 October, Headquarters 1 ATF Main closed at Nui Dat, along with the radio bunker on Nui Dat hill, and reopened at Vung Tau.[51] The bulk of the Squadron prepared for embarkation on HMAS *Sydney* on 6 November 1971, for the return trip to Australia.[52] 547 Signal Troop was to follow a short while later, flying out of Vietnam on 23 December.[53]

With the impending closure of the base at Vung Tau, studies had been initiated in mid-1971 aimed at providing a viable alternative to the Royal Australian Signals provided and manned HF rear link to Australia.[54] The solution involved a combination of submarine cable, HF radio links and radio relay systems from Saigon to Clark Air Force Base in the Philippines, to Singapore,

48 Frost *Australia's War in Vietnam*, pp. 150-151.

49 Ibid., p. 144.

50 AWM 98, R 193/1/8, 'AFV Communications', 9 Nov 1971; AWM 95, CD, So Sigs, HQ AFV, 'Report of Interfuse Communications', 31 Jan 1972; and AWM 95, CD 110 Sig Sqn, Monthly Report, Sep 1971.

51 AWM 95, CD, 104 Sig Sqn Commanders Diary Narrative, Oct 1971; and AWM 95, CD 110 Sig Sqn, Monthly Report, Oct 1971.

52 AWM 95, CD, 104 Sig Sqn, Final Report - Sep/Oct 71.

53 RASCM, '547 Signal Troop: June 1971 to December 1971.'

54 RASCM, DSigs, AAFV Installations, 5-LC-0, 'Communications to AFV - Hire of a Commercial Channel', 21 Jul 1971.

then HMAS *Harman*, Naval Radio Station in Canberra, and on to Melbourne.[55] Testing of the circuit commenced in early October and was not completed until the end of November. Eventually, the Major Relay Station at Vung Tau was also to be relocated to Saigon and installed there as the communications centre for the Headquarters Australian Army Assistance Group Vietnam (AAAGV), which was to remain behind. It did so until December 1972 when the newly elected Labor Government was to have the Group returned to Australia.[56]

A special troop was raised to provide tactical communications for the residual element of the Task Force.[57] The troop was to form part of the Headquarters AAAGV. This Group remained in charge of the 23 officers and 120 other ranks (consisting of AATTV personnel in Phuoc Tuy and the Headquarters AAAGV staff in Saigon) that made up the remaining force.[58]

As these events were unfolding, an element of 127 Signal Squadron was deployed to Vietnam to assist the 110 Signal Squadron packing team in the de-installation, dismantling and packaging of fragile and expensive electronic equipment for its return to Australia. This team was augmented by a specialist element of the US 1st Signal Brigade, Communications Assets Recovery Agency (CARA), who were to remain until early January 1972.[59]

On 16 December 1971, the HF rear link from Vung Tau to Melbourne was finally closed, under the control of 6 Signal

---

55 AWM 98, R193/1/8, HQ AFV, 'AFV Communications', 9 Nov 1971; and AWM 102, AAAGV R193/1/8, 'Standing Operating Procedures for HQ AAAGV - Aust Rear Link', 13 Apr 1972.

56 AWM 95, CD, So Sigs, HQ AFV, 'Report of Interfuse Communications', 31 Jan 1972; and AWM 98, R193/1/8, HQ AFV, 'AFV Communications', 9 Nov 1971.

57 AWM 95, CD, 104 Sig Sqn, Final Report - Sep/Oct 71.

58 Hon D. Fairbairn, *Defence Report,* 1972, p. 32.

59 AWM 95, CD, So Sigs, HQ AFV, 'Report of Interfuse Communications', 31 Jan 1972; and AWM 95, CD, 110 Sig Sqn Monthly Report Dec 1971.

Regiment at the Melbourne end.[60] The Vung Tau Major Relay Station was subsequently closed on 19 January 1972.[61]

One of the final comments made by the Squadron Commander of 110 Signal Squadron was that

> The theme for INTERFUSE on the [signals] net could very well be too much too soon and too little too late, too much equipment and personnel out too soon, and too little information on remaining and new requirements too late to plan effectively for them. This has been a major cause of the heavy workload now existing, and a lesson for future withdrawals is to resist strongly any attempt to thin out [Signals] personnel early.[62]

The last message was transmitted on the circuit in the morning of 17 December 1972, noting that 1,790,000 messages had been transmitted over the circuit since it had been established on 30 June 1965.[63]

Vietnam - Reflections

Australian soldiers had been committed to the defence of the South Vietnamese regime for over ten years by the time the last Australian soldiers left Vietnam in December 1972. The major role of the Task Force, that of destroying the main force communist Vietnamese units in Phuoc Tuy Province, had been partly accomplished, but the success was dependent on the South Vietnamese Government officials effectively eradicating

60 AWM 95, CD, 110 Sig Sqn Monthly Report Dec 1971; and AWM 102, 437/1/8, Cable: 'Interfuse Rear link Comms', 6 Dec 1971.of.

61 AWM 95, CD, HQ AFV Monthly Report, Jan 1972, including 110 Sig Sqn Monthly Report.

62 AWM 95, CD, HQ AFV Monthly Report, Jan 1972, including 110 Sig Sqn Monthly Report.

63 RASCM, CGS 50583 RAYWFF of 162355Z DEC 72 to AUSTG Vietnam for COMAAAGV from CGS.

the Viet Cong structure from the villages, and this never happened.[64] The implications of this trend had been discerned as early as the beginning of 1967, when Major General Mackay, the first Commander AFV, expressed grave doubts about the War's outcome to his superior, General Sir John Wilton, because of the chronic inefficiency and corruption of the Vietnamese Government and because there appeared to be no understanding Vietnamese military or civilian leaders and there seemed to be no will to win.[65]

For the Royal Australian Corps of Signals, the Vietnam experience provided several high points which served as rays of hope in a War in which a sense of overall purpose had been lost. The assistance rendered through the Civic Action programme was, for many, a way of making up for the seemingly senseless death and destruction that was such a feature of this war. On the other hand, from a professional point of view, the time in Vietnam was a satisfying experience for many, as skills were put to the test in providing, under trying circumstances, the vital communications link that acted as the lifeline for many men isolated in dispersed fire support bases and on patrols. Moreover, the exposure to American equipment and the Americans' lavish scale of operations was also, for many, an unforgettable experience that left a lasting impression.

Their efforts were rewarded by the praise of men such as Brigadier S.P. Weir, the Task Force Commander from September 1969 to May 1970, who stated that

> The secure, ever-reliable communications links made my task easy, and the generally reliable signals intelligence greatly supplemented my 'intuition'. During my tour of

---

64 McNeill, 'An Outline of the Australian Military Involvement in Vietnam July 1962 - December 1972', pp. 18-19.

65 Horner, 'A Complex Command: the Role of the Commander, Australian Force, Vietnam', p. 22.

> command of 1 ATF, the Signal Corps component could not have been more effective.[66]

Similarly, Brigadier O.D. Jackson, the first Task Force Commander in Vietnam, had this to say

> My first recollection...is of the excellence of the Australian Task Force tactical communications...I always knew that so long as I was at my headquarters, in my helicopter or with my pack set, I had certain and clear communications by radio with commanders at literally all levels in the Force. I had not known this before in World War II, in Korea, or in training exercises at home in Australia or abroad... It certainly spoke volumes for signals personnel, planning, training, operations and equipment.[67]

Apart from serving to demonstrate the abilities of the Australians to their allies, the experience in Vietnam also served as a catalyst for development within the Signal Corps. Radio relay had featured predominantly for the first time in the campaign, providing the Force with its triangulated links between Saigon, Vung Tau and Nui Dat as well as out to the fire support bases whilst the Task Force was out on operations.

Limitations in the field of trunk telephone operating were also highlighted as were deficiencies in the area of field operable automatic telephone switchboards and line and cable laying skills, but the most significant deficiency highlighted during the campaign was that of communications security. It was quite clear that the enemy had been effectively monitoring the airwaves throughout the campaign, and the Australian response to this had at times been patently inadequate.

---

66 RASCM, Letter Weir to Lawrence, 1977.

67 Letter from Brigadier O.D Jackson to Lieutenant Colonel G.J. Lawrence, 1977.

The lavish scale of American operations also served to hide many of the deficiencies experienced by the Australian Force throughout the campaign, commencing with the experience of Captain Twiss and 709 Signal Troop in 1965. It was clear that in any future deployment, adequate, modern, air-conditioned and well-maintained equipment would have to be available if the fragile high-technology equipment of the Signal Corps was to be able to carry out its purpose. The January 1972 Monthly Report highlighted this when it concluded saying 'the Australian Forces in Vietnam have relied to a tremendous extent upon...unofficial assistance, and upon the goodwill of the American Forces.'[68]

Major General Graham observed that, probably for the first time in history, communications in Vietnam were able to match in dimension and efficiency the great advances in mobility. The helicopter, allied with the more traditional means of tracked vehicles, aircraft, small ships and boats, made possible the application of concepts which were not new in principle but which previously could never be more than partially implemented. These concepts depended primarily on their success in communications, because without them, the many complex activities that had to proceed simultaneously and in coordination could not have been controlled. Mobility, therefore, supported by communications, enabled the troops to seize fleeting opportunities which units of similar size in World War II or Korea could never have hoped to meet.[69]

On the other hand, the combination of a complex command structure, combined with modern communications capabilities as experienced during the Vietnam War, resulted, particularly at the higher levels, in decision cycles being longer than those of the enemy, who had a far less sophisticated hierarchy and communications system. Commanders appeared to be confronted with an excess of information with little ability to distin-

---

68 AWM 95, CD, HQ AFV Monthly Report, Jan 1972, including 110 Sig Sqn Monthly Report.

69 RASCM, Letter Graham to Lawrence, 1977.

guish between the essential and peripheral. It is evident that the ability to interpret information had not yet matched the technological capability to provide vast quantities of information that required interpreting. This problem was also compounded by the poor communications security, which often gave the enemy the opportunity to stay one step ahead, despite the odds. In fact, the poor communications security played a critical role in undermining the benefits of better communications technology in Vietnam. Moreover, the massive quantity of new and high-technology equipment, including communications technology, had a mesmerising effect on commanders and their staffs. This, in part at least, resulted in the US and allied forces being unable to appreciate the fundamental flaws in the campaign strategy.

On a broader front, communications enabled our side of the war to be brought into living rooms around the world. As General Graham observed

> This was as marvellous technologically as it was damaging militarily, but however one views it, it is a factor which those responsible for conducting warfare will always in the future have to contend with.[70]

For many, the memories of Vietnam linger on, and for the Corps, the experience was a watershed as it was for the Army as a whole. No longer was Australia reliant on working closely with British Forces on operations. In the future, Australia would also seek to reduce its reliance on American support. In the meantime, the veterans would have to cope with the effects of the war on their own lives.

70 Ibid.

CHAPTER NINE

THE ARMY'S STRATEGIC COMMUNICATIONS NETWORK

Introduction

THE DRAMATIC REDUCTION in size of the Australian Army, after World War II, meant that decisions had to be made as to whether or not the Army's strategic telegraph message network that had developed during the war years should be maintained thereafter. The network had become quite extensive by war's end to include overseas and internal links. The overseas HF radio links consisted of transmitter and receiver sites, connecting the relay stations and signal centres. These formed part of the British Empire's Army Wireless Chain which had links connecting the Army Headquarters of the participating nations. The main Australian station was at Army Headquarters Signals in Melbourne.[1]

The internal Australian links consisted of interstate telephone, teleprinter and radio links as well as intrastate and district telephone links. In Melbourne, wartime HF transmitters (including the Marconi SWB 8 and SWB 11) were used in a galvanised iron building known as the 'transmitter shack' (later to become the Gymnasium) at Diggers Rest, while a mixture of HF receivers

1 RASCM L. Moore, 'Outline and Resume of the Development of Army Fixed Communication Network', p. 1; RASCM Directorate of Signals 'A Short History of the Royal Australian Corps of Signals (809); RASCM Royal Australian Signals, 'An Introduction to 402 Signal Regiment', p. 3; and AWM 121, 9/C/5, 'Manning Problems in the AUSTCAN Network'.

(including some dual diversity and single channel equipment) were used at nearby Rockbank. Antennas were rhombics and horizontal dipoles. For dual diversity reception, two antennas were used to overcome fading.[2]

Wartime experience had shown that Australia should never again allow itself the 'technological nakedness' that had left it so vulnerable in 1942 and which had required such effort to overcome. As a result, the PMG was active in developing a civil communications infrastructure in Australia in the early post-war years. However, the role of the Army's strategic communications network was not so clear.[3]

The choice had to be made between three options: firstly, retaining and upgrading the network; secondly, integrating it with the Navy and Air Force's networks; or thirdly, abolishing it and relying solely on civil means of communication. This chapter outlines how the decision to retain the network as a separate identity was made; the steps that were then taken to modernise it; and the technological developments that made it possible.

Developments in Communications Technology

The first post-war Director of Signals, Colonel A.D. (Archie) Molloy, was a stocky, muscular man, with piercing blue eyes, who was pugnacious, aggressive and bounding with energy - a man renowned as a clever operator and a visionary who laid the foundations for the peacetime Signals Corps. He observed that one of the most remarkable features of World War II was the growth in the scope and quality of military communications. The scale of essential communications was greatly increased, as was the demand on the traffic carrying capacity of the system. With the increased distances involved and the greater mobility of forces, it was normally not possible to provide line channels. Consequently, radio had become the primary and often the only

---

2 RASCM Moore, op.cit., p. 1.

3 Moyal, Clear Across Australia, pp. 174-179.

means of communication.[4] In the immediate post-war years, long before the advent of satellites, long haul wireless communications world-wide were still by HF skywave which, when reflected off the ionosphere, could achieve world wide coverage given sufficient transmitter power and sophisticated antenna design.[5]

Increased traffic demands meant that there were more wireless circuits and consequently congestion in the usable frequency bands. So a logical step was the development of high-speed multi-channel radio circuits which enabled more traffic to be handled on a given frequency in a given period. The high-speed technique employed in the early stages of the war was usually the high-speed morse system but eventually the introduction of the teleprinter or teletypewriter was perfected and this conferred on a radio circuit much the same traffic handling capacity as a line circuit employing similar terminal equipment.[6] The trend during World War II, in the sphere of the higher powered radio sets serving the rear areas, was toward high speed machine telegraph either manually-retyped or employing a tape relay system for both wireless and line.[7]

The US-developed teletypewriters were considered the best type of machine telegraph for use with radios and considerable interest in them was shown by the Australian Army's Directorate of Signals. They provided greater ease and efficiency of operation than the old high speed morse (Wheatstone) system. The system developed by the US Army Signal Corps which employed teletypewriters and other similar instruments, using a 5-unit (Baudot) code in their operation, was termed the 'Semi-Automatic Tape Relay System'. In this method, the information

4 RASCM Address by the Director of Signals at Albert Park on 14 May 1947 - Appendix B to the Directorate of Signals' 'Monthly Information Bulletin', No 17.

5 Warner, *The Vital Link*, p. 239.

6 Teleprinters had first been introduced on line circuits in Australia in the mid-1920s. See Moyal, *Clear Across Australia*, pp. 187-188.

7 RASCM Address by the Director of Signals at Albert Park on 14 May 1947.

to be transmitted was converted initially into the form of perforations in a paper tape, using combinations of a 5-unit code for each character. The message was received and resent in tape form until it reached its ultimate destination. At all intermediate offices it was necessary only to transfer the tape from one automatic instrument to another.[8] This process eliminated the slow and costly manual, letter by letter, reprocessing of traffic at intermediate offices that had been the case up to that point, although it still retained the use of operators to relay the tape between message circuits (hence its description as 'semi' automatic).[9]

This new equipment had a number of advantages over morse. These advantages included a reduced risk of failure or transmission interference, greater transmission accuracy, a simplified message handling procedure for the teletypewriter operators, a manpower saving as a result of the development of automatic or 'on line' encrypting facilities, greater speed of transmission and flexibility of use due to its increased ability to be integrated into other military or civilian communication systems.[10]

As this technology was developing, the British Army Wireless Chain had come into operation, using HF skywave as its mode of operation. In 1939, the Army Wireless Chain had connected with Egypt, India and Hong Kong. Australia was connected shortly thereafter and, by 1947, the Chain reached its peak when East Africa, West Africa, Austria, Burma, Ceylon, Egypt, Germany, Greece, Hong Kong, Delhi, Iraq, Italy, Japan, Malaya, Australia, New Zealand and Canada were connected.[11] A similar world-wide semi-automatic network had been established by the Americans stretching eastward and westward from

8 Directorate of Signals, 'Modern Concepts of Intercommunication', in *Australian Army Journal,* No 11, Feb-Apr 1950, p. 28.

9 'Equipment for the AMF Semi-Automatic Telegraph Tape Relay System', in *'Signals Bulletin,* Vol 1, No 2., 14 December 1951, p. 37.

10 MP 742/1, 323/6/425, 'AMF Communication System'.

11 Colonel R.M. Adams, *Through to 1970,* Royal Signals Institution, London, 1970, p. 107; and Nalder, *The Royal Corps of Signals,* p. 479.

the Pentagon in Washington D.C. known as the Army Command Administrative Network (ACAN).[12]

The British War Office proposed, in June 1945, that the Army Wireless Chain be converted from a chain of high-speed morse stations to a chain of radio teleprinter stations, using the semi-automatic tape relay system. This would link the British War Office with the various British overseas commands and with the military headquarters in the various dominions and colonies. As part of this conversion, it was envisaged that the long distance radio links would be converted to either CFS or SSB operation, employing high-powered space diversity reception.[13] CFS was applicable to circuits where one teletypewriter alone could cope with the traffic offering and was usually not a multi-channel facility. CFS was also known as frequency shift working or frequency shift keying. This technique enabled teleprinters to be used over radio circuits with greater efficiency.[14] Unlike CFS, SSB carrier operation was applicable to circuits where there was a heavy volume of traffic to be passed. It allowed up to six teletypewriter circuits to be operated simultaneously over the one radio circuit, thereby achieving economy in the use of both equipment and frequencies.[15] These upgrades in the US and UK demanded a response from the Australian Army. Would the Signal Corps update as well to enable it to be inter-operable internationally or would it cut its losses and abolish the wartime network?

---

12 Coker & Rios, *A Concise History of the US Army Signal Corps*, pp. 23-24.

13 MP 729/8, 67/431/48, 'Conversion of the Army Wireless Chain to Radio Teleprinter Working'. This method involved the use of two separate or spaced antennas for the one receiver so as to overcome poor reception of radio signals.

14 MP 729/8, 67/431/48, 'Conversion of the Army Wireless Chain to Radio Teleprinter Working'.

15 MP 742/1, 323/6/425, 'AMF Communication System'.

Abolish, Integrate or Upgrade? The Army Wireless Chain to 1950

Major General C.H. Simpson, the senior wartime Signals officer in the Australian Army, submitted a request in early 1945 to purchase 60 radio teletypewriter conversion kits for use between Army Headquarters and subordinate formations. Approval from the Commander-in-Chief, General Sir Thomas Blamey, was not obtained before the cessation of hostilities and resulted in the order being reduced to two equipments - which meant that they would be used only to conduct trials.[16]

As a consequence, the need for morse code operators continued in the immediate post war years. For instance, signal operators posted to the Signal Centre at Army Headquarters Signals in Melbourne quickly absorbed the esoteric art of 100 words per minute morse code which used an ink morse code writer on tape or a morse printer for receiving. For transmission at high speed, the messages were punched on a Wheatstone perforator and transmitted at high speed through a sending head. When morse printers could not keep up with the volume of incoming traffic the paper tape of dots and dashes were sent to the transcription hut where Australian Women's Army Service (AWAS) Signalwomen converted the code to digestible English.[17]

Despite the equipment shortcomings experienced at the end of the war, radio circuits between Army Headquarters in Melbourne and the BCOF in Japan were converted to radio-teletypewriter operation in July 1946, using make-shift equipment and an outmoded technique. This was done on the advice of the British War Office, which had instigated similar stop gap measures while the design and production of equipment was finalised for the operation of radio-teletype, multi-channel, SSB cir-

16 MP 729/8, 67/431/48, 'Conversion of the Army Wireless Chain to Radio Teleprinter Working'.

17 RASCM 'The History of the Army Wireless Chain, the COMCAN and AUSTCAN' manuscript by Colonel R.G. Lawrence, 1977, p. 1.

cuits.[18] The British War Office determined that their conversion process would take place in three stages. Stage One involved the modification of stations for carrier frequency shift working (single channel). Stage Two involved the conversion to single side band working SWB 11 equipment (multi-channel, 10 kilowatt HF transmitters). The third stage involved the replacement of some of the SWB 11 equipment by the newer and more advanced S15 or E10 (multi-channel 30 kilowatt HF transmitter) equipment.[19] Physically, the Marconi-built E10 transmitter installed at Diggers Rest was the size of a small two bedroom house![20] Apprentices were known to regard the essential tools to be a broom stick for earthing and a file or heavy glass paper for cleaning the thick copper contacts.[21]

For strategic reasons, Australia was committed to operating the Australian terminals of the Empire Army Wireless Chain circuits in the post-war period and parallel action was at least temporarily required to maintain compatibility. The channel to Kure was established when the force was sent to Japan in early 1946. It existed as a complementary requirement for both the Empire Wireless Chain and the Australian Army.[22] Australia provided and maintained this channel for military and higher level traffic of the countries involved in the BCOF. The channels to the islands of Morotai and to Rabaul were only maintained until 1947, whilst Australian Army commitments in those areas continued to exist. Circuits internal to Australia were also maintained so that efficient command and control could be maintained over the elements of the Army retained for the post-war period'.[23]

18 MP 742/1, 323/6/425, 'AMF Communication System'.

19 CRS A2653, Vol 3 1947, Military Board Proceedings, Agendum 100/1947, 9 July 1947, 'Conversion of the Army Chain to Radio Teleprinter Working'; and Nalder, *The Royal Corps of Signals*, p. 479.

20 Letter from Major General R.P. Woollard, September 1988, p. 5.

21 Notes from Lieutenant Colonel N.R. Churches, 1991.

22 RASCM 'The History of the Army Wireless Chain, the COMCAN and AUSTCAN' manuscript by Colonel R.G. Lawrence, 1977, p. 3.

23 CRS A2653, Vol 2 1947, - 32/1946 'Acquisition of "Grosvenor" as a Permanent Signal Office to Serve AHQ'.

Radio links were maintained until May 1947 on the Australian Army's communications network from the Army Headquarters Signals in Melbourne to Adelaide, Hobart, Perth, Darwin, Duntroon in Canberra and Rabaul in New Britain. The links to Sydney and Brisbane included both teleprinter and telephone channels but voice channels on strategic HF circuits were difficult to provide and rarely used. The maintenance of the BCOF commitment in Japan centred on Sydney and there was a significant traffic load arising from the commitment. Overseas radio links, maintained in Australia as part of the Empire Army Wireless Chain at the time, were from Melbourne to Ceylon, Singapore, Delhi, Wellington and Kure in Japan.[24] Direct links to London were found to be spasmodic and unreliable, so messages for London and for further relay via London were relayed via the Colombo or Delhi stations.[25] With cross connection, it was possible to route messages to any part of the British Empire.[26] All overseas links were connected from Army Headquarters Signals, Melbourne, and traffic from all the other states to each other had to be relayed through the Melbourne complex.[27]

Army Headquarters Signals (renamed 403 Signal Regiment upon being relocated to Watsonia in January 1961, and then 6 Signal Regiment in 1965)[28] was formed on 28 May 1946 when the two wartime units known as Land Headquarters Heavy Wireless Group and Land Force Headquarters Signals were amalgamated.[29] At the end of World War II, Army Headquarters Signals was based at Albert Park in Melbourne. The Signal Centre was nearby at 'Grosvenor' on Queens Road and the transmitting and receiv-

24 CRS A2653, Vol 2 1947, - 59/1947: 'Summary of Proceedings of the Military Board' 19 May 1947 'AMF Communication System'; RASCM 32/1946 'AMF Communication System'; and *Signals Bulletin,* Vol 1, No 3, March 1952, Annex G.

25 RASCM 'The History of the Army Wireless Chain, the COMCAN and AUSTCAN' manuscript by Colonel R.G. Lawrence, 1977, p. 2.

26 Ibid., p. 1.

27 RASCM Moore, op.cit., p. 3.

28 Barker, *Signals,* p. 141; and RASCM 6 Signal Regiment Unit History, prepared in 1987 by Lieutenant M. Oliver-Weymouth, p. 4.

29 Barker, *Signals,* p. 141; and RASCM 6 Signal Regiment Unit History, p. 2.

ing outstations were at Diggers Rest and Rockbank respectively. It wasn't long before personnel were reduced by 200 from 700 other ranks and 43 officers.[30] (by August 1950, the total remaining reached 25 officers and 219 other ranks.[31]) Until 1947, the Signal Officer-in-Chief and the Director of Signals were accommodated at Grosvenor. Both the Signal Officer-in-Chief (until the position was disbanded in 1947) and the Director of Signals' Staff were members, and used the messing facilities, of Army Headquarters Signals at its hutted barracks on the opposite side of Albert Park Lake (a five to ten minute walk away).[32]

As these events were unfolding, the Australian Defence Committee[33] was evaluating the Services communications networks from a different perspective. In April 1947, the Defence Committee recommended, on the advice of the Joint Planning Committee,[34] that the Defence Communications Committee:

- keep itself informed of the receipt and use of modern equipment by the individual services;
- produce a plan for an integrated system;
- monitor equipment purchased by the individual Services to ensure maximum economy; and
- integrate Service communications from Australia to other Empire countries in accordance with the decisions of the UK authorities when their decisions were made known.[35]

30 CRS A2653, Vol 2 1947, Military Board Agendum 32/1946.
31 CRS A2653, Vol 4, 1950, Military Board Agendum 60/1950.
32 Woollard, letter, September 1988.
33 The Defence Committee consisted of the Chiefs of the Naval, General and Air Staffs (CNS, CGS & CAS) as well as the Secretary of the Department of Defence.
34 The Joint Planning Committee consisted of the Deputy Chief of Naval Staff, the Army's Director of Military Operations and Plans, and the Director of Air Staff Policy and Plans.
35 CRS A2031, Defence Committee Minutes, No 114/1947, 1 April 1947; and MP 729/8, 53/431/15, 'Introduction of Radio Teletype for the Communications Organisation in the RAAF'.

Integration meant combining with the separate but smaller communication networks operated by the Royal Australian Air Force and the Royal Australian Navy. These three networks had developed independently of each other, although they were interconnected by single channel links at strategic locations.[36]

Soon afterwards, in July 1947, the Defence Committee concluded that the following four principles should be applied as the post war policy in respect of Service communications:

- they should employ the most modern techniques;
- with the exception of Station Terminal equipment (which, was considered unlikely to be produced economically in Australia), efforts be made to produce items which may reasonably be so produced in Australia;
- the provision of an automatic telegraphy network be on a scale adequate for the immediate requirements of an emergency, and not merely for peacetime; and
- the three Services' systems be integrated as far as practicable consistent with efficiency and economy.[37]

The last part of this proposal was viewed by the Minister for the Army at the time, Cyril Chambers, as a means to implement cost-cutting measures. Chambers had toured Army installations in late 1946 in South Australia and had visited the Signals Section at Warradale, near Keswick. Afterwards he wrote that:

> the whole business appeared to be uneconomic and over manned. It is probable that the retention of this system is costing the taxpayer somewhere in the vicinity of £200,000 per annum.

36 RASCM Moore, op.cit., p 6.

37 CRS A2031, Defence Committee Minute No 256/1947, 29 July 1947; and MP 742/1, 240/7/405, 'Coordination of Service Signal Communications.'

Chambers then tasked the Military Board, in December 1946, to examine the Signal service Australia-wide, including its cost and benefits, and to submit its views to him for further consideration.[38] According to the then Lieutenant Colonel Tim Vincent in the Directorate of Signals, prior to this point, it had been accepted doctrine that Australia would participate in the UK's Army Wireless Chain and have its own internal strategic communications network as well. Apparently, the driving force behind the opposition to the network was the Secretary for the Department of the Army, Mr F. R. Sinclair, who allegedly hid key policy documents on the Army Wireless Chain to undermine the efforts aimed at maintaining and upgrading the network![39] The Vice Chief of the General Staff, Lieutenant General S.F. Rowell, relying on the advice given to him by the Director of Signals, Brigadier Molloy, replied to the Minister on 14 January 1947. He argued that the retention was justified for four reasons. Firstly, the first essential in any military organization was the maintenance of the chain of command and the maintenance of the means to exercise command and convey orders - the Signal Service; secondly, the 73 men then working on the strategic communications network was the minimum nucleus for rapid expansion in an emergency; thirdly, the complexity of the communication equipment meant that it could not be installed and operated at short notice; and fourthly, the cost of maintaining the 73 men was a small premium to pay to ensure a degree of communication preparedness which could not otherwise be obtained for a state of emergency.[40]

The difference between paying for 73 men and sending all the traffic via the PMG would only have been £14,000. Rowell went on to argue that the deficiency of telephone channels between Melbourne, Sydney and Brisbane was such that the return of these trunks to the PMG would not discernibly improve

38 CRS A2653, Vol 2 1947, Minute from Cyril Chambers, Minister for the Army, 19 November 1946.

39 Vincent, interview, September 1988.

40 CRS A2653, Vol 2 1947, Letter from the VCGS, Lieutenant General S.F. Rowell to the Minister for the Army, Cyril Chambers, 14 January 1947.

the service to the public whilst it would have greatly hampered the conduct of Service business. The Army Minister responded stating 'I am not impressed by this case'.[41] Rowell then went further, in April 1947, by drawing on the recommendations of the Defence Communications Committee, which argued that no economy would result for inter-state communications by the introduction of a Joint Service system until certain modern equipment could be procured. Rowell also highlighted the views of the British Joint Communications Board arguing that until the existing strategic radio network could be replaced, any unilateral alteration to the existing international system would upset the Imperial network.[42]

Mr Chambers, accompanied by Molloy, visited the Army Signals establishment based at Grosvenor, Albert Park, on 18 May 1947 as well as the Army Wireless transmitting and receiving installations at Diggers Rest and Rockbank.[43] This tour served to reinforce to Mr Chambers the opinion that each Service was doing a job which should have been done collectively.[44]

The Army's highest body, the Military Board, met to discuss the matter further on 1 August 1947. There, the Minister approved the retention of a fraction of the inter-state communications links although the international links were to be maintained.[45]

This meant that many of the links in use at the end of World War II were closed down. What remained of the internal communication system was a speech channel and a single teleprinter channel between Army Headquarters in Melbourne and Eastern Command Headquarters in Sydney, a radio circuit to Northern Command, based in Brisbane and Western Command based in Perth. The remainder of the circuits were handed over to the

41 Ibid.

42 CRS A2653, Vol 2 1947, Letter from the VCGS, Lieutenant General S.F. Rowell to the Secretary for the Department of the Army, 2 April 1947.

43 CRS A2653, Vol 2, 1947, Military Board Agendum 32/1946.

44 CRS A2653, Vol 1, 1947.

45 MP 729/8. 67/431/48, 'Empire Army Wireless Chain'; and CRS A2653, Vol 1 1947, 'Retention of Signals Service'.

PMG.[46] Signals offices were closed in Queensland, Western Australia, South Australia and the Northern Territory, and all teletype maintenance was carried out from then on by the PMG, with the exception of the teletypes at Army Headquarters in Melbourne and in Sydney. The manning of switchboards and Special Despatch Services in all states (except Victoria and South Australia), which had been major Signals tasks since the Corps was formed, ceased to be a Signal Corps responsibility with effect on 14 August 1947.[47]

The strategic communications side of the Australian Signal Corps had, at this stage, reached its lowest ebb since before World War II. Had it not been for Brigadier Molloy, during his reign as Director of Signals at this time, there may not have been a fixed communication system at all for the Army in Australia. He jeopardised his career to ensure that the Army did not revert to PMG communications only, as had been the case prior to World War II.[48] With commendable foresight, he later also managed to gain staff endorsement for a modern fixed communication system linking all commands with multi-channel HF radios, housed in modern buildings. The completion of this task was left for his successors to accomplish.[49]

Meanwhile, life at AHQ Signal Regiment continued following a routine that adjusted to changes as they unfolded. The then Signalman B.C. (Bernie) Saunders (later Major), posted to Traffic Squadron, AHQ Signal Regiment in 1948, recalls that the Officer Commanding, Major Ted Maclean, was an ex-operator who led by example. 'Possibly his greatest attribute was that he acted as a buffer between the Directorate and the operators, as his knowledge, along with that of his Chief Duty Signals Officer was so good, many staff hare-brained operational ideas never got past

---

46 MP 729/8, 53/431/15 (S), 'AMF Internal Communications', Jun '47, drafted by VCGS Maj Gen Rowell; and MP 729/8, 67/431/48.

47 'Signal Information Bulletin', No 20, 15 August 1947, p. 7.

48 RASCM Moore, op.cit., p. 9.

49 'A Short History of the Royal Australian Corps of Signals', written for publication in the RA Signals Newsletter Vol 10, No 11, November 1981.

them.'[50] The operators had their own job to do and he let them get on with it.

The operating system then in use was a one-to-one effort with an operator at each end of the teletype, or hand-speed morse circuit, who transmitted and received signal traffic. Operators had to be reasonably skilled and, when competent, gained a wonderful sense of achievement. Each circuit was personally 'owned' by a specific operator on shift and new operators had little chance of working one except when the designated operators answered the call of nature or had a tea or meal break. The system was tried and true and engendered encouragement to learn one's trade.[51]

From a technical viewpoint, massive advances in the commercial electronic sphere created a high demand for good engineers and technicians. Consequently, those so qualified tended not to stay in the Army for long. This created an on-going problem of retaining the best people and maintaining the technical standards. Despite these difficulties the technicians and operators adequately coped with the work at hand. This was due to a number of factors one of which was the small traffic load.[52] It could conversely be argued that the Army has served as an excellent trainer, providing a steady influx of trained manpower to civil life which has helped to maintain the economic growth of Australia.[53]

Faulty wiring was a common problem and on occasions Bill Potts, the unit electrician, was thrown out of the ceiling when jolted by touching live wires. Although he was a small, wizened, shaky man, the story has it that he was only 24 years old.
Saunders quipped - 'this I doubt but I do know that there was no way to get him into the traffic area ceiling unless he had previously fortified himself with a number of whiskies.'[54] On the subject of alcohol - the air conditioning coolers in the transmitter

50 Saunders, letter, August 1990.
51 Ibid.
52 Ibid.
53 Churches, notes.
54 Saunders, letter, August 1990.

stations at Diggers rest and Acacia Ridge were know to be able to keep a lot of beers cold![55]

Despite the initial setback referred to earlier at the hands of the Minister, Mr Chambers, with the cutback of the network, further steps were taken to ensure the network's viability. In February 1948, a request was submitted to the Minister with the support of the Military Board, for the purchase of the wartime Signal Centre in Melbourne, 'Grosvenor', from the owner who had been leasing it to the Army since early in World War II. A further request was made to improve the circuits connecting it to the receiving and transmitting outstations at Rockbank and Diggers Rest, outside Melbourne. The Minister approved the purchase and the circuit improvements on 3 March 1948.[56]

Following the Minister's efforts to reduce the Army Communications network, the Military Board had not given up and considered that there were numerous advantages to be gained from converting the network to the radio teletype method of operation. These included the capacity for rapid expansion in emergencies; the possibility of integration with the US and UK; greater speed, accuracy, reliability and economies would be effected, training would be simplified, and the system would incorporate the latest cipher techniques and allow for future developments. The Military Board also envisaged that if radio teletype was not introduced, Australia's system would be obsolescent and would become a bottleneck for communications emanating from BCOF in Japan. It would also not be able to participate in any combined plans involving modern communications until such time as the delay in modernising the system could be made good.[57] The more cynical view held by operators at the keyboards was that efforts were being made to scrap the system of high speed morse operation because it was a highly

55 Churches, notes.

56 MP 729/8, 67/431/48; 'Signals Information Bulletin', No 28, 15 April 1948, p. 3; and CRS A2653, Military Board Minutes, Vol 2 1947, 8 March 1948.

57 MP 742/1, 323/6/425, 'AMF Communication System'.

complex system that was difficult to maintain and operate with a continual floating population of inexperienced staff at AHQ Signal Regiment.[58]

If agreed to, it was clear that implementation of this conversion programme would involve significant costs.[59] For instance, the procurement of the transmitting and receiving apparatus for conversion to SSB multi-channel operation was estimated at £63,141. Also, a small quantity of the 5-unit (Baudot) code telegraph equipment (enough to meet the peace time traffic requirements) was estimated to cost £30,553 and the required training (conducted in Britain and Singapore) for the conversion teams was estimated to cost £4900.[60] By 1949, this estimate was increased by the inclusion of further replacement and expansion costs to £236,750.[61] The Military Board also considered that the Army's inter and intrastate network should be converted to radio teletype operation with single channel frequency carrier shift circuits. The carrier shift conversion kits were estimated to cost £40,000. The telegraph equipment was estimated to cost a further £11,960.[62] Given the Government's budgetary constraints at the time, these expenses were significant.

As the Royal Australian Navy and Royal Australian Air Force also had a need for similar equipment, joint discussions were held and all three Services agreed to adopt a plan whereby, given the constraints in purchasing in US dollars as a result of Australia's continued membership of the sterling economic area, only the automatic telegraph machines would be imported from the US. The remaining equipments would be manufactured in Australia. The automatic telegraph machines were made by the US Teletype Corporation of Chicago. Most of the remaining equipment was designed and assembled by Standard Telephones and

58 Saunders, letter, August 1990.
59 CRS A2653, Military Board Proceedings, Agendum 100/1947.
60 MP 742/1, 323/6/425, 'AMF Communication System'; and MP 742, 323/6/425, 'Conversion of Army Wireless Chain to Radio Teletype Working'.
61 MP 729/8, 67/431/48, 'Empire Army Wireless Chain'.
62 Ibid.

Cables in Sydney.[63] The introduction of this equipment into service was planned to follow the same three stages outlined by the British. That is, the modification of stations for frequency shift signalling operations, the introduction of tape relay and the conversion to SSB working, followed by the introduction of the E10 transmitters.[64]

As the three Services were deliberating over the need for common equipment, the Defence Committee continued to debate the issue of tri-service integration of the networks. They argued that Australia's international military links should be integrated in accordance with the pending decisions of the UK authorities.[65] In contrast, the Minister for the Army, Mr Chambers, found it difficult to appreciate why a decision on this matter had to be given by the UK Government before amalgamation could be effected in Australia.[66] The position was clarified by the then acting Minister for Defence, J.B. (Ben) Chifley, who stated that 'one of the most important principles for co-operation in British Commonwealth Defence... is standardisation to the greatest possible extent in organisation, equipment and training. Uniformity is of vital importance [and] the present trend in these matters is toward integration on an Empire-basis subject to sovereign control of policy of each part of the Empire by its own Parliament and Government.' He went on to point out that 'the United Kingdom... would be the hub and Australia one of the links, and it would be impracticable, for communication technical reasons, for Australia to take unilateral action to effect integration'.[67] The Australian Government still

63 'Equipment for the AMF Semi-Automatic Telegraph Tape Relay System', in '*Signals Bulletin*, Vol 1, No 2., 14 December 1951, p. 37.

64 MP 729/8, 67/431/48.

65 CRS A2031, Defence Committee Minutes, No 114/1947, 'Signal Communications: Requirements for Reorganisation Period and Post War Army', 1 April 1947.

66 CRS A2653, Vol 2 1947, 'Amalgamation of Overseas Signal Services', Minute from Cyril Chambers to the Secretary Department of the Army, 31 October 1947.

67 MP 742/1, 240/1/405, Letter to the Minister of the Army, C. Chambers from J.B. Chifley, 9 December 1947.

relied heavily, in 1947, on developments in Britain to determine the way ahead for Australia.

The results of the UK Chiefs of Staffs' study on network integration had been passed on to Australia by early December 1948. The British considered that neither economy nor efficiency would be achieved by full integration, that essential freedom of action would be jeopardised and that, in any case, it would be impracticable at that time. In response, the Australian Joint Communications Committee and the Defence Committee determined, in December 1948, that whilst technically feasible, there were a number of factors against tri-service integration. The main factors given were: firstly, the lack of flexibility and the loss of independence and freedom of action; secondly, integration would decrease the facilities for alternate routing in emergencies; and thirdly, the facilities were essentially incapable of integration. For instance, 75 percent of the RAAF and 50 per cent of the RAN shore-based communication facilities at any given centre were purely for these Services' requirements and therefore incapable of integration. Similarly, between 70-80 percent of the effort of the Army's major communication centre related to overseas communications requirements, which were affected by the UK decision on integration.[68]

With these points in mind, as well as the fact that the networks were organised with the UK as the hub, the Defence Committee considered (and the Minister for Defence agreed) that it would be impracticable for the Australian Services to take unilateral action

---

68 MP 729/8, 53/431/20, Letter from J.J. Dedman to C. Chambers, February 1948; MP 729/8, 53/431/34, Memorandum: 'Services Communications Policy', 15 December 1948; and Nalder, *The Royal Corps of Signals,* pp. 479-480.

to affect integration at this time.[69] Prime Ministerial approval was given to this decision on 16 December 1948.[70]

In the light of the decision not to integrate, the British Joint Communications Board recommended a policy of partial integration to take advantage of the many good points, whilst avoiding the perceived disadvantages.[71] The Australian response was, likewise, to recommend mutual collaboration as the best and most acceptable measure of coordination. The Defence Committee also supported the formation of an Advisory Committee on Australian inter-departmental radio and communications matters to help bring about a greater measure of rationalisation at the Australian terminals.[72]

In line with the move to encourage tri-service collaboration, the Defence Committee endorsed the Joint Communications Committee policy to provide the maximum degree of standardisation in communications equipment. This involved the circulation of user requirement specifications in the initial planning stages for new equipments. It also involved the adoption of Joint Service standards and specifications to which manufacturers would be required to conform. The Defence Committee also recommended that these standards and specifications be submitted to the Australian Inter-Departmental Telecommunications Advisory Committee (AID-TAC) for consideration of adoption by all Government Departments and instrumentalities.[73]

---

69 MP 729/8, 53/431/20, Letter from J.J. Dedman to C. Chambers, December 1948; MP 729/8, 53/431/34, Memorandum: 'Services Communications Policy', 15 December 1948; and MP 729/8, 53/431/27, 'Outline of Army's Case for the Retention of A Signals Organisation'.

70 MP 729/8, 67/431/48. This position was reaffirmed in the UK and Australia in October 1953 - See MP 719/8, 67/432/12, 'Integration of Communications Systems'; and CRS A2653, 1954, Vol 3, 'COMCAN-AUSTCAN', Military Board Agendum 59/1954.

71 CRS A816, 48/301/118, 'Integration of UK Fighting Services Strategic Wireless Networks'.

72 MP 729/8, 53/431/20, Letter from J.J. Dedman to C. Chambers, December 1948.

73 CRS A2031, Defence Committee Minutes, No 18/1949, 'Policy in Respect of Standardisation of Equipment', 3 February 1949.

A further nine months were to pass before 7 September 1949, when the Minister for Defence approved that the 'plan for the Army Wireless Chain be endorsed subject to the commitment of £236,750 being covered by the Five Year Plan'. The decision to re-establish the interstate radio circuit links on the grounds of financial economy and improved efficiency followed seven months later in April 1950. Nine months after the outbreak of the Korean War, on 22 January 1951, approval was finally granted for the reinstatement of the radio links (initially using manual morse) to Queensland, Tasmania and the Northern Territory which had been closed in 1947.[74] It was argued that the military telegraph circuits from Army Headquarters in Melbourne to the various Commands and Military Districts would be re-established by radio and auto-telegraph to promote financial and economic efficiency increases - the same reason used by Minister Chambers, in 1947, for their closure.[75] It had become obvious that the economies spoken of in 1947 were, at this stage, illusory.[76] From here on, the services had an ongoing requirement to maintain the network as a consequence of Australia's continued military commitments in regional conflicts through to the end of Australia's involvement in the Vietnam War. By that stage the value of the network would be established beyond doubt.

In order for the Army to make up for lost ground in this field, men had to be trained to install and operate new equipment. The then Corporal G.J. Mapson, recalled the following:

> It was now February 1949 and Captain George Sonnenberg from [the Directorate of Signals] visited us to explain an ambitious project involving the Army Wireless Chain and the adoption of new high-powered HF radio techniques including

74 MP 742, 323/6/425, 'AMF D Sigs Instruction No 36, Melbourne, January 1952.'

75 MP 729/8, 67/431/48; and MP 742, 96/8/248.

76 G. Greenwood & N. Harper (eds), *Australia in World Affairs 1961-1965*, Cheshire, Melbourne, 1968, p. 267.

multi-channel transmission. Of the half dozen Line Mechanics left in the Corps, 4 were to go to a special course in the UK with a similar number of Radio Mechanics...Corporals 'Slim' Boyd, Ian 'Blue Nelson, 'Mick' Rooney and I were the Line Mechanics selected and so we happily enlisted for 6 years in the Regular Army, a requirement for overseas training in those days.

We knew precious little about the Army Wireless Chain and so we were attached to AHQ Signal Regiment, before departure for the UK... Those weeks were a blur of activity and novel experiences. We visited Diggers Rest and Rockbank, discovered radio transmitters the size of pocket battleships called SWB 8s and SWB 11s. There were rows of these all with glowing, white-hot power amplifiers, obscenely high kilovolt readings and there was an ozone nip in the air reminiscent of a violent thunderstorm.

This was very much a Commonwealth Army network, coordinated by a Joint Communications Board in the UK. Technically it used on-off keying, (CW) mostly with teletype start-stop Murray code at 50 baud. As conditions deteriorated the circuits would revert to machine morse and finally hand-speed morse. The cipher system was wholly off-line which of course added considerably to message handling times, both incoming and outgoing. There were occasional tests of Frequency Shift Keying and there was dual diversity reception but all circuits were limited to single channel operation.

> We were impressed by the scale of the task performed by AHQ Signal Regiment exemplified in the acres and acres of rhombics, dipoles and assorted long haul HF aerials at Diggers Rest and Rockbank. We spent some time at the [signal centre] at Grosvenor in Queens Road, Albert Park and were inducted into the mysteries of System Control, Wheatstone variable speed morse transmitters... and now for the first time I understood the proud boast that the Corps was on operational duty 24 hours a day, 365 days a year. I learned that a late message was a catastrophe and a lost message was at least a hanging offence and virtue was an empty washboard.[77]
>
> [Then] in mid 1949 we travelled to the UK on the P&O liner *Stratheden* which was back to its pre-war elegance after service as a troopship. A 4 week cruise is the only way to approach an Army Training Course. I remember thinking at the time it should be made compulsory. We found the wartime austerity of the UK, which was still very evident, an enormous contrast. Rationing and ration books were still in being and the scars of the blitz were showing in many parts of London.[78]

At the School of Signals in Catterick, Yorkshire, the trainees were joined by Captain (later Major General) R. P. (Bob) Woollard and Staff Sergeant Joe Budge who had just completed long term courses there. The course was also well attended by British servicemen. There was a common core of theory for some weeks

77 The 'washboard' was where outstanding tasks were placed for actioning by duty personnel.

78 Letter from Lieutenant Colonel G.J. Mapson, December 1989.

and then the course was split. The radio technicians studied the new E10 (30 Kilowatt) SSB transmitters, while the line mechanics were coached in every aspect of the new voice frequency telegraph system. This equipment represented a significant advance on the less powerful and slower morse telegraph system which it would replace.

After qualifying on the course, the men were given a series of attachments to round off their training, including three months at the Signals Research and Development Establishment (SRDE). Mapson recalls:

> At SRDE... we were forced to live in abject luxury in the Grand Hotel, Bournemouth. We were [also] invited to participate in trials of the first on-line cipher equipment (the 5UCO), which was also destined to come into service on the Army Wireless Chain. This was a great experience and placed us at the leading edge of communications technology in those days.
>
> We rounded off our training with attachments to the War Office Signal Regiment, Richmond Park, London and to the outstations at Bampton (receivers) near Oxford and Droitwitch (transmitters) in Worcestershire. It is now 40 years since we did that...conversion training and I marvel that the techniques used today in the long haul HF radio business are essentially unchanged from those we studied, although the equipments are now smaller, more reliable and more automated...[79]

The group returned to Australia in March 1950 and were set to work to introduce the new equipment and techniques into the Corps. Most of them were posted to the then newly established 1

79 Ibid.

Signal Project Squadron (mentioned in Chapter Two) to plan and implement the installation of the new system.[80]

Consolidation and Growth in the 1950s

During this period, from 1947 to 1950, when the future of the interstate network was in doubt, the Army's strategic signal communications in New South Wales had been provided by transmitters and receivers located at Middle Head, connected by land line to the signal centre at Victoria Barracks, Paddington. In 1950, the transmitters were moved to the Base Signal Park at Kingswood, NSW. A VHF link was installed to carry transmissions from Victoria Barracks in Sydney to the transmitters at Kingswood.[81]

The interstate links had been provided to supplement the civil system throughout Australia but, in the early 1950s, the installations were somewhat makeshift. With the outbreak of the Korean War in April 1950, a great strain was placed on Army Headquarters Signals and the other Signals sites carrying 'operational' circuit links as traffic was heavy due to the build-up for the commitment to Korea. At this stage, the commitment to the Woomera Rocket Range was also a drain on the Signal Corps' limited resources.[82]

AHQ Signal Regiment was organised into three squadrons during this period. These were Wireless Squadron, Operating Squadron, and Administrative Squadron.[83] Communications links working from Grosvenor included radio teletype to Japan and Singapore (with a morse backup to Singapore), scheduled

80 Ibid.

81 RASCM 134 Signal Squadron Unit History, prepared in June 1986.

82 Weir, letter, July 1988.

83 Wireless Squadron included a Headquarters and Central Office Troop at Grosvenor, Transmitting Centre Troop at Diggers Rest and Receiving Centre Troop at Rockbank. Operating Squadron included a Headquarters, four Operator Troops, a Cipher Troop and a Despatch Rider troop. Administrative Squadron included storemen, general dutymen and drivers. CRS A2653, Vol 4, 1950, Military Board Agendum 60/1950.

radio teletype links to Suez, London, and Sydney; and a morse link to Wellington, New Zealand. The other Australian capital cities had continuous wave (CW) morse links working day schedules only. Included in the network was a land line to the various other tributary stations, such as the Central Army Records Office (CARO) at Albert Park, Headquarters Southern Command (Victoria), RAAF Base Frognall (in suburban Melbourne) and Puckapunyal Army camp. A metropolitan Signals Despatch Service, using mainly WRAAC drivers, was also provided. A temporary circuit was worked from Grosvenor to Maralinga during the bomb tests.[84]

The Army Wireless Chain underwent significant development as the Korean War and the Malayan Emergency emphasised the need for the Army to acquire new equipment.[85] In June 1952, the term 'Army Wireless Chain' was replaced by 'Commonwealth Communication Army Network', abbreviated to 'COMCAN'.[86] In order to standardise the nomenclature with the UK-based international network and the United States' Army Command and Administration Network (ACAN), the Australian Army's inter-state network became known as the Australian Communications Army Network or AUSTCAN. This term applied to the Army's radio and line network linking Army Headquarters and the British Commonwealth Force in Korea.[87] As agreed to in the late 1940s, it was intended that Australian practice continue to follow as closely as possible that laid down by the War Office in Britain in regard to principles, engineering techniques and procedure.[88]

The Australian hub of COMCAN was the Army Headquarters Signal Centre at Grosvenor. This hub was classed as one of the

84 Weir, letter, July 1988. According to Weir, the link to Vientiane was operated in either 1955 or 1956 by operators at Grosvenor.
85 Woollard, letter, September 1988.
86 *Signals Bulletin,* Vol 1, No 4, 21 July 1952, p. 30.
87 *Signals Bulletin,* Vol 3, No 1, January 1954, p. 36; and Letter from Lieutenant Colonel G.J. Lawrence.
88 *Signals Bulletin,* Vol 3, No 1, January 1954, p. 38.

six international 'Primary Relay Stations' in the world-wide network.[89] A large traffic load passed through this hub as it was also the main station for AUSTCAN. Consequently, in March 1953, a new Signal Office was opened in Melbourne's Victoria Barracks. This office became a 'tributary station' to the Primary Relay Station at Grosvenor. Its opening eased the load at Grosvenor and obviated the need to maintain a regular Signals Despatch Service mail run between Army Headquarters, in Victoria Barracks, and Grosvenor. The two sites were linked by two hired PMG lines using the 5-unit code and tape relay technique. Sufficient teletype equipment (including the EE97B teletypewriters) were installed to operate two duplex channels.[90] The operating facilities, in use at Grosvenor until 1955, were based on a crude tape relay system using banks of Transmitter Distributor (TD) heads and TG7B printers and perforators.[91]

Operators included men and women on day and night shifts - and sometimes the night shifts were sufficiently quiet to allow for a little 'kip', although, of course, the women always slept in a separate room - usually the cipher office or 'Rockex room'.

> On one hot and humid night, having retired to the Rockex room and pulled down the curtain, the girls decided to protect their difficult to iron uniform khaki frocks by removing them and kipped down in their underwear. Later, one of the girls awoke to go to the toilet and, not wishing to go alone, woke her companions who decided to go together. Knowing they were the only awake occupants of the building, they decided against donning their frocks for the quick trip. So, in their underwear, they tiptoed through the main office. Afterwards, on the spur of the moment, they decided to go

89 'History of the Post War Army', September 1953 precis, p. 6
90 *Signals Bulletin,* Vol 2, No 2, 30 April 1953, p. 27.
91 Weir, letter, July 1988.

> up to the roof to get a breath of cool night air. Suddenly they were bathed in torchlight and a Commonwealth Police officer demanded to know 'who's there'. Panic ensued. Three females in a state of undress jumped back from the parapet saying nothing, ran downstairs and, having locked themselves out, rang the bell, woke the men and, on entering, brushed aside the Sergeant opening the door gasping 'quick, we were spotted on the roof, the guard will be here any minute'. The quick thinking duty cipher officer turned the lights and cipher machines on to give the appearance of busyness. Shortly after the guard appeared but was told that they had been too busy to notice any goings on outside. The incident was eventually reported as a false alarm but the girls were given a fatherly admonition - the men had their reputation with their wives to maintain![92]

Antics aside, the stations on the network played an important role in providing the communications necessary for the effective command of the diverse elements of the post-war Army. As demands continued to grow, so the imperative for the network to grow became stronger.

The plan to upgrade the strategic communications network to form the AUSTCAN and Australian component of COMCAN involved the procurement and installation of new equipment and the development of sites. In the period from 1949 to 1953, according to Major Les Moore, a Staff Officer at the Directorate of Signals in 1953, miscellaneous equipments had been ordered for the 'network' but this had been done on an ad-hoc basis without overall system planning - some things never change!

92 Powell, letter, August 1990.

By 1951, the equipment ordered for the partial modernisation of the international network had arrived in Australia, but it could not be used until new buildings were erected. Delays in obtaining suitable designs had retarded this part of the project.[93] In 1953, Lieutenant Colonel de Courcey Browne and his team in the Directorate of Signals were tasked to examine what was held and what was on order from the overall system viewpoint. They discovered that most items on order would not inter-operate and, not surprisingly, as a result a number of the equipment orders had to be cancelled.[94]

The War Office, in 1953, deciding that it could wait no longer, introduced the SSB transmission method to Australia, via the Singapore and Nairobi routes (using modified SWB11 equipment on the Melbourne end of the circuit), and to Wellington, while leaving the Melbourne-London direct circuit on carrier frequency shift working. This proved unsatisfactory due to mismatched equipment on some of the links.[95] The channel between Australia and Kenya (Nairobi) was wanted as an alternative to the Melbourne-Singapore-London link, and to replace the direct route between Melbourne and London. The circuit to Africa was activated on 22 July and the channel was available for traffic from then on.[96] On the AUSTCAN links, the conversion to carrier frequency shift, radio-teletype working took place gradually. It had commenced with Sydney in 1947 and was followed by Perth

93 RASCM Barker, unpublished manuscript, p. 385; (S) *Signals Bulletin*, Vol 2, No 1, 15 December 1952, pp. 33-37; and Vol 3, No 2, May 1954, pp. 44-54.

94 RASCM Moore, op.cit., p. 4.

95 RASCM Barker, unpublished manuscript, p. 385; *Signals Bulletin,* Vol 2, No 1, 15 December 1952, pp. 33-37; and Vol 3, No 2, May 1954, pp. 44-54.

96 RASCM Barker, unpublished manuscript, p. 385; and *Signals Bulletin,* Vol 2, No 1, 15 December 1952, pp. 33-37. On 21 June 1960 a temporary link was established with the United States' Strategic Army Communication System (STARCOM) with a Frequency Shift Keying circuit between Melbourne and Fort Davis, San Francisco. This was the 100th birthday of the US Signal Corps and formal greetings were exchanged by the CGS, Lieutenant General Sir Ragnar Garrett and the US Army Chief of Staff, General Lemnitzer as well as between the two Corps Directors. See '*Signals Bulletin,* Vol 9, No 2, July 1960, pp. 64-66.

in late 1951, Adelaide in 1955, with Hobart and Darwin remaining to be converted by early 1956.[97]

Another task undertaken by the Directorate of Signals was to examine and acquire proper sites for transmitter and receiver buildings in all states except Victoria. The sites at Rockbank and Diggers Rest, near Melbourne, were adequate. Progress was slow and no actual approval for any construction to commence had been given by mid-1952. This resulted in all portions of the project being reconsidered under a significantly curtailed works programme in the following financial year.[98]

A construction programme finally got underway in the mid-1950s but took a number of years to reach completion.[99] The Army Minister at the time, Mr Josiah Francis, was initially hesitant about committing the required £1,288,587 that had been proposed.[100] This figure soon reached £2.9 million on equipment and £.9 million on works. At Diggers Rest, work commenced on the new transmitter building in August 1955.[101]

Construction of the new buildings at the receiver site at Rockbank commenced in 1956, three years after an additional 888 acres of land had been bought in anticipation of the greater need.[102] The Tape Relay Centre at Grosvenor was also renovated as part of the upgrade program. In Sydney, sites were bought at Wallgrove (for transmitting), Bringelly (for receiving) and Dundas (for the Major Relay Station site).[103] Sites for the state AUSTCAN links were also selected and developed during this period.

---

97 RASCM Barker, unpublished manuscript, 179 SL 6/16.

98 *Signals Bulletin,* Vol 2, No 1, 15 December 1952, pp. 33-34.

99 'History of the Post War Army', September 1953 precis, p. 6

100 CRS A2653, 1954, Vol 3, Military Board Agendum 52/1954, 'COMCAN-AUSTCAN', p. 7.

101 RASCM Directorate of Signals, 'Post War Developments in R Aust Sigs, as at November 1955', pp. 5-6.

102 MP 729/8, 67/431/48, 'Empire Army Wireless Chain'; and CRS A2653, Vol 1, 1953, Military Board Proceedings, Note 175, 15 May 1953.

103 RASCM Directorate of Signals, 'Post War Developments in R Aust Sigs', as at November 1955, pp. 5-6.

- Queensland: Lytton (for transmitting), Redbank Plains (for receiving) and Victoria Barracks, Brisbane (for the signal centre).
- South Australia: O'Hallorans Hill, St Kilda and Keswick Barracks.
- Perth: Jandakot, Westfield and Swan Barracks.
- Tasmania: Brighton (for transmitting and receiving) and Anglesea Barracks.
- Northern Territory: 16 Mile, 23 Mile, and Larrakeyah Barracks.[104]

The project progressed slowly through the late 1950s and was not completed until 1961. This occurred only after the Military Board agreed, in July 1959, that the COMCAN/AUSTCAN installations should be completed and put into use at the earliest possible date; and subsequently approved an increase in the establishment of 1 Signal Project Squadron, in order to help fulfil this commitment.[105]

For the individual state links into AUSTCAN, the transmitter and receiver sites were usually separated and connected to the relay station by VHF FM or UHF PTM (pulse-time modulation) radio bearers. The UHF PTM bearers were used to connect only the Melbourne Primary Relay and Sydney Major Relay stations to their transmitters and receivers. These bearer systems were 2000 MHz, 23 channel, 2.5 watt equipments made by STC in Australia. There were quite a few 'teething problems'. The Amalgamated Wireless of Australasia (AWA) 150 MHz, five channel, 10 watt bearers were extensively used. A number of line circuits were also hired from the PMG but the basic philosophy underlying AUSTCAN was for the system to be as independent as possible, whether it be for circuits, power supply or personnel. On one

104 RASCM Directorate of Signals, 'Post War Developments in R Aust Sigs', as at November 1955, p. 6; and (S) RASCM 'Minutes of D Sigs Staff Conference held at Grosvenor, 6 Sep '55'.

105 CRS A2653, Vol 3 1959, Military Board Agendum 154/1959, 17 July 1959.

occasion, the PMG lost its line circuits across the Nullabor and the AUSTCAN Perth-Melbourne HF radio circuits carried a considerable amount of commercial traffic on their behalf.[106]

The Director of Signals from 1954 to 1958 was Colonel D (Tim) Vincent. The then Colonel Vincent was a tall, angular, Duntroon graduate from 1938, who looked the part of a soldier. He was a dedicated and capable officer with a dry sense of humour and regarded as being a man who always had matters of the Corps at heart. He was also an aviator with a mechanical bent who later rose to the rank of Major General, serving as Commander Australian Force Vietnam from January 1967 to January 1968.

Major General Vincent gives the Director of Staff Duties in 1953, Brigadier H.G. Edgar, much of the credit for the staffing and planning of the development of the Australian component of the COMCAN and for AUSTCAN. He took personal control of the situation because he viewed the network as a national standby communications system.[107] Until this time, the decisions concerning the communications network, made in 1948, had not received the planning and financial support to enable them to be carried out. Moreover, AUSTCAN represented the peacetime requirement for interstate links which were considered essential for effective functioning of the higher command of the Army in the event of an emergency.[108]

1954 to 1956, incidentally, was a period that saw a large number of Service marriages between members of AHQ Signal Regiment. According to Major Kel Weir, this situation was due to 'probably so many bachelors coming home from Japan and Korea like myself finally felt the urge to settle down.'[109]

During this period, changes were also occurring in the area of codes and cryptographic equipment. Off-line encryption,

106 RASCM Lieutenant Colonel G.J. Lawrence, 'History of AUSTCAN', pp. 2-3.
107 Vincent, letter, September 1988.
108 CRS A2653, 1954, Vol 3, Military Board Agendum 52/1954, 'COMCAN-AUSTCAN', pp. 8-9.
109 Weir, letter, July 1988.

using methods adapted from the UK War Office, Inter-Service and Army manuals were in use including 'Typex' and 'Rockex'. These were labour intensive, slow and often prone to error. The first on-line cipher equipment in use in Australia was BID (British Inter-Departmental) equipment which used a 'one-time' tape on a line circuit between Sydney and Melbourne. It was introduced in the mid-1950s and was operated by 'Operators Keyboard and Cipher' (OKC) rather than technicians and was used to 'cover' traffic up to secret level. The tapes came on a wooden spool and were issued by 1 Cipher Distribution and Research (CD & R) Troop at Grosvenor.[110]

At about this time, the BID-610 on-line cipher equipment was also introduced to the radio teletype circuits. This equipment used one-time tapes and was operated by technicians. The first of these were installed in Grosvenor in 1955. The decision to use British equipment was dictated by the overseas circuits which all went to either British or New Zealand stations such as Singapore, Port Said in Egypt, BCOF in Japan, and Wellington.[111]

Later in 1955, the first tape relay consoles were installed by 1 Signal Project Squadron under Major Topp. A four shift roster was worked at Grosvenor with a Sergeant on each shift and newly graduated Lieutenants were posted as Duty Signals Officers (DSOs).[112]

A new cipher machine, called 'Derby', was introduced in early 1959 for use within Australia on selected AUSTCAN circuits. Army Headquarters was relocated from Melbourne to Canberra in 1959 and this equipment was used on the circuits established between Melbourne and the new Army Headquarters. The Derby was an 'electronic teleprinter cryptographic regenerative repeater mixer' that could be used in connection with all

110 'Signals Information Bulletin', No 21, 15 September 1947, p. 2. CD&R Troop is now 700 Signal Troop.

111 Weir, letter, December 1988.

112 Included among these were Heather MacGregor, Steve Hart, Bob Jones, Warren Meredith, Alex Hanley, J.P. Nugent, and Trevor Phillips. Weir, letter, July 1988.

standard 5-unit code teleprinter or teletype equipment to provide a teleprinter secrecy system.[113] Plain text or automatically encrypted text could be transmitted to or received from a distant station. Incoming text in cipher could be automatically or manually deciphered and made to appear on the associated teleprinter in plain text.[114]

Later developments in the early 1960s saw the change to US on-line and off-line equipment. This move reflected the growing ties with the US, marked by the Military Standardisation Agreement signed by Australia and the US in 1957.[115]

When Army Headquarters relocated from Melbourne in 1959, a Signal Troop was established in Canberra. The Troop was raised on 15 January 1959 and placed under the command of Eastern Command Signal Regiment (which became 402 Signal Regiment in 1960 and then 5 Signal Regiment in 1965). A fire, on 5 September 1960, virtually destroyed the signal centre and a British Empire Medal was subsequently awarded to Corporal G. Haddon for saving the life of Corporal Jim Flanagan who was trapped in the Signal Centre vault during the fire.[116]

At Hobart, the new signal centre and tape relay station (built as part of the upgrade to semi-automatic tape relay working), was handed over by 1 Signal Project Squadron, on 4 August 1959. The building included new equipment such as the modular, fully transistorised, voice frequency telegraph channelling equipment with eight channels (TCA Type 1851A). At the time, this was an

---

113 Derby was a compact, electrically operated unit, weighing 60 pounds and it used 'one time' key tape for the code. It operated at 45 bauds or 61 words per minute (normal US speed), or 50 bauds, that is, 67 words per minute (normal European speed).

114 *Signals Bulletin,* Vol 3, No 2, July 1959, p. 20.

115 NSC 7513/2 'Long Range US Policy Interests in Australia and New Zealand', 23 August 1957, cited in G. Pemberton, *All The Way,* p. 67.

116 RASCM 'History of 135 Signal Squadron', prepared in May 1986; *Signals Bulletin,* Vol 9 No 4, December 1960, chapter 5.

ultra-modern device and Hobart was the first terminal in Australia to receive it for military purposes.[117]

Shortly afterwards, in January 1961, AHQ Signal Regiment relocated to Watsonia, north-east of Melbourne and changed its title to 403 Signal Regiment.[118] The regiment continued to operate and maintain the primary relay station and provide signal office facilities for the Victorian region.[119]

At the official opening of the barracks trees were planted by dignitaries and brass plaques installed. This marked the commencement of a practice that continued for some years. A detailed plan was needed to record where and what types of trees had been planted and by whom. Thus it was that the Regimental Sergeant Major, Warrant Officer Class One Bill Blair (later Major), was made (via the unit's Routine Orders) the unit's arboriculturist. The typist's misreading of the draft resulted in him being appointed the unit abortionist - with much hilarity as a series of phone calls from the unit's females making pseudo appointments followed. The order was amended![120]

There had also been celebrations at the departure from Albert Park with bon fires on the parade ground fuelled by barrack room and office furniture, and numerous items of unwanted stores. By contrast, Watsonia provided permanent living quarters and a semi-rural atmosphere. The brand new operations building had been the subject of years of planning to provide trouble free torn-tape relay communications for over a hundred circuits anticipated to be needed in the event of war. The only hitch was that the powers-that-be had procrastinated for so long as to where the building would be erected that torn-tape systems were in decline and the era of fully automatic operations was

117 RASCM Barker, unpublished manuscript, p. 388. See also *Signals Bulletin*, Vol 5, No 2, November 1958, pp. 33-36; Vol 8 No 1, April 1959, pp. 37-38; Vol 8 No 2, July 1959, p. 23; and Vol 8 No 3, October 1959, p. 19.
118 RASCM '6 Signal Regiment Unit History', p. 4.
119 RASCM 'Notes by Director of Signals for visit by Brigadier Mohammed Suleman, Feb 1962', p. 13.
120 Powell, letter, August 1990.

beginning. Consequently, the twenty to thirty torn-tape circuits then in existence were banished to one end of the enormous traffic hall.[121]

The delays encountered with the upgrade of the network to nation-wide semi-automatic tape relay were attributed primarily to the financial constraints imposed. Other constraints included the availability of certain suitable items of equipment, the progressive equipment purchasing policy (based on the anticipated communication requirement at the time), and the short timeframe for procurement action after financial authorisation had been granted.[122]

In Sydney, new transmitter, receiver and relay stations were completed in 1960 and 1961; and, in July 1960, 402 Signal Regiment was raised at Dundas, under Lieutenant Colonel D.A.C. Griffiths. It was made up from 3 Squadron, AHQ Signal Regiment, Eastern Command Inter-Communications Troop and Eastern Command Signal Regiment (which had one Regular Army and one CMF Squadron).[123]

402 Signal Regiment's facilities were officially opened on 21 August 1961, by the Minister for the Army, Mr J. O. Cramer, who transmitted an 'around the world' message which took 12 minutes to complete via Nairobi, Boddington, O-Hawa, Washington, Fort Davis, Melbourne, Wallgrove and Dundas. The transmitter station was opened at Wallgrove and the receiver station at Bringelly.[124]

The Network goes Automatic: AUSTCAN in the 1960s

As the network upgrade programme continued, technological developments began to show up the limitations of the semi-au-

121 Saunders, letter, August 1990.
122 '*Signals Bulletin,* Vol 5, No 2, November 1958, pp. 28-33.
123 RASCM 134 Signal Squadron Unit History, prepared in June 1986; RASCM Barker, unpublished manuscript, p. 388. See also *Signals Bulletin,* Vol 5, No 2, November 1958, pp. 33-36; and Vol 8 No 1, April 1959, pp. 37-38.
124 *Signals Bulletin,* Vol 10, No 1, October 1961, Chapter 5; and RASCM 134 Signal Squadron Unit History, prepared in June 1986.

tomatic tape relay system that had been adopted after World War II. The first attempt to build a practical fully-automatic tape relay system was made in the US in the 1950s. This machine was an improvement but it was noisy, costly and not really automatic as it required supervision. Amongst engineers, it was agreed that automatic tape relay centres of the future would have either electronic computers or specially designed electronic components.[125]

In the field of civil telecommunications, similar developments were occurring. In 1959, the PMG had instituted a message switching system because of the growing needs of post offices to communicate with each other in order to handle the growing telegraph traffic demands. By this system, a series of automatic switching centres directed each message to its destination for printing out on the receiving teleprinter. 1959 also saw the first broadband microwave radio telecommunications link in Australia between Melbourne and Bendigo, mentioned in Chapter Two. In 1960, the PMG introduced an electronic (or electro-mechanical) telephone exchange equipment which offered space savings and increased reliability. Then, by early 1962, coaxial cables were installed providing the first broadband link joining Sydney, Canberra and Melbourne.[126]

In 1957, the Directorate of Signals, under Colonel Vincent, deliberated over designs for the 'gateway' stations in Melbourne and Sydney. It was clear that a mechanism superior to 'torn tape' was required.[127] The specifications were raised and computer controlled systems were analysed in both the US and UK, but nothing was found in military circles that could do the job with reliability.[128] Then, early in 1958, the Directorate of Signals was informed that the US National Security Agency (NSA) in conjunction with STC (a subsidiary of ITT), had developed and installed a

125 RASCM Barker, unpublished manuscript, p. 386.
126 For details on the developments in civil telecommunications in Australia at this time see Moyal, *Clear Across Australia*.
127 Vincent, interview, September 1988.
128 RASCM Moore, op.cit., p. 8.

wired-logic, electronic-relay, computer-controlled switch at Fort Meade and that it apparently worked.[129] The Fort Meade facility was then used as a sales exercise with discount inducements and promises of further refinements than those found in the Fort Meade prototype.[130] The machine also told the operators if they did something wrong, and it recorded all the messages. All this represented a quantum jump from the manpower intensive method of torn tape operation.[131]

An order for two of these '**S**imultaneous **T**ransmit **R**eceive **A**nd **D**istribute' (STRAD) fully-automatic electronic relay systems, to be installed in the Melbourne and Sydney Relay Stations, was placed with Standard Telephones and Cables Pty Ltd UK (STC-UK). Orders were also placed for Kleinschmidt Automatic Telegraph machines with AWA. The capacity of the STRAD systems was to be 35 duplex circuits in Melbourne and 24 in Sydney. Both systems were designed so that these capacities could be increased when necessary.[132]

On the face of things the Signal Corps took a bold step in placing a double order of STRAD when it still did not have a proven track record. Rumour had it, however, that 'some palms had been greased' and fears were not allayed when a member of the Directorate allegedly resigned his commission and was appointed, perhaps coincidentally, to an executive position with STC.[133]

The initial purchase price of £300,354 was high but the justification was considered valid on the grounds of manpower saving.[134] Previously a full squadron of operating personnel plus technical and maintenance support was required. Thereafter, one 'shift' of staff for STRAD consisted of only about 12 people.

129 Vincent, interview, September 1988.
130 Saunders, letter, August 1990.
131 Vincent, interview, September 1988.
132 *Signals Bulletin,* Vol 8, No 4, December 1959, p. 16; and RASCM Barker, unpublished manuscript, p. 386.
133 Saunders, letter, August 1990.
134 RASCM Address by the Deputy Chief of General Staff, Major General L. G. O'Donnell, 9 May 1984, on the occasion of the cutover from STRAD to DINTS.

Further advantages included the speed in cross office switching, reduction of human error and ease of repair and maintenance.[135] STRAD was designed to perform a number of functions automatically including: transmission of received messages; numbering, detection of missing serials, logging, storage of delayed messages, taping and printing of messages for transfer to civil systems, alerting staff for high priority messages and prioritising the despatch of messages.[136]

The equipment was installed by the contractors and 1 Signal Project Squadron, using hundreds of miles of wiring and over a quarter of a million transistors and diodes in each site.[137] STRAD was opened in Sydney in 1962 and in Melbourne on 31 October 1963, thus providing the Australian Signals Corps with a world 'first' in terms of Army's strategic communication network as the US, UK, Canada and other European countries were still using electro-mechanical devices.[138] Moreover, the PMG did not receive a fully automatic message exchange until 1966.[139] At the time the powers-that-be were bursting with pride and took the opportunity to boast about their cutting edge technology, but perhaps the others delayed opting for the new system for good reasons.

For in spite of STRAD's features, it had a number of disadvantages: it was susceptible to human error, it was expensive and bulky, and it required a continuous power supply, necessitating expensive emergency generators with 'no-break' facilities.[140] It was common, because of circuit schedules and outages, to have message back-logs in excess of the 100 messages the integral magnetic overflow drum was designed to cope with, so it was essential to have another storage facility. It took a few

135 RASCM Barker, unpublished manuscript, p. 387.

136 Ibid.

137 RASCM Barker, unpublished manuscript, p. 387; and RASCM Address by the Deputy Chief of the General Staff, Major General L.G. O'Donnell, 9 May 1984 on the occasion of the cutover from STRAD to DINTS.

138 RASCM Barker, unpublished manuscript, p. 386; and RASCM Moore, op.cit., p. 8.

139 Moyal, *Clear Across Australia*, p. 189.

140 RASCM Barker, unpublished manuscript, p. 387.

months for the increased overflow mechanism to be installed and this, unlike the original fully electronic one, had mechanical parts subject to wear and tear. Modifications had to be made to the transfer equipment before compatibility was achieved and, by the time STRAD was installed, hundreds of modifications had been completed which in turn compounded the maintenance problem. Further problems were experienced with fluctuating temperatures, which required the air-conditioning thermostats to be set low - necessitating the wearing of winter clothing in midsummer.

Eventually STC despatched a specialist to provide on-going maintenance assistance. He was reputed to have exclaimed eventually in exasperation 'we have created a monster. Its size and complexity are too great for the human mind to fully understand.' Perhaps many of the problems would have been overcome with less difficulty had STRAD's budget not been pruned. As a result of cutbacks, the error detection and correction facilities that were considered by STC to be necessary, had not been purchased.[141]

Despite the difficulties, Major Les Moore's calculations in support of the purchase showed that finances would break even after approximately two years operation of STRAD. From then on, substantial savings would be accrued.[142]

Captain P.J.A. ('Tubby') Evans was serving with 402 Signal Regiment at Watsonia in Melbourne at the time that STRAD was introduced. Captain Evans was a very technically capable and gregarious pipe smoker. He later went on to become a Brigadier. Evans recalled that a difficulty existed in trying to make the 'old and bold' accept STRAD, as there was much mistrust of this electronic equipment.[143] After a lengthy period of installation, the signal centre supervisors and sergeant operators attended a course under STC technicians. Four months of trials on the

141 Saunders, letter, August 1990.
142 RASCM Moore, op.cit., p. 8.
143 Letter from Brigadier P.J.A. Evans

equipment followed, resulting in a relatively smooth transition to the new system.[144]

The 'White House' at Dundas, where STRAD occupied several rooms, was fitted with a temperamental fire alarm system. Initially, the fire brigade responded with two units, then one and finally after the 'umpteenth' call, by telephone.[145] Early difficulties were also found with the on-line cipher equipment, which did not have sufficient alarms when it went out of phase thus causing traffic losses.[146] Operators and technicians were beginning to reconcile themselves to the new system acknowledging 'we have it! There will be no replacement for many years! We have to learn to live with it! - and they did, by adapting procedures and countering its weaknesses.[147]

Teleprinters were also undergoing developments at this time. In 1960, tape machines could punch at the rate of 75 words per minute and overseas there was a prototype page printer capable of 3000 words per minute. High-speed operation was combined with more use of electronics so that, by the early 1960s, improvements in automatic telegraphy made obsolete the old Teletype EE-97 and fixed station models.

The Australian Army embarked upon a machine-telegraph re-equipment programme and, after a detailed study, Kleinschmidt products were adopted for both combat Signals units and COMCAN/AUSTCAN units. The Kleinschmidt machines had fewer parts, were more reliable and could work at speeds of up to 100 words per minute.[148]

By the time the two STRADs were installed, AUSTCAN had semi-automatic tape relay telegraph installations in the minor relay stations at all the other capital cities around Australia con-

---

144 Course participants included Warrant Officers Class Two Doug Miller, Joy Curran, Kel Weir, Sergeants Neville McDonald and Appleby. Weir, letter, March 1988.

145 etter from Lieutenant Colonel B.G. Swan.

146 Weir, letter, March 1988.

147 Saunders, letter, August 1990.

148 RASCM Barker, unpublished manuscript, p. 388.

nected by one HF radio to either Sydney or Melbourne, both of which were connected to Canberra via a line link. They also had one base station capable of operation to a fixed or mobile station operating in their state-based command area.[149]

The Sydney station was capable of taking over only some major circuits from Melbourne in an emergency. This limitation was imposed by the restricted antenna space available. However, the Melbourne station was capable of taking over all of the Sydney message traffic responsibilities. The COMCAN network was connected to AUSTCAN by means of three HF radio circuits from Singapore to Melbourne, Nairobi to Sydney, and Wellington to Sydney.[150]

Integration: The Debate Continues

While the network was being upgraded, a number of reviews had been instigated in the years following the 1950 decision to maintain the three separate Services' strategic communications networks. In 1954 and 1955, the Joint Communications Committee (subordinate to the Defence Committee, and including representatives from the three Services and the Department of Defence), as well as the Board of Business Administration, had concluded that 'effective progress has been made and the technical basis for coordination and cooperation is sound.' However, they argued that a margin of capacity must be available to each Service separately. They then concluded that amalgamation was not feasible and that maximum efforts were already being made.[151]

A further review was initiated in November 1960 and conducted by the Joint Communications and Joint Planning

149 RASCM 'Notes by Director of Signals for visit by Brigadier Mohammed Suleman, Feb 1962', p. 5.

150 Ibid.

151 AWM 121, 6/A/1, 'Internal Communication Networks of the Services in Australia', Brief for CGS, 8 June 1955; and 'Board of Business Administration - Report on Internal Communications Networks of the Services in Australia', August 1954, p. 3.

Committees. The review, completed in 1963,[152] also concluded that integration would nullify any savings because of the rebuilding and alterations that would be required around the nation. It was also decided that no unilateral action should be taken in Australia on this matter without firstly coordinating this aspect with the UK.[153] Reliance was still placed on compatibility with the British Services' networks, as Australian troops continued to serve alongside British troops in Malaya and Singapore, and shortly thereafter, in Borneo as well.

The 1963 review had also stressed the need to ensure 'maximum economy.' In response, the decision was made on 30 November 1964 to raise a Joint Communication Centre at the recently completed Defence Headquarters at Russell Hill in Canberra which would result in an annual saving of £14,000. There would also be a reduction in space requirements of 3,290 square feet at Russell Hill, and additional capital savings. The Joint Communications Centre was drawn from the old AHQ Signal Troop, under Army control as an independent squadron.[154] (In 1969, the Navy and Air Force components were withdrawn and the newly renamed 135 Signal Squadron became responsible for providing the same functions for the three Services Headquarters in Canberra.)[155]

In the following years, between 1965 and 1970, the size of the Australian Regular Army rose from 28,000 to 40,000 as a result of the introduction of selective National Service and the commitment of troops to Vietnam. This was an increase of 55 per cent and, in the same period, the Signals fixed communica-

---

152 AWM 121, 6/M/1, 'Rationalisation of Communications - Army Fixed Communications Network', Enclosure 3.

153 AWM 121, 6/A/2, 'Australian Services' Communications Policy', Annex A to JPC Report No 38/1963.

154 AWM 121, 6/A/2, 'Minutes of the Chiefs of Staff Committee, No 5 1963', 23 January 1963; Lieutenant Colonel Ian Willoughby, letter, August 1990; and RASCM 'History of 135 Signal Squadron'.

155 RASCM 'History of 135 Signal Squadron'. In May 1972, the Squadron received its first WRAAC Administration Officer the then Captain V.A. Graylin.

tion system handled traffic which grew from 127 million groups in 1965, to 561 million groups in 1970. This was made possible by the developments in faster and more reliable communications technology and represented an increase close to 450 per cent![156]

The major contributing station to this increase in message traffic was the signal squadrons in South Vietnam (discussed in previous chapters). For most of the time that Australian forces were in Vietnam, the torn-tape relay station in Saigon provided the national circuit link to Australia. This station was served by a transmitting station at Nui Dat and a receiving station at Vung Tau.[157]

The traffic volume was also contributed to by the opening, in 1963, of a Major Relay Station in Papua New Guinea (as discussed in Chapter Two). A fourteen channel torn-tape relay and transmitting station was located in Murray Barracks, Port Moresby. The station was supported by a receiving station at the nearby Taurama Barracks. Outer links were maintained as combined transmitting, receiving stations and communication centres at Vanimo, Wewak and Lae. These stations were connected to Port Moresby by multi-channel radio circuits.[158]

As a result of this rapid growth, it was difficult for the Corps to maintain AUSTCAN stations efficiently and a large part of this problem could be traced to personnel shortages. Consequently, in 1971 the Director of Signals, Colonel J. I. Williamson, requested assistance. He outlined where the deficiencies lay, indicating a deficiency of between 42 and 66 percent for each rank level in AUSTCAN units around the nation. He concluded that a greater allocation of the more intellectually gifted soldiers was required by Signals so that the higher technical trades could be filled, pro-

156 AWM 121, 9/C/5, Colonel J.I. Williamson, Director of Signals, 'Manning Problems in the AUSTCAN Network', 26 May 1971, p. 1.

157 AWM 121, 6/M/1, Annex A to AHQ DCGS 204/70, dated 22 May 70.

158 Ibid.

viding the operating trades with higher quality people than they were then receiving.[159]

A further communications rationalisation study was conducted in 1970 following the 1966 integration of the British Army's COMCAN stations with the Royal Air Force network.[160] This time, the reasons for avoiding integration that had been used in earlier studies were reduced. However, priority was now given to compatibility with the equipment of US forces. The need to have a degree of survivability against sabotage, military action and natural disaster was also emphasised.[161] In arguing for the retention of the status quo, the Army was able to point to the large traffic volume handled by its strategic facilities.

Despite the reticence to integrate, measures had been taken to streamline the links (apart from the Joint Communications Centre in Canberra) prior to the study in 1970. In Tasmania, for instance, by 1959, all traffic was passed via the Army radio link from Hobart to Melbourne. Elsewhere, however, the three Services continued to provide duplicate facilities. In Sydney, the Army and RAAF had HF transmitters twelve miles apart at Wallgrove and Londonderry, and receivers 24 miles apart at Bringelly and Marsden Park, while all three armed Services had their own tape relay centres at Potts Point (Navy), Dundas (Army) and Penrith (Air Force). In Melbourne, a similar situation could be found, although, after 1959, the Navy had commenced sharing the Air Force's facilities at Frognall, in Melbourne, through a communication centre in Victoria Barracks, Melbourne.[162]

The 1970 rationalisation study was commissioned at a time when the civil telecommunications facilities had been

159 AWM 121, 9/C/5, Colonel J.I. Williamson, Director of Signals, 'Manning Problems in the AUSTCAN Network', 26 May 1971, p. 1.

160 Warner, *The Vital Link*, p. 242.

161 AWM 200, R193/1/5(a), 'Guiding Principles for Planning Communications for the Australian Services', July 1970.

162 AWM 121, 6/A/1, 'Report by the Joint Communications Committee - Integration of Services Communication Facilities', Report No 3/63, and 2/63.

significantly expanded by the introduction of microwave radio links, stretching from Perth to Sydney to Cairns. These links were connected with the international submarine coaxial cable routes, introduced around the globe in the 1960s, which provided 80 voice channels, linking Australia with Asia and America. Australia was also now linked into the Intelsat worldwide satellite network. Civil satellite communications were becoming an integral part of the international services on offer in Australia and this presented the armed Services with a challenge to their rationale for existence.[163]

The main reason for the 1970 review, however, was the increasing expense of trying to keep up with the advances in communications technology. The three Services could not afford to do it if resources were not being shared. This problem was accentuated by the introduction of costly automatic message switching facilities (of which STRAD had been a forerunner) and the dawning of the era of satellite communications. Duplication of effort needed to be eliminated wherever possible. Thus, a far reaching overhaul of the Services communications networks took place.[164]

The preparatory work for the overhaul had been completed by August 1971 and work on the substantive study had begun. Each Service and Department in the Defence Group was required to contribute to the study which was conducted by the 'Defence Communications Rationalization Steering Committee'.[165] This

163 Moyal, *Clear Across Australia,* pp. 239-243; and AWM 121, 6/M/1, 'Rationalisation of Communications- Army Fixed Communications Network', 1970, Enclosure 3 - 'Reasons for Rationalisation Studies'.

164 AWM 121, 6/M/1, 'Rationalisation of Communications- Army Fixed Communications Network', 1970, Enclosure 3 - 'Reasons for Rationalisation Studies'.

165 'Defence Services Communications Study: Question No 3138', *Commonwealth Parliamentary Debates,* 20 Eliz II, Vol H of R 73, 17 August 1971, p. 176; and AWM 102, 749/1/2, Col J.I. Williamson (D Sigs), 'Defence Communications Rationalization Studies, 7 April 1971'.

committee had been set up in 1970 and reported to the Chiefs of Staff Committee.[166]

The development of military applications for satellite communications technology had a significant bearing on the proceedings of the Rationalization Steering Committee. By 1970, military satellite communication systems were being introduced in the US, UK and other NATO countries.[167] Thus, in Australia, it was clear that consideration had to be given to a system in which satellites were one possible means of providing bearers. In the case of the Territory of Papua and New Guinea, it was noted that if a need arose for a deployment of land forces there, satellite communications were essential, as no other system was economically feasible or practicable. The then Director of Signals, Colonel J.I. Williamson, further argued that 'we should always maintain control over the system that we need for command and control of our own forces' and that an attempt should not be made to use the satellite communications systems of other countries for purely internal purposes.[168]

While the Rationalization Steering Committee continued to deliberate over the fate of the three Services' communications networks, the Directorate of Signals sought the support of the Army's Director of Operations, in December 1972, in order to separate the functions of the Army Communications Agency (ACA) from the Directorate of Signals. The ACA was responsible for the technical oversight and direction of AUSTCAN or the Army Fixed Services Communication System (ACS), as it had become known. The reasons given for this desired change were that ACA worked separately from the Directorate and had day-

---

166 AWM 121, 6/M/2, 'Brief for CGS: COSC Agendum No 34/1973 - Defence Communications Rationalisation Studies Steering Committee Report'.

167 T. Gander, *Encyclopaedia of the Modern British Army,* (3rd Ed), Stephens, Willingsborough, 1986, p. 76; and T. M. Rienzi (Major General), *Vietnam Studies: Communications Electronics 1962-1970,* Department of the Army, Washington D.C., 1972, p. 18.

168 AWM 121, 6/F/1, 'D Sigs Comments on a Study of Military Communications Satellites for Defence Communications', Feb 1972.

to-day technical control of the network. The final, and perhaps most significant, reason given by the Directorate of Signals was that the formal identification of ACA would strengthen Army's position to ensure its fixed services communications requirements were met.[169] It was clearly in the Directorate of Signals' interest to make this change in order to exert as much influence on the deliberations as possible.

By July 1973, the Rationalisation Studies Steering Committee prepared its report and recommended that the separate Army, Navy and Air Force networks, as well as those of the Defence and Supply Departments, be integrated into a single network. They also recommended that, by 1985, the network be upgraded to provide secure teletypewriter, data, voice and facsimile communications.[170]

Economic considerations weighed heavily on the deliberations leading to this recommendation. For instance, it was anticipated that $50 million would be required for capital expenditure with an annual recurring expenditure of about $14 million in 1973 dollars up to 1985 - an amount not significantly different from predicted expenditure on the existing systems. Moreover, a real benefit was seen to be gained by being able to provide improved grades of service plus savings in personnel, facilities and recurring costs. The Committee looked to the experience of both the US and UK and observed that their network integration had also resulted in substantial personnel economies.[171]

The Defence Communications Network (DEFCOMMNET), that was to emerge in 1975, and its successor in the late 1980s, the Defence Integrated Secure Communications Network

---

169 AWM 121, 9/C/5, Minute from Major General S.C. Graham, Chief of Operations to ARPS: 'Command and Control of Fixed Signal Units', Dec 1972.

170 These goals were realised with the introduction of the Defence Integrated Secure Communications Network (DISCON) by 1989. AWM 121, 6/M/2, 'Brief for CGS: COSC Agendum No 34/1973 - Defence Communications Rationalisation Studies Steering Committee Report'.

171 AWM 121, 6/M/2, 'Brief for CGS: COSC Agendum No 34/1973 - Defence Communications Rationalisation Studies Steering Committee Report'.

(DISCON), owed much to the work that had gone into the operation of the Army Wireless Chain and its successors the COMCAN and AUSTCAN. The development of and introduction into service of improved communications technology resulted in significant economies and improvements in the service provided. Messages could be on-line encrypted and transmitted at a far greater rate and with substantially improved reliability than had been possible in 1947. These improvements helped ease the personnel requirements necessary to maintain the network.

Reflections

The contrast in procedures from 1947 to 1972 was stark. In 1947, for instance, the process of sending a security classified message to London, involved a number of manually processed steps. Firstly, the message would be hand written and passed to the signal centre where it would be manually encrypted off-line and then transmitted using high speed morse over single-channel HF skywave radios to Ceylon. There it would be relayed manually by an operator (ie re-keyed in morse code), again using HF skywave, to the receiver station in London. There the message would be received in morse, decrypted manually (off-line) and typed out for reading by the intended recipient. By 1972, the message could be sent more reliably and quickly. It would be typed into a teleprinter at the communication centre, encoded on-line and transmitted using a single-sideband, carrier frequency shift multi-channel HF transmitter. The relay would occur automatically, probably in Singapore, and would be automatically onforwarded to the receiving station in London where it would be automatically decrypted and printed out for onforwarding to the intended recipient.

These improvements altered both the method of operations and the actual network configuration to the point where integration with the Navy and Air Force networks became not only possible, but in their best interests. The three Services' network controllers saw that, corporately, they stood a better chance of keeping up with technological developments, given the financial constraints under which they had to operate.

Since the end of World War II, the maintenance of a strategic international and interstate communications network had been validated. The network went through numerous changes culminating in integration, but along the way it had provided Australia with an independent means of communicating with and commanding and controlling its contingents deployed both overseas and around Australia.

The network was shaped largely by developments in communications technology but it was also responsive to the personalities in charge of the Corps' destiny and the needs of the Army within the financial constraints. In turn, the network shaped the structure and nature of the Royal Australian Corps of Signals. The following chapter demonstrates how the Corps responded to these and other challenges. It responded by evolving to meet the requirements generated by the network - requirements for direction and policy formulation (to guide its development) and training (to operate and maintain it effectively).

CHAPTER TEN

TRAINING, THE CORPS IN THE COMMUNITY AND THE DIRECTORATE OF SIGNALS

HAVING VIEWED THE development of the field force and specialist signal units since World War II, as well as the operational commitments of the Corps both overseas and in Australia (with the strategic communications network), it is appropriate to now reflect on the training and policy direction aspects of the Signal Corps. Here, developments in the field of communications technology made the job of setting policy and training an ongoing challenge. The arrival of new technologies meant that trade streams had to be regularly updated and new policy guidelines had to be introduced and managed. Interoperability with allies was also an important consideration as technical and operating skills often had to be tailored to accomodate the direction set by the UK and US. The equipping of our soldiers with the right skills and equipment was the responsibility of the School of Signals in conjunction with the Directorate of Signals.

It is also appropriate that a view be presented here of some of the Corps domestic aspects. The involvement with the Australian civil community in the years from 1947 to 1972 is also looked at.

Training

Signals training in the post Word War II era was to be markedly different from that conducted previously. Prior to World War II, it had centred around training parades in local drill halls, weekend bivouacs, night and morning courses, annual camps and special exercises for officers. These were usually run by Permanent Military Force adjutants and non-commissioned officers on behalf of the militia (CMF) units being served. When CMF instructors were used, they were often only one step ahead of those being trained; besides, for many the camps and bivouacs were social, rather than military, occasions. Moreover, the technology then available was not as complex and so did not require the level of training that its post-war successors would require. Eventually, the Army School of Signals was established at Victoria Barracks in Melbourne, on 1 October 1936. The School moved to Georges Heights in Sydney, in August 1939, and a number of further moves followed during World War II.[1]

By the end of 1942, there were sixteen Signals training units of various types throughout Australia and New Guinea. This number was rapidly reduced at the end of the war. The Central Signals Training Depot moved to Balcombe, Victoria, in February 1946, where it was replaced by 1st Signal Depot Battalion, on 25 August 1946. The Army Headquarters School of Signals had moved to Balcombe in 1945 and the two units worked together until October 1947, when they combined. The School of Signals then emerged, with a new establishment, to become the major training unit for the Signals Corps in the post war period.[2]

The School at Balcombe

During these early post war years, from 1945 to 1947, Balcombe was Signals territory exclusively. The place was later shared with the School of Survey, the Army Apprentices School and, later still, the School of Music. The camp consisted of a sprawling conglomerate of huts built for the US 41st Division and constructed of corrugated iron with cement sheet roofing; unlined, unheated

1 Barker, *Signals*, pp. 116-118.
2 Barker, *Signals*, pp. 211-212; and 'Signals Information Bulletin', No 21, 15 September 1947, p.2.

and sparsely furnished, with external ablution blocks open to the elements and reliant for hot water on wood-fired boilers stoked by the roving picquet during the night.

Balcombe used tons of wood which, upon delivery, had to be split and distributed to the messes, kitchens and boilers. According to G.J. Mapson, that problem was solved with typical Army flair through extra training.

> A minor misdemeanour (like accidentally rattling a mess tin whilst marching to mess) would earn an invitation from 'Jimmy the One' [Warrant Officer Class One Jim Eastlake, the Regimental Sergeant Major of the School of Signals] to attend his woodchopping parade at 1630 hours. Most signalmen could handle an axe with rare skill in those days!
>
> Bugle calls on 78 rpm records were piped over an extensive public announcement system and these regulated activities from Reveille to Lights Out. It was not Hi-Fi but it balanced any shortcomings in quality with a formidable power output. It penetrated all over camp and a mile or two down wind. The Corporal of the Guard played reveille at 0630 and we began the new day with one of half a dozen recorded marches and a country ditty or two. Those of us sprung by the Orderly Officer for being in bed after reveille supplied the bulk of the woodheap detail for the day. The more sympathetic Orderly Officers rattled the corrugated iron with a cane before entering the huts so giving the smart movers just enough time to get their feet on the floor.
>
> Moreover, we got Saturday afternoon and all Sunday off if we were not rostered for

> duty. We were paraded in uniform at the bus stop at noon on Saturday, inspected by the Orderly Officer, handed a leave pass (only if we passed inspection) and marched onto the bus. We caught the last bus back to camp from Frankston station at about 2200 hours and leave expired at 2359 hours on Sunday. Missing that last bus meant a brisk 9 mile walk along the Nepean Highway which was rarely travelled by cars in an era when petrol was rationed and cars were scarcer than hens teeth![3]

The Signal Corps owes a great deal to the residual war-experienced Officers, Warrant Officers and Non-Commissioned Officers who stayed on and held the Corps together in that immediate post-war era when the future shape of the Australian Army was uncertain and before the new crop of Royal Military College 'regulars' emerged. These were men with high standards who made sure that their students knew what was expected of them and the Corps. They made training realistic, urgent and valid as they had 'been there' and could recite hair raising anecdotes from personal experience whenever the trainees failed to heed a lesson or tried to take a short cut.[4]

Although there was this continuitity immediately after the war, Signals training, after World War II, was significantly different from what it had been like prior to 1939. There were now four general types of Signals training available: unit training, courses at the School, civil technical training and overseas training. Unit training included courses of practical instruction and exercises. Courses available included attendance at the School of Signals and at courses run by other Arms' and Services, such as infantry. For civil schooling, selected personnel attended schools and universities and technical colleges for further technical training.

3 Letter from Lieutenant Colonel G.J. Mapson, December 1989.
4 Mapson, letter; and Powell, letter.

Overseas training included a Long Tele-communications Course for officers and a Foreman of Signals course for other ranks at the School of Signals at Catterick, UK, the Military College of Science in UK. It also included attachments to British Signal Units and exchange tours between Australian and British Signals officers in the UK and the British Army on the Rhine (BAOR) in Germany.[5]

A further difference was introduced, in 1947, when trade titles were replaced. The old Electrician Signals became Radio Mechanic, Lineman Mechanic became Line Mechanic, Instrument Mechanic became Telegraph Mechanic, Fitter Signals became Vehicle Mechanic and two new trade titles were introduced namely Operator Keyboard and Cipher (OKC) and Line Test Clerk.[6] Bernie Saunders, for instance, recalls that while at Bricosat Signals Troop, Tokyo, he was promoted to provisional Warrant Officer.

> When I queried why provisional I was told I did not have a trade although I was fully qualified, which was rather peculiar seeing I held a top rank in a trade position. The upshot [of the trade changes] was that I had to be trade tested as an Operator Keyboard Class 1 to substantiate my rank. Later... I qualified as a cipher operator but eventually had to do a conversion course to qualify as an Operator Radio. Thus 13 years after I came into the Army, RA Sigs granted me the distinction of of being Class 2 in a trade in which I was fully qualified when I enlisted [having transferred from the RAAF with the qualification].[7]

In the years following its formation in 1947, the School of Signals, had to make a number of adjustments, within budgetary constraints, to cope with demand. Most of these were due

5 RASCM, 'Post War Developments in R Aust Sigs as at November 1955'; *Signals Bulletin*, Vol 1, No 1, September 1951, pp. 16-18; and Murphy, 'History of the Post War Army', p. 236.

6 'Signals Information Bulletin', No 22, 15 October 1947, p. 1.

7 Saunders, letter.

to the advances in communications technology and changes in military organisation. The 1947 amalgamation left a total of 202 personnel.[8] Then, in 1952, the School absorbed 1 Technical Maintenance Troop, 1 Line Carrier Troop and 1 Wireless Carrier Training Troop when the size and responsibilities of the School expanded following the introduction of the National Service Training Scheme.[9] A considerable re-arrangement of courses, scheduled for 1952 and later, was also required.[10]

The end result was that the School reorganised into a headquarters and two training wings. The first was Corps Training Wing which provided career type courses for officers, non-commissioned officers and post graduates from both the Royal Military College (RMC) Duntroon, and the Officer Cadet School (OCS) Portsea. Other courses included promotion qualification courses up to the rank of major, equipment and procedures refresher courses, and signals tactics courses. Students for these courses were drawn from all arms of the Regular Army and the CMF[11]

Tactical procedures training was one of the responisibilities of Corps Wing. One such procedure was known as 'step-up' which was adopted from the UK in 1952. The step-up system involved part of a divisional or brigade headquarters in the field, including the Signals component, being sent forward to the new location, to set up and take over control of the formation from the old headquarters. This obviated the need to maintain communications whilst on the move.[12]

---

8 MP 742, 96/8/270, 'School of Signals Reorganisation, 27 March 1952'; Barker, unpublished manuscript, p. 397; MP 897, 106/7/3114, 'School of Signals Reorganisation'; and CRS A2653, Vol 2 1947, Military Board Agendum 59/1947,"AMF Communication Sysytem': Memorandum from CGS dated 16 May 1947, p. 2.

9 MP 742, 96/8/270, 'School of Signals Reorganisation'; *Signals Bulletin,* Vol 1 No 2, December 1951, p. 14; and Barker, unpublished manuscript, p. 397.

10 Barker, unpublished manuscript, p. 397.

11 'Post War Developments in R Aust Sigs as at November 1955'; and *Signals Bulletin,* Vol 1, No 3, pp. 10-12.

12 *Signals Bulletin,* Vol1, No 3, March 1952, p. 6.

The other major component of the School was Trade Training Wing which conducted trade training from basic to advanced courses in signal trades. It was later divided into Operator Wing and Technician Wing. Basic trade courses ranged from about 18 months for radio technicians to about six weeks for technical storemen. Advanced trade courses for promotion ranged from about four months for technicians to two months for operators.[13] Most of the tuition was given at Balcombe but, in 1953, a supplementary Mobile Wing was formed to visit all Command areas to conduct refresher courses primarily for CMF members.[14]

The Mobile Wing supplemented the CMF evening parades, weekend bivouacs and annual camps. These events were aimed at training potential officers and NCOs, continuing the basic training for National Servicemen allotted to Signals units, and continuing advanced and basic training to volunteer recruits in signal subjects.[15] In contrast to their pre-World War II counterparts, CMF Signals units were in a better position, as they had more modern training equipment. The greater numbers attending camps and parades also made it possible to carry out more comprehensive exercises than before.[16]

In addition to Corps and Trade Wings, the Infantry Regimental Signals Wing was located at the School of Signals in 1955, providing basic and refresher courses for regimental signalling officers and other ranks.[17] The first Officer Commanding was Captain Laurie Watts (Infantry Corps). This Wing was raised because the then Director of Infantry, Colonel 'Bunny' Austin, was not satisfied with the standards achieved by regimental signallers during

---

13 Woollard, notes, September 1988.

14 RASCM, 'Post war Developments in R Aust Sigs as at November 1955'; *Signals Bulletin,* Vol 5, No 2, November 1958, p. 20; *Signals Bulletin*, Vol 2, No 2, April 1953, p. 12; Barker, unpublished manuscript, p. 398; *Signals Bulletin,* Vol 10, No 1, October 1961, p. 2.03; and RASCM, 'History of the Post War Army: R Aust Signals', p. 3.

15 RASCM, 'Post war Developments in R Aust Sigs as at November 1955', p. 3.

16 RASCM, 'History of the Post WarArmy: R Aust Signals', p. 4.

17 RASCM, 'Post war Developments in R Aust Sigs as at November 1955'; and Barker, unpublished manuscript, p. 398.

operations in Malaya so he sought help from Signals. Instructors were provided by Corps and Trade Wings and the results achieved meant that regimental signallers were better able to meet the operational demands of the Malayan Emergency (as attested to in Chapter Three), the 'Confrontation' crisis and the early stages of Australia's involvement in Vietnam.[18]

In May 1947, the Director of Signals, Brigadier A.D. Molloy, had predicted that there would be no place for any equipment which employed morse code. This prediction followed advice from the British Army's Directorate of Signals. In 1948, the Australian Army faithfully followed British precedent in abandoning morse training but, by April 1953, a policy reversal occurred.[19] At this time, the Australian Directorate of Signals enunciated that 'morse remains the basic means of communication and in all new signal equipment, this facility is retained so that communications can be maintained under the worst possible conditions'.[20] Technological developments had been significant but not so much as to negate the need for a working knowledge of morse code. Thereafter live morse circuits would be retained on AUSTCAN until 1961.[21]

National Service

Compulsory military training was introduced for service in Australia in 1950. Fears of a possible global war, foreshadowed by the Malayan Emergency and the Cold War in Europe, and an existing 'limited' war in Korea had been the catalyst. The idea was to form a trained reserve of manpower from which the nation could raise forces to be put into the field with the shortest possible delay. According to Mr Holt, the then Minister for Labour and

18 Woollard, notes,September 1988 (He was Chief Instructor at the School of Signals from 1954 to 1957).
19 Nalder, *The Royal Corps of Signals,* pp. 478-9; and 'Signals Information Bulletin', No 26, 15 February 1948.
20 RASCM, Barker, unpublished manuscript, pp. 391-2
21 Saunders, letter.

National Service, it was also 'to improve the physical fitness of our young manhood'. The Scheme provided for the call up of all male British subjects on their reaching the age of 18 years, with some provision for deferments.[22]

In August 1951, the scheme was introduced, with National Service training battalions raised in all states. At the commencement of the scheme, Corps training was to be carried out as part of the second half of the 176 days training to which each trainee was committed. A period of five years was required to complete this part-time training. In 1953, this requirement was reduced to 140 days training, to be completed in three years.

The experiences of Lieutenant Colonel B.G. Swan illustrate the induction procedures. His introduction to service life started with an interview with a public servant in the Department of Labour and National Service Office in Newtown, Sydney.

> When asked which 'Corps' I preferred I nominated 'Signals' because it sounded interesting. Then, on 24 April 1952, I reported to Marrickville depot with cut lunch and tooth brush and was bussed to 13 National Service Training Battalion Ingleburn. B Company was 'Signals' with the rest of the battalion being infantry. Our platoon of 40 men was the last of 1300 men to go through the Q store in three days. All items on the list were therefore issued regardless of size (they were corrected within a fortnight). It was tan boots and raven oil and maybe that's why I have disliked tan ever since.

---

22 T.B. Millar, *Australia's Defence*, MUP, 2nd edition, 1969, p. 132-133; G. Greenwood & N Harper (eds), *Australia in World Affairs 1961-1965*, Cheshire, Melbourne, 1968, p. 262; N.T. Shields in R. Forward & B Reece (eds) *Conscription in Australia*, QUP, 1968, p. 66; F. Alexander, From Curtin to Menzies and After, Nelson, 1973, p. 137; and Murphy, op.cit., p. 68.

> The next day we were marched to the camp theatre for an ANZAC day address by Lieutenant Colonel I.B. Ferguson, the Commanding Officer of 3 RAR during the battle of 'Kapyong' in Korea in 1950. Our platoon was in marquees for 98 days. The sides had to be rolled up by day for ventilation and many a cold winter's night they stayed up because no one was prepared to take them down and put them up again. We were issued with the wartime woollen service dress, and, while it was not very glamorous, it was certainly nice and warm to change into of an evening. At the time (it was the third intake) National Service was 98 days full time with no corps training. Later it was reduced to 77 days and included three weeks corps training.

Swan was subsequently posted to 1 Infantry Divisional Signal Regiment.[23]

National Service recruits were allotted to CMF units by residential areas rather than on the basis of suitability for the Corps to which they were posted. Some of those posted to Signals were semi-literate, and few had any inclination or capacity to undertake trade and technical training.[24] Experience also showed that it was better to disregard civil experience and to give all recruits aptitude tests to indicate a mechanical or electrical inclination. These problems were rectified when changes were introduced in 1953. After the changes took effect, Signals personnel carried out four weeks of Corps introductory training as a final stage in the training programme of the training battalion. For the average tradesman posted to a training battalion as a 'Regimental Duty' non-commissioned officer (NCO), this presented quite a

23 B.G. Swan, letter, p. 1-2.

24 MP 897, 171/1/223, Colonel Roseblade (Director of Sinals)'Selection Procedure for National Service Personnel Allotted to RA Sigs CMF Units', Jan '52;

challenge as he had to revise his own basic military training to be able instruct in weapons and 'fieldcraft' while also having to revise his basic Corps subjects.[25]

The structure of the School of Signals meant that in addition to a school, it also performed the functions of a Signal Training Battalion. Regular Army other-rank entrants to the Corps went to the School after recruit training at the Recruit Training Battalion. Allocation of recruits to Corps was based on an arbitrary percentage allocation. Thus, depending on the success of the Army's recruiting, other ranks entered the School at a rate varying from about 30 per month to two or three. The system was not efficient but it was difficult to suggest any alternative which could cope with the inflexible policy of percentage allocation to Corps from each recruit intake.[26]

Trainees were held at the School until about 20 had arrived. They were then placed on a one week 'Corps Indoctrination' course. This gave them an insight into the Corps' history and traditions, its units and equipment and the nature of duties, skills and training required for the various Corps trades. Aptitude tests were also conducted and at the end of the course each trainee was allocated to a trade. Trainees were then held at the School until sufficient members were accumulated to commence the particular course. For 'common' trades such as clerks and drivers, arrangements were made for individuals concerned to be sent to the Command Training Centre for instruction and qualification.[27] When trainees were trade qualified, the Directorate of Signals was advised to arrange their posting to a Signal Corps unit. Trainees awaiting training at the School performed all the

---

25 Molloy, 'Post War Developments In R Aust Sigs', p. 3; Barker, unpublished ms, p. 402 (c); *Signals Bulletin,* Vol 3, No 3, November 1954, pp.18-19; *Signals Bulletin,* Vol 2, No 3, October 1953, p. 13; and Weir, letter, March 1988, p. 4.

26 Each intake would contain a percentage of individuals of the various levels of 'I.Q.' Woollard, notes, September 1988.

27 *Signals Bulletin,* Vol 3, No 1, January 1954, p. 16; and Woollard, notes, September 1988.

'general' duties for the School, but once an individual commenced a course he was exempt from duties. Females (WRAAC) underwent a similar process, except that they were quartered at the WRAAC Barracks at Mount Martha about 1.5 miles distant from Balcombe.[28]

As a result of the re-organisation in 1960 which saw the demise of National Service training, the manpower of the Corps rose considerably, although the proportion to trades did not remain the same. Consequently, the output of the School of Signals had to be re-directed towards those trades in which the greatest deficiencies existed. It was envisaged that it would take a number of years before the manpower by trades would match the establishment structure of the Corps.[29]

Each to His Own: Corps Training Policy

The question of technical versus military skills and the proportions in which they were needed arose frequently during the 1950s and 1960s. Increasingly complex machinery required highly skilled technicians to maintain it; but signalmen were soldiers as well as tradesmen and there was a constant necessity to ensure that both aspects of the recruit's military education were sufficiently developed.[30] Nevertheless, a popularly held view within the Corps, in the 1950s and 1960s, was that there was neither the time nor the need for the communicators to be soldiers first - a view strongly contested within the Corps by the regimental and squadron sergeants major.[31]

From 1948 until 1953, this problem was aggravated by the shortage of tradesmen and recruits suitable for advanced training. Consequently, the Corps had to resort to posting men to units before they were fully qualified. This process was known as 'absorption training' and it required the signalmen con-

28 Woollard, notes, September 1988.
29 *Signals Bulletin*, Vol 9, No 1, April 1960, p. 3.
30 Barker, unpublished manuscript, p. 392.
31 'To Each His Own', in *Signals Bulletin*, Vol XIII, No 1, March 1964, p. 3.16.

cerned to work toward full qualifications while serving in a unit. By 1953, it was estimated that approximately 25 per cent of the total strength of tradesmen were in this category.[32] According to Bernie Saunders, 'the Army equation appeared to be that two half trained men was equal to one who was fully trained, whereas anyone connected with a trade will tell you that an unqualified tradesman was as good as two men short.'[33] This programme proved unsatisfactory and was eventually abandoned. Those remaining unqualified signalmen already serving in units had to return to the School to complete their training. Another difficulty was the poor standard of technical knowledge of a number of the qualified tradesmen. There were also a number of qualified tradesmen who had received their qualifications either without having been tested or having been given only perfunctory tests. Consequently, in 1953, tradesmen were ordered to be tested in all Commands. This measure served to raise and maintain a uniform trade standard for the Corps.[34]

Part of the problem was that because of problems associated with tape relay operation, a situation arose whereby poorer operators were sent on courses as the better ones had to be retained to cope with the work. Consequently, some sub-par tradesmen were produced.[35]

Another problem faced by the Corps was that there was a severe shortage of senior technicians known as 'Foreman of Signals'. Prior to 1954, higher level training for technicians had been provided by sending personnel to civilian establishments or overseas. The first Foreman of Signals course conducted at the School of Signals commenced in August 1954 and lasted for 63 weeks. This was designed to provide the Corps with a regular source of approximately twelve Foremen of Signals each year

32 *Signals Bulletin,* Vol 2, No 2, April 1953, pp. 9-11.

33 Saunders, letter.

34 *Signals Bulletin,* Vol 2, No 2, April 1953, pp. 9-11; *Signals Bulletin,* Vol 2,No 1, December 1952, pp. 13-14; and T. Barker, unpublished manuscript, p. 394.

35 Saunders, letter.

from then on. Previously, only one a year had returned from training in the UK.[36] However, this course was only conducted once at the School of Signals, Balcombe. Because of the difficulty of finding the comparatively large number of highly qualified instructors and potential Foremen required for the 18 months course, any repetition of this course was considered impractical by the Directorate of Signals.[37]

During this period, the Director of Signals obtained approval for the rank structure of Foreman of Signals to be rationalised. Previously, all Foremen were Staff Sergeants. To advance beyond that rank, individuals were required to leave the technical 'stream'. The new arrangements eliminated that restriction by providing positions in the ranks of Staff Sergeant, Warrant Officer Class Two and Warrant Officer Class One. This move improved their morale and did much to attract and retain personnel with potential into this stream.[38]

The question of technical versus military skills was also raised in relation to the recruitment of officers. The Directorate of Signals realised that some officers had both high technical qualifications and the capacity for competent military leadership but, as there were not enough so endowed, emphasis was on recruiting potential officers who were well educated technically or capable of becoming so. Meanwhile, others likely to be good commanders were accepted even though they may have lacked engineering skills. An officer recruited in either category was not excluded from joining the other, but the categorisation provided a useful base from which to start career training. Generally, those allotted to the technical group were given advanced training at universities or on long telecommunications courses with a view to their being appointed to a technical staff or to units which

36 *Signals Bulletin,* Vol 3, No 2, May 1954, pp. 18-19.
37 Woollard, notes, September 1988.
38 Ibid.

used complex equipment.[39] Use was also made of the facilities of the PMG and Australasian Wireless Amalgamated (AWA) facilities for training officers who could be attached for liaison and instruction, as well as for research and development.[40] The second group of officers was directed to Staff College training and service with tactical and field units.[41]

An article in the Signals Corps' quarterly professional technical publication *Signals Bulletin*, pointed out the difficulties of this scheme and stated that both categories of officer were needed and were inter-dependent and that no invidious distinctions should exist.[42] Nevertheless, effort was put into those with the best potential rather than spreading the effort across the Corps. According to Major General Vincent, this led to 'a race of superior officers, far ahead of most other officers in many respects.' This process of selection had started back in Brigadier Molloy's time as the Director of Signals in the years immediately after World War II. These selected officers included Clarke and Burnard (later Colonels), Roseblade, Honeysett, Morel and Evans (later Brigadiers), as well as Williamson, Woollard and Hockney (later Major Generals).[43]

One such selected officer was the then Captain P.J.A. Evans who wrote that on completion of his Master of Engineering Science at the University of New South Wales he was interviewed by the Corps Director:

> I was ushered into the presence of the Director of Signals and must admit to [have been] feeling

---

39 *Signals Bulletin,* Vol 1, No 1, pp 12-16; MP 742, 323/6/423, Military Board Agendum 103/1947, 'Scientific Technical Industrial Trade Trade requirements'; and RASCM, Barker, unpublished manuscript, p. 393.

40 'Signal Information Bulletin', No 17, 15 May 1947, p. 2; MP 742, 323/6/423; and CRS A2031, Defence Committee Minute No 350/1947 Research and Development of Defence Telecommunications Equipment.

41 *Signals Bulletin,* Vol 1, No 1, pp 12-16; MP 742, 323/6/423, Military Board Agendum 103/1947, 'Scientific Technical Industrial Trade Requirements'; and RASCM, Barker, unpublished manuscript, p. 393.

42 *Signals Bulletin,* Vol 1, No 1, p. 15; and Barker, unpublished manuscript, p. 393.

43 Vincent, interview, September 1988.

> a little proud of my academic achievement. The Director of Signals said 'What I'm about to say Peter, I want you to take seriously. This degree of yours will not be held against you in any way! What's more I have a posting that will use your new found knowledge - you are to be Station Commander at Diggers Rest.' To say I was a little shattered would be an understatement until I realized that I was being told I would not be treated as a boffin and would be allowed to follow a normal career.[44]

In the immediate post-war period, steps were taken to commence selecting and training the officers who would have the technical skill required by the Corps in the then Interim Army. To this end, Officers Telecommunication Courses of six months duration were commenced at the School of Signals in 1947. The purpose of this course was 'to teach signals subalterns the basic technical and tactical requirements of a junior officer in R Aust Sigs.'[45] It provided advanced study for selected officers of the Corps of Signals as well as the Corps of Electrical and Mechanical Engineers, in the operation, maintenance and design of radio and line telecommunications equipment. The first one of these was conducted between 25 August 1947 and 7 March 1948.[46] Many of these students had no difficulties with the course and went on to complete University courses in Science or Electrical Engineering. Others found the technical aspects of the course fearsome and were inclined to lose interest when they found that other students were far ahead of them. The course covered subjects such as the fundamentals of electricity and magnetism,

---

44 Notes on Corps History from Brigadier P.J.A. Evans, 1988.

45 *Signals Bulletin,* Vol 5, No 1, June 1956, p. 19; and Woollard, letter, 1990.

46 Included on this first course were Captains L.D. Maclean & H.G. Shackloth, Lieutenants D.R. Douglas, H.J. Matthews, J. Morgan, A. Jackson, M.C. McVeity and L.G. Moore. MP 742, 323/6/382, 'Officers Telecommunications Course (Short) No 1, Aust, 25 Aug-7 Mar '48'.

mathematics, electronics theory, electronics design and analysis, field carrier principles and equipment, radio theory, transmission line theory, Fourier Analysis and advanced aerial theory. Included in the course was an attachment to laboratories, as well as to the Army Headquarters Signal Regiment's transmitting and receiving stations, to conduct specific research projects.[47]

The Long Telecommunications Engineering Course was held at Catterick in the UK. Major General Woollard recalls that RMC graduates who had not done post RMC university courses because of the war were being posted for tertiary training.

> I had graduated from RMC in 1941 (after 2 ½ years) and had not attended university. However, as I had passed the first Australian Staff College Course (at Seymour, Victoria) in 1946 and had managed to do well in a Mathematics/Physics refresher course held by D Sigs in early 1947, I was selected to do post graduate training in the UK [as a Captain along with Lieutenant Colonel, later Brigadier, R.K. (Rex) Roseblade].[48]

The Long Telecommunications Engineering Course at Catterick was a post-graduate course in applied military telecommunication engineering and, for the Australian participants, was followed by attachments to various establishments which were designed to provide special training to meet a specific Australian requirement. General Woollard further recounts:

> My first attachment was to the War Office Signal Regiment in London and then to the Signals Research and Development Establishment (SRDE) at Christchurch in the

47 MP 742, 323/6/382, 'Officers Telecommunications Course (Short) No 1, Aust, 25 Aug-7 Mar '48'; and D Sigs Technical Liaison Letter No 33, 25 May 1948, pp. 5-7.

48 Woollard, notes, September 1988.

> 'New Forest'. These were concerned mainly with circuit engineering and antenna design. I was then joined by a team of seven sent from Australia to be instructed in the installation, operation and maintenance of the newly ordered equipment.[49] The training was in the form of a series of 6 to 8 week courses conducted by: School of Signals Catterick (Single Side Band theory and installation practices), Marconi Wireless Company (on transmitters), ST&C (receivers), Ericsson (Single Side Band Drive and Monitor Units), and SRDE Christchurch (on apparatus 5UCO - the first on-line encryption device to come into service).[50]

Some time later, in 1969, the Telecommunications Engineering Course was abolished and then replaced, in 1970, by the Long and Short Telecommunications Engineering Management Courses. The long course was a two year course designed to train selected officers, not holding degrees, in systems planning, engineering and management; and the short course was a 12 month course designed for engineering and science graduates.[51] Non-Commissioned Officers also went overseas on training courses. The first main contingent was with the then Captain Sonnenberg for the Army Wireless Chain Project in 1949, as mentioned in Chapter Nine.

The United States also provided excellent facilities for overseas training as it was a place where some of the most advanced research in telecommunications was occurring. The Australian Army was keen to send personnel and there

49 Mentioned under Project & Construction Signals in Chapter Two, this team consisted of men allocated to 1 Signal Project Squadron and was led by Captain G.T. (George) Sonnenberg with six Sergeants.
50 Woollard, notes, September 1988.
51 *Signals Bulletin,* Vol XVII, No 2, April 1970, pp. 7.1-7.3; and AWM 98, 842/1/11, 'Telecommunications Engineering Courses: UK'., Feb 1970.

was considerable interest in the courses offered by the United States. One of the first officers to study in the USA was Major L.W. (Len) Cumpston, who was attached to the Development and Equipment Laboratories at Fort Monmouth, New Jersey, in February 1946. Australia's severe dollar exchange shortage limited the number of servicemen who could go. For example, in 1949, the Director of Signals attempted to enrol two Australians in the US Signal Corps Officers' Advanced School but this was not approved on the basis of the currency issue.[52] Nevertheless, the policy of training in the US remained an aim and in later years Signals Corps personnel were able to be sent to places such as Fort Monmouth, Camp Davis California, the Electronic Proving Grounds in Arizona and other American establishments. The currency problem did not affect, to the same extent, travel to Europe as there was a steady stream of Australians to Royal Signals Schools in the UK or for exchange duty postings with the British Army on the Rhine in Germany.[53]

On the tactical side, Signals officer courses were conducted by the School in Australia. Many officers found difficulty in applying their theoretical knowledge to specific situations on the ground and it was contended that this was due to the specialised nature of the duties of Signals officers in peace time.[54] The Directorate of Signals expressed concern that many officers appeared to simply attend without making any preparation for the work that they were required to do.[55] In 1954, the Directorate went so far as to say that 'recent observations have shown that a disturbing lack of knowledge is sometimes apparent amongst officers concerning the functioning of their own Corps and its equipment.' To counter this ignorance, one month officer refresher courses were conducted at the School of Signals.[56] By

52 Pemberton, *All The Way,* pp. 2-3; and RASCM, Barker, unpublished manuscript, pp. 394-5.

53 RASCM, Barker, unpublished manuscript, pp. 394-5.

54 *Signals Bulletin,* Vol 2, No 3, October 1953, p. 8.

55 *Signals Bulletin,* Vol 3, No 2, May 1954, p. 15.

56 Ibid., p. 17.

1963, the School was running four formal courses for the officers, including: a four-week indoctrination course on receiving their commissions; a junior officers course of eight weeks duration, conducted about four years after commission; an advanced officers course, for officers with about five years of seniority as captains; and a senior officer refresher course for majors to lieutenant colonels.[57]

Also in 1963, a new trade structure was introduced into the Corps, reflecting the requirements of the Corps to match the changes in equipment used. The new trades were as follows:

- Technician Signals, replacing the old driver electrician, for employment largely on radio relay equipment;
- Technician Electronic, incorporating the roles of the previous Line Mechanics and Radio Mechanics;
- Technician Electronic (Cipher), made into a separate trade from the Technician Electronic due to the growing requirement for automated and specialised encryption equipment;
- Operator Keyboard and Radio, combining the previous roles of the Operator-Wireless-and-Line (OWL) and the Operator-Keyboard; and
- Operator-Radio and Technician-Telegraph, replacing, in name only, the Operator-Wireless and Telegraph-Mechanic.[58]

Whenever the Royal Australian Corps of Signals adopted new technology, the School adapted by developing suitable courses. An example of this was the introduction of semi-conductors (transistors). In December 1960, the only transistorised equipment in service in the Australian Army was Voice Frequency (VF) telegraph. These equipments were the responsibility of the line

57 'ARA Officer Training', *Signals Bulletin,* Vol XII, No 1, March 1963, p. 2.02.

58 'New Trade Structure', in *Signals Bulletin,* Vol XII, No 2, June 1963, pp. 2.02-3.

mechanic but, from 1959, all mechanics on basic courses were taught elementary semi-conductor theory. It soon became clear that knowledge of this subject would have to be increased as new transistorised equipment, such as the AN/PRC-25 portable radios and the STRAD automatic message switch, were entering service with the Army.[59] At the time, the School was teaching the same course on transistor theory to both trainee and advanced mechanics because the topic was as new to tradesmen returning to the School as it was to beginners. A shortage of facilities meant that practical instruction on equipment had to be limited to line mechanics who spent about 50 periods of their course studying VF Telegraph circuitry.[60] Others were encouraged to study publications in their own time.[61] However, the School designed and built a number of basic circuits for laboratory experiments and with these, it was eventually able to give practical experience to all mechanics. As the number of transistorised equipments in service increased and the more experienced tradesmen were trained, the standards for tradesmen in the different classes were raised progressively. This training was considered to be very important and, by 1964, all appropriate tradesmen were being taught transistor as well as electron tube circuitry.[62] By way of contrast, the Army Apprentices School were still solidly behind electron tube theory with only an occasional foray into semi-conductors being attempted in the mid 1960s.[63]

Another example of adaptation to changing need was the introduction of cipher mechanics courses when 'Derby' machines were introduced, in 1959. In 1948 and later, cipher operators were trained at the Signals Directorate as the need arose, but establishment of the trade Operator Keyboard and Cipher (OKC) led to this

59 *Signals Bulletin,* Vol 9, No 2, July 1960, p. 13; and RASCM, Barker, unpublished manuscript, p. 398.
60 RASCM, Barker, unpublished manuscript, p. 398.
61 *Signals Bulletin,* Vol 9, No 2, July 1960, p. 14.
62 RASCM, Barker, unpublished manuscript, p. 398.
63 Notes by Lieutenant Colonel N.R. Churches, 1990 (an Apprentice Radio Mechanic from 1967 to 1969).

training being passed to the School. Under existing arrangements the need for cipher mechanics solely for maintaining cipher equipment was so small that special courses to train them could not be justified. However, as the new machines were to have 'on-line' and speech secrecy facilities, the necessity for specialists to care for them was forecast as an essential requirement. For this reason, the training of cipher mechanics as well as operators was made a responsibility of the School of Signals.[64]

Despite the perennial shortage of qualified instructors and other difficulties encountered in the 1950s and 1960s, the School of Signals managed to cope well with the demands made upon it. Perhaps the worst setback came in 1960, when the Army was reorganised onto the 'pentropic' organisation and National Service training was abolished. Although the overall establishment of the Corps was increased at the time, that of the School was decreased by approximately one third, from 117 to 85 all ranks. As a result, some of the scheduled courses for 1961 were cancelled. The School had to limit itself to aptitude testing of all recruits and providing basic courses for Mechanic, OWL and OKC trades, advanced courses for Mechanics and OKCs, plus officer and NCO equipment and tactics courses. The courses for advanced Operator Keyboard and advanced OKC were delegated to Southern Command, based in Melbourne and that for advanced OWL, to Eastern Command in Sydney. The task of training tradesmen other than those mentioned had to be given to the units of the Corps.[65]

By the middle of 1965, changes in establishments, following the general Army reorganisation and the revival of National Service training, together imposed a trade training commitment that was, once again, beyond the resources of the School. Several measures were taken to relieve the situation. The establishment of the School

64 *Signals Bulletin*, Vol 2, No 3, October 1953, p. 11; and RASCM, Barker, unpublished manuscript, p. 399.

65 *Signals Bulletin*, Vol 10, NO 1, October 1961, pp. 2.02-3; and RASCM, Barker, unpublished manuscript, pp. 399-400.

was increased to provide additional instructors at Balcombe; an Eastern Command Wing was detached to Ingleburn, New South Wales; and a contract was signed for the Marconi School of Wireless to operate a Services Wing at East Hills, New South Wales. The main duties of the Eastern Command Wing were: to administer the trainees at the Marconi School; to conduct basic courses for Signals Corps trainees, both Regular Army and National Servicemen; and to give trade training in some Signals trades particularly Lineman Field, Rigger Signals, Operator Switchboard, Despatch Rider and Technician Signals. The Marconi School of Wireless taught elementary courses for operators and technicians, and a radio course for junior officers. Both Wings in New South Wales were controlled from Balcombe by the Chief Instructor at the School of Signals so far as syllabuses and standards were concerned. The Director of Signals from 1963 to 1967, Colonel R.P. Woollard, pointed out that although the Marconi School was controversial, no one could advance a practical alternative. The point at issue was the lack of Corps indoctrination and lower quality instruction at the Marconi School. This problem was overcome to an extent by requiring all Marconi students to complete their trade courses at Balcombe before becoming qualified.[66]

Meanwhile at Balcombe, the reorganisation resulted in Trade Wing being split into Technician and Operator Wings and a system of specialist instructors was introduced. Courses were scheduled between specialists subjects which meant that instructors now had adequate time to prepare and improve their lessons and aids. In Corps Wing, the production of the Corps 'Blue Book' reference manual was carried out under the direction of Major Peter Mudd and included Captain (later Lieutenant Colonel) Ian Willoughby and the attached British Royal Signals officer, Captain Mike Collins, who later immigrated and joined the Royal Australian Signals Corps.[67]

---

66 RASCM, Letter from Major General R.P. Woollard to Lieutenant Colonel G.J. Lawrence, Feb 1977; 'Training Policy' in *Signals Bulletin,* Vol XIV, No 1, July 1965, pp. 2.01-2.03; and RASCM, Barker, unpublished manuscript, p. 400.

67 Letter from Lieutenant Colonel I. Willoughby.

After five years of reliance on a dispersed array of teaching establishments, a more streamlined state of affairs and an increase in the number and variety of courses became available in 1970 with the concentration of Corps training in Victoria. Funding had finally been allocated for the construction of new facilities, so the School of Signals finally settled in Watsonia Barracks at Macleod, a suburb to the north-east of the city of Melbourne. Work on the buildings for a new School had started in 1968, the School occupied them in June 1970 and the official opening was performed by Mr Andrew Peacock, the then Minister for the Army, on 9 October 1970.[68]

At the new site, there were three main buildings: one for administration and training, another for technical maintenance and stores, and a third which was a covered training area. This was a welcome improvement on facilities at Balcombe, where the School had used 80 buildings (most of them only temporary huts built in World War II) spread over seven hectares. Furthermore, at Watsonia there was an increase in the supply of up to date teaching aids, so that it was possible to look forward with confidence so far as accommodation and equipment were concerned.[69] The Corps finally had the facilities to adequately cope with the stresses of training in the fast changing field of military communications technology.

The Directorate of Signals, Corps Domestic Matters and Aid to the Civil Community

The Corps of Signals finally had a home base with the new School at Watsonia Barracks. This was to become a place regarded as the home for the 'grand fellowship' that is the Royal Australian Corps of Signals. The Corps' heritage, safeguarded by the Signals Corps Committee and the Colonels Commandant of the various states, now had a focal point at the School of Signals and the

68 RASCM, Barker, unpublished manuscript, p. 401.
69 Ibid.

Watsonia Barracks area, so commemorations were to follow. As for the future direction of the Corps, the Director of Signals and his staff were given the responsibility of guiding the Corps through an era of immense technological change. The task was daunting and their efforts deserve credit.

The Director of Signals and his Staff

After World War II, the wartime position of Signal-Officer-in-Chief had been abolished following the retirement of Major General C.H. Simpson. From then on the technical direction and control of all signals personnel was vested in the Director of Signals. He acted as Head of Corps and Signals Adviser to the Chief of the General Staff and members of the Military Board. Under him at the Signals Directorate were a limited staff who were responsible for the implementation of signals policy. This responsibility included the detailed work in relation to signals training, equipment, organisation, and 'staff duties', security, liaison with overseas headquarters and other governmental departments. Each command also had a Signals staff officer attached to the General Staff. These officers acted as Signals advisers to the General Officer Commanding and staff of his respective command. In addition, they dealt directly with Signals units in the Command on all technical matters, thus providing a direct link between the units and the Signals Directorate.[70]

The execution of these responsibilities was divided between various commands within the army, but the technical oversight and control lay with the Directorate of Signals at Army Headquarters. The Director of Signals (along with the various Signal staff officers throughout the Army at various headquarters), had oversight over the following areas:

70 J.E. Murphy, 'History of the Post War Army', unpublished manuscript, 1955, p. 241.

- the operation and maintenance of exisiting telecommunications;
- the development of communications policy and procedures;
- the provision of advice on communications requirements to commanders at all levels;
- the training of Signals personnel;
- liaison with other authorities concerned with telecommunications;
- the provision of assistance in the maintenance of communications security; and
- the sponsoring, development and purchasing of telecommunications equipment required to fulfill the Corps' responsibilities.[71]

In 1951, the Directorate was organised into five sections working under the Director and a Staff Officer Grade 1 (Lieutenant Colonel) for co-ordination. The five sections were commanded by Staff Officers Grade 2 (Majors):

- Section One concerned organisations and 'staff duties';
- Section Two concerned training (this was changed to also incorporate 'liaison' duties in 1953);
- Section Three concerned communications equipment (this was changed to 'Field Force' communications in 1953);
- Section Four was for liaison with the Department of Defence (which, until 1974, was a separate entity from the Department of the Army); and

71 See *Signals Bulletin,* Vol 1, No 1, September 1951, p. 1; and 'The Royal Australian Corps of Signals', Handout to Students at RMC, Duntroon 1970; and *Signal Training,* Vol 1, *Signal Organisation and Tactics,* Pamphlet No 1, The Higher Organization of Signals, War office, 1950, p. 1.

- Section Five concerned the operation and maintenance of existing telecommunications (this was changed in 1953 to 'static signal services').[72]

In 1956, the Directorate was again reorganised into an operations 'division' and an engineering 'division', as well as a Defence liaison section, each under the direction of a Lieutenant Colonel.[73]

The Directorate of Signals was also involved in the procurement and development of equipment for use in the Signals Corps. In the early 1950s this was a simple procedure. The Directorate of Signals was the sponsor and represented the users. A section in the Master General of the Ordnance (MGO) Branch of Army Headquarters was responsible for translating the user requirement as expressed by Operations Branch into development or production specifications with the aid of the Army Design Establishment (ADE). A section of the then Department of Supply provided a limited research and development facility but most of the development work was sub-let by contract to manufacturers such as AWA. There were Signals officers in both the MGO and the Department of Supply. On the completion and acceptance of development work, the MGO raised purchase orders with the developing contractor for manufacture, to the extent of funds allotted and available. The new equipment was then introduced into service under arrangements made by the Directorate of Signals and distributed according to the Directorate's recommendations.[74]

During this period, the policy concerning the responsibility for the repair and maintenance of Signal equipment, held by all arms of the Army, was developed beyond the stage reached in World War II. The division of responsibility for the repair of Signal

---

72 *Signals Bulletin,* Vol 1, No 1, September 1951, p. 1; and *Signals Bulletin,* Vol 2, No 3, October 1953, p. 5.

73 *Signals Bulletin,* Vol 5, No 1, June 1956, p. 22.

74 Woollard, notes, September 1988 (a staff officer in the Directorate of Signals' Equipment Section from 1950 to 1953).

equipment had been a contentious subject dating back to the formation of the Corps of Electrical and Mechanical Engineers (RAEME) during World War II. The RAEME view was that all levels of repair were its responsibility with the users being responsible for operator maintenance only. This view was strongly contested by Signals on the grounds that it was responsible for communications and that responsibility could only be exercised if it had control of the key element of the telecommunications function - equipment. The General Staff had supported the Signals Corps in this view until the early 1950s when it became clear that there was a duplication of RAEME and Signals electronic repair facilities at the base or 'third echelon' level, due to the introduction of electronic equipment in other areas such as radar. As the Signals Corps could not accept responsibility for non-communication repairs, it was reluctantly agreed that RAEME would be responsible for third echelon repair for all Signal equipment except for that incorporated in fixed (strategic network) Signal systems including cryptographic equipments. Signals then retained responsibility for first and second 'echelon' repair of Signals equipment held by all arms. This was done via the technical maintenance troops in major Signals units.[75]

The Directorate Representing the Army

While the Royal Australian Corps of Signals was developing its Field Force, strategic communications network, signals intelligence capability and its training functions, the Directorate of Signals was overseeing a number of other Service, inter-departmental and international developments. These included the establishment of inter-departmental committees, Corps conferences, official visits and other projects.

The Signals Directorate represented the Army on a number of Australian Joint Services and Inter-Departmental Committees. One such committee was the Joint Communications Committee

75 Woollard, notes, September 1988.

which answered to the Joint Planning Committee of the Defence Committee. The Joint Communications Committee consisted of the Director of Naval Communications, the Director of Signals (Army), and the Director of Radar and Telecommunications (RAAF). Its function was to advise on all aspects of telecommunications (except research and development), continuously review communications boards and conferences, and co-opt representatives from other departments to assist.[76] It also arranged, on an interservice basis, for the preparation of 'plans for communication arrangements to meet a state of emergency'.[77]

Inter-Departmental Communications Committees were managed by the PMG and the Director of Signals had representation on the Australian Inter-Departmental Frequency Allocation Commitee (AIDFAC), the Telecommunications Advisory Committee (AIDTAC) and the Radio Frequency Allocation Review Committee (FARC).[78] The AIDFAC met to determine the break-up of the frequency spectrum for its different uses by the various interested Departments. The post-war internationally recognised break-up of the frequency spectrum had previously been agreed upon, in 1947, at the Atlantic City, USA, Telecommunications Conference. This conference provided only for the sub-division of the frequency spectrum into specific bands and did not provide for the detailed re-adjustments of Armed Services which, under the new regulations, found themselves 'out of band'.[79] This was followed, in 1951, by an 'Extraordinary Administrative Radio Conference (EARC) in Geneva, Switzerland, where plans

76 MP 927, A65/1/332, 'Main Communications Committees on which the Directorate of Signals has Representation.'

77 RASCM, Department of the Army, 'Signal Communications', in The Army War Book, November 1959', p. 36.1.

78 AWM 121, 6/A/4, 'National Management of the Radio Frequency Spectrum'.

79 Departments concerned included the following: PMG's Department, Department of the Army, Department of the Air Force, Department of Civil Aviation, Overseas Telecommunications Commission, Department of the Navy and the Department of Supply and Development. Director of Signals Technical Liaison Letters, Nos 29, 34, & 36, 1948; and *Signals Bulletin,* Vol 1, No 3, 17 March 1952, pp. 32-34.

were drawn up for the compilation of the new international frequency list.[80] The AIDTAC was concerned with the coordination of radio and cable communications between the various governmental agencies in Australia and it dealt with matters of major policy.[81] For instance, where a Department considered that economic or mutual advantages might have accrued, or if its plans had inter-departmental ramifications, it could submit the matter to the AIDTAC.[82]

Provision was also made for links with other Commonwealth nations by means of the Joint Service Representatives of both Britain and New Zealand to foster co-operation in Commonwealth Defence matters.[83] In order for the Australian Armed Services to be able to take part in discussions concerning communications and electronics matters affecting Commonwealth Defence, the Navy, Army and Air Force's Communications Directors (including the Army's Director of Signals) were directed to participate in the inaugural meeting of the British Joint Communications-Electronics Branch (BJCEB) meeting in Britain, in May, 1951.[84] The meeting discussed the linking of British and Commonwealth Services functions to the North Atlantic Treaty Organisation (NATO), the policy for interoperability of automatic telegraphy, and existing and proposed global cable and wireless networks from their Defence aspects, in relation to possible theatres of war.[85]

---

80 *Signals Bulletin,* Vol 1, No 3, 17 March 1952, pp. 32-34.

81 MP 729/8, 67/431/54, 'Formation of Telecommunications Advisory Committee'; MP 729/8, 67/431/48; and MP 927, A65/1/332, 'Main Communications Committees on which the Directorate of Signals has Representation.'

82 MP 729/8, 53/431/26, 'Collaboration between Services and/or Civil Departments in Sharing Communication Plant and Services'.

83 AWM 123, Box 95/6, op.cit., p1.

84 CRS A2031, Defence Committee Minutes, 41/1951, 22 February 1951, p1-2; and CRS A2031, Defence Committee Minutes No 214/1950 'British Commonwealth Co-operation in Defence Communications Matters', 9 November 1950, & No 264/1950, 14 December 1950.

85 CRS A2031, Defence Committee Minutes, No 41/1951, 22 February 1951. In attendance were Colonel A.D. Molloy, Commander I.H. McDonald (RAN), and Wing Commander J.W. Reddrop (RAAF). *Signals Bulletin,* Vol 2, No 1, December 1952, p. 48.

Efforts aimed at standardising procedures between allies were stimulated by this conference and were well advanced by March 1952, when a list, indicating the extent of procedural standardisation within the Australian Army, was published in the *Signals Bulletin*. This was important as British War Office procedure publications were used by the Australian Army through the 1950s including on operations in Malaya.[86] Subsequent Commonwealth meetings with the BJCEB, held bi-annually from 1952 to1960, were represented by only one of the three Services' Directors. Australian Army representation at subsequent meetings was usually by the Signal Corps officer of the London representative staff and not from the Directorate in Australia.[87]

The Radio Frequency Allocation Review Committee (FARC) was formed in 1960, in reponse to the 1959 Ordinary Administrative Radio Conference of the International Telecommunications Union. The conference had agreed to revised international radio regulations and it was important for the Defence group of Departments to meet with other Departmental and civil group representatives to settle the new arrangements. This was successfully achieved in 1961 after a 16-day conference.[88] However, due to the spectacular advances in electronics, further meetings were convened to cope with the frequency congestion and over-saturation in parts of the radio frequency spectrum. The Defence group of Departments' main interest in this matters was to ensure, as far as possible, global operational compatibility with Australia's allies in both the military and scientific fields, in accordance with the strategic outlook of the Government, which placed emphasis on operational deploy-

---

86 *Signals Bulletin,* Vol 1, No 3, March 1952, pp. 37-38, 'Allied Communication Procedures.'

87 CRS A2031, Defence Committee Minutes No 169/1952,'British Joint Communications Electronics Board Conference', 25 June 1952; *Signals Bulletin,* Vol 2, No 1, December 1952, pp. 49-54; AWM 121, 6/A/1, Defence Committee Minute No 68/1960; and Report of the Joint Planning Committee No 7/1954.

88 AWM121,6/A/1,JointCommsCommitteeReportNo12/61,'Recommendations of the Radio Frequency Allocations Review Commitee.

ments alongside major allies.[89] In the FARC's report, tabled in Parliament in 1961, it was made clear that

> as far as the Defence Group of interests are concerned the frequency allocations recommended by the Committee are, in the main, in close agreement with the Geneva Table of allocations [for the region]; cater for estimated expansion as far as [could] be forseen; and [would] not have an adverse effect on the Defence Group's activities.[90]

Project Mallard: A Lame Duck?

Standardisation of procedures and equipment was further propelled by the inclusion of Australia, in January 1963, as a member of the America-Britain-Canada (ABC) Army Standardisation Programme, thus extending the title to ABCA. Under this programme, effort was directed toward effecting economies among the member armies through the use of combined technological and scientific resources. It was hoped that reciprocal arrangements for funding, exchange of personnel, materiel, information and visits, as well as the joint use of facilities, could be arranged through this forum. The programme was administered by the Washington Standardisation Officers (WSO), including representatives of the four armies involved. The WSO had oversight over a number of permanent Quadripartite Standardisation Committees which met to establish, maintain and review lists of various projects.[91] Australia's participation in this programme was consistent with the views expressed by the Australian Prime

89 AWM 121, 6/A/4, 'National Management of the Radio Frequency Spectrum'.

90 'Report of the Radio Frequency Allocations Review Committee', in Commonwealth of Australia, *Parliamentary Papers,* General, 23rd Parliament, 3rd Session, 1961, Vol 4, pp. 13-14.

91 K.G. Groom (Lieutenant Colonel), 'ABCA: The American-British-Canadian-Australian Standardisation Programme', in *Australian Army Journal,* No 177, Ferbuary 1964, pp. 30-32.

Minister, Robert Menzies, who, in April 1957, stated that 'common sense dictates that...we should pay considerable attention to the logistic aspect of war, and standardise so far as we can with the Americans.'[92]

The Australian Army became involved in 'Project Mallard', in line with this policy. Project Mallard had its origins in a 1965 proposal that the interoperability of US and UK communication systems be sought 'through the use of interface equipment and exchange of communication liaison detachments'.[93] Project Mallard was a major co-operative communications development programme aimed at developing a secure tactical trunk communications system for the four respective ABCA Armies by about 1975.[94] This new concept was based on the understanding that future conflicts would involve less clearly defined battlefronts and therefore less secure trunk communications. This, in turn, meant that armies required communications systems based on interlocking nodes for alternate routing in case of a breakdown at any one particular node. It was intended to provide for the tactical communications requirements of armies at brigade, division, and corps level.[95] The research and development, which ensued through the project proved to be valuable in assisiting the NATO countries of Western Europe to standardise their tactical communication systems as well.[96] The system was to be automatically switched, completely interoperable between the four armies, together with the associated air formations and, where applicable, the navies. It was also intended to be capable of handling all modes of communications from simple writ-

92 'Statement by the Prime Minister', Commonwealth of Australia, *Parliamentary Papers,* General Session, 1957-1958, Vol 5, p. 647.

93 AWM 121, 12/B/23, 'TEAL XV - Agenda Item 6: Interoperability of Future Communications Systems'.

94 *Defence Report,* 1967, p. 33; and AWM 121, 282/B/3, Memorandum of Understanding for Project Mallard.

95 AWM 121, 12/B/23, 'Interoperability of Future Communications Systems'.

96 AWM 121, 282/B/3, 'Project Mallard, Release of Information to NATO'.

ten messages and voice radio links to teletype and facsimile information.[97]

In a brief for the Chief of the General Staff, Australia's objectives in participating in Project Mallard were outlined. The 'fundamental' objective was to obtain a secure tactical communications system which was completely interoperable with the other ABCA nations. The second objective was to advance technology in Australia by participation in the research, development and production of parts of the system. By doing this it was considered that Australia could obtain equipments within the same time frame as they were becoming available overseas while reducing its research, development and production costs, and stimulating Australian industry at the same time.[98]

On 15 September 1966, the Defence Committee made the recommendations that Australia participate in Phase II and III of Project Mallard, and that Australia's contribution (to be met from the Army and Supply Department's allotments), should be limited to three per cent of the then estimated cost of $20 million for Phase II. The Australian Government saw this share as a 'meaningful allocation' of the scientific and technical content of the task. These recommendations were subsequently approved by the Minister for Defence, on 22 September, 1966.[99]

The Washington Standardisation Officers subsequently set up a working group, with Australian and Canadian representation. The working group was replaced, in 1967, at Fort Monmouth New Jersey by an international management board known as the Project Management Board (PMB). The PMB consisted of the US, Canada and Australia with Britain joining shortly afterwards.[100] The Project was intended to be four phased:

97 *Defence Report,* 1967, p. 33; and AWM 121, 282/B/3, Memorandum of Understanding for Project Mallard.

98 AWM 121, 282/B/2, Project Mallard.

99 Ibid.

100 AWM 121, 12/B/23, 'TEAL XV - Agenda Item 6: Interoperability of Future Communications Systems'.

- Phase I comprised of a study of three systems and technique investigations in support of the system studies, to be completed by early 1969.
- Phase II involved the development of functional models, from the best component of the three systems studies of Phase I, and the defining of contracts, to be completed by November 1970.
- Phase III was intended to concern the engineering development of test models, and
- Phase IV was intended to be the production of the system to satisfy the requirements of the Governments concerned.[101]

Australia's contribution to Phase I and IIA cost Australia $600,000. During Phase I, four studies were completed, two by the Weapons Research Establishment and one each by Plessey and AWA. The results of the studies were forwarded to the Joint Engineering Agency, ADE and the major system contractors. The PMG also showed interest in some of the Mallard work performed.[102]

In 1968, while Phase I was continuing, Australia joined the Canada, United Kingdom, United States (CANUKUS) forum, upon invitation. The forum dealt with the formulation of agreed national positions of mutual interest in the military communications electronics field.[103] Australia's participation was precipitated by its involvement with Mallard, as numerous CANUKUS documents pertaining to Mallard had not been releasable to Australia without special authority.[104]

The negotiations for Phase II were conducted in Washington D.C., in December 1968, and these concluded 'very satisfacto-

---

101 AWM 121, 282/B/3, Memorandum of Understanding for Project Mallard; and AWM 121, 282/B/2.

102 CRS A2653, Military Board Proceedings, Vol 8, 1970, 'Project Mallard'.

103 AWM 121, 6/Q/1, 'Brief for DMO&P: AUSCANUKUS - MCEB.'

104 AWM 121, 6/H/2, 'Brief for CGS: Proposal for AUSCANUKUS Forum on Military Communications Electronics.' C of S Agendum 56/68.

rily' from Australia's point of view.[105] In Phase IIA, Australia was responsible for two hardware projects and a simulation study and a proportion of the costs for administrative and engineering support. The ADE, which was responsible to ensure that weapons and equipment were effectively and economically engineered to meet Army requirements, became intimately involved in the development of the functional models for Phase IIA of Project Mallard.[106] In particular, ADE's involvement with the Mallard tasks included: a Loop Multiplexer/Combiner, in conjunction with STC in Sydney; and a Tactical Facsimile Machine, in conjunction with Plessey Telecommunications Pty Ltd. The ADE was also involved in several other Mallard tasks such as assessing the communication traffic requirements for the Australian Services for potential deployments in South East Asia.[107]

In February 1970, the Project began experiencing a delay as a result of a US embargo on the Phase IIA work packages. A meeting of Australian, Canadian and British Programme Managers was held in London, early in August 1970, to consider Phase IIA schedules in the light of the US embargo. It was becoming clear that the Americans would proceed with a Mallard type system on a uni-lateral tri-service basis. It also became clear that the UK intended to proceed with Mallard with or without US support.[108]

Although there had been strong support within the US Department of Defence for the Mallard system, it failed to gain Congressional support. This was because of the reduced 1970 US Defence Vote and because of the recommendation, from a Congressional Joint House Committee, that the US terminate its membership from the project.[109] Official notification of the

105 CRS A 2653, Military Board Proceedings, Vol 3, 1969, No 143/1969, 'Project Mallard'; and Vol 8, 1970, 'Project Mallard'.

106 CRS A2653, Military Board Proceedings, Vol 7, 1970.

107 CRS A2653, Military Board Proceedings, Vol 7, 1970; and Vol 8, 1970, 'Project Mallard'.

108 CRS A2653, Military Board Proceedings, Vol 7, 1970; and CRS A2653, Military Board Proceedings, Vol 2, 1970.

109 CRS A2653, Military Board Proceedings, Vol 3, 1970.

United States' termination of arrangements was given on 25 September 1970.[110] It was clear that vested interests in the area of US electronics manufacturers and Congress could not accept other countries manufacturing equipment at the expense of their electronics companies. Therefore, a co-operative project such as Mallard was not possible.[111] Nevertheless, despite fears about the well being of the international standardisation efforts, the ABCA standardisation agreement remained and the four countries continued to standardise where possible.[112] So, although vested economic interests in the United States were too powerful to allow Mallard to proceed, the interests of all four nations involved in developing inter-operable systems endured as this remained consistent with their strategic outlook and perceived combat roles.

With the closure of Project Mallard, the Australian Army found itself seriously affected. It was back at the position it had been in some years previously, lacking automatic telephone and message switching systems and secure voice facilities within and between tactical headquarters.[113] In the meantime, the American and British Armies eventually managed to produce their tactical communications networks in the late 1970s and early 1980s (known as Tri-Tac and Ptarmigan respectively), using much of the knowledge gained from the research conducted for Project Mallard. The determining and procuring of an integrated tactical communications system for the Australian field Army would take a while longer.[114]

110 CRS A2653, Military Board Proceedings, Vol 7, 1970.

111 Hockney, interview, August 1989.

112 Hockney, interview, August 1989; and CRS A2653, Military Board Proceedings, Vol 3, 1970, 'Project Mallard'.

113 AWM 121, 12/B/23,

114 AWM 121, 12/B/23, Colonel J.I. Williamson, 'Provision of Tactical Communications Pending The Introduction of the post 1980 System', October 1971; and Gander, *Encyclopaedia of the Modern British Army,* pp. 62-64.

The Corps at Home: Domestic Corps Matters

The Royal Australian Corps of Signals has been a 'grand fellowship' for many people, providing an extended family. Thus, the 'domestic' side has provided an important introspective facet to the Corps. It covers Corps-wide sporting competitions, Corps Committees, commemorative activities and Corps Conferences. It is not, however, intended to provide here a comprehensive listing of items found in the Corps Memoranda.

The Directorate of Signals conducted Corps Conferences at the School of Signals in Balcombe and, from 1970 onwards, in Watsonia. These were four or five-day events with administrative and technical support being provided by the School. They were attended by the commanding officers of major Signals units, officers commanding independent Signals squadrons, chief signals officers and Signal staff officers of the various Army commands as well as senior staff from the Directorate of Signals. Occasionally the Director of the Royal New Zealand Signals Corps, and the British Chief Signal Officer of the Far Eastern Land Forces (FARELF), based in Singapore, were also in attendance. These conferences were significant events for the Corps, as the Director used them to brief his senior officers and vice versa. Topics discussed included policy aspects of equipment (usage and acquisition), organisations, personnel, training, contingency plans, and domestic Corps matters. Officers attending were required to state the implications of these matters within their respective spheres of influence and to advise on relevant happenings in their commands. Sometimes portions of these briefings were achieved by means of indoor exercises. Each conference was concluded with a 'Corps Conference Dinner'. These were often memorable occasions.[115]

115 Woollard, notes, September 1988; *Signals Bulletin,* Vol 6, No 1, December 1957, pp. 7-18; *Signals Bulletin,* Vol 8, No 4, December 1959, p. 30; and *Signals Bulletin,* Vol 5, No 1, June 1956, pp. 35-36.

In 1955, approval was granted for the appointment of six Colonels Commandant to the Royal Australian Corps of Signals. By November 1958, five had been appointed and these were: Major-General R. Kendall, CBE (the Representative Colonel Commandant in 1958); Major-General Sir Jack Stevens, KBE, CB, DSO, ED (who retired as Colonel Commandant in 1960); Brigadier H.H. Edwards, CBE, DSO; Colonel J.H. Thyer, CBE, DSO; and Major-General C.H. Simpson, CBE, MC, VD.[116] Other representative Colonel Commandants following included Brigadier H.H. Edwards, from April 1959 to September 1959; Major General C.H. Simpson from September 1959 to December 1960, Brigadier J.H. Thyer in 1961 and 1964, and Brigadier C.H. Kappe in 1962 and 1965.[117]

The Directorate of Signals always had a close association with the School of Signals, as the School provided a focus for the Corps and most members of the Corps often returned there on a number of occasions. Consequently, it was fitting that the Officers' and Sergeants' Messes associated with the School of Signals should come to be regarded by many, although not all, as the Corps Messes for the officers and 'seniors'. In the Officers' Mess at Balcombe, it was the practice for officers courses at the School to donate a piece of silver to the Mess at their final formal dinner. Alternatively, a sum of money was donated to the 'Silver Fund'. Corps Conferences and other major activities at the School followed this custom. Also, individual officers on the School staff did the same thing on being posted away from the School. The principal items held by the Officers' Mess in the mid 1950s were a number of fine silver candelabra, a silver statuette of Mercury (presented by the Royal Signals Corps to the Royal Australian Signals Corps when the Royal prefix was granted in 1948), and a silver statuette of a KATCOM (Korean Attached

116 *Signals Bulletin,* Vol 5, No 2, November 1958, p. 26; *Signals Bulletin,* Vol 9, No 3, October 1960, p. 3.03; *Signals Bulletin,* Vol 8, No 3, December 1959, p. 17; and CRS A2653, Vol 2, 1955, Supplement No 2 to Military Board Agendum 24/1955, 'Appointment of Honorary Colonels - Royal Australian Corps of Signals'.

117 'Representative Colonel Commandants', Annex B to Corps Memorandum No 3

Troops - Commonwealth). These personnel had been attached to the British Commonwealth Divisional Signal Regiment during the Korean War. Four such statuettes were made and paid for from the residual Officers' Mess funds of that regiment when it was disbanded, one each being presented to the Royal Signals, Royal New Zealand Signals, Royal Canadian Signals and Royal Australian Signals.[118]

Gifts were also presented to the Corps from the United Kingdom and the United States. In 1946, the then Major (later Brigadier), K.R. (Keith) Colwill, received, on behalf of the Australian Corps of Signals, a silver salver from Her Royal Highness, The Princess Royal, Princess Mary, as a memento of the co-operation that had existed between the Royal Corps of Signals and the Australian Corps of Signals throughout World War II. In 1960, the United States Signal Corps presented a wooden plaque to the then Director of Signals, Colonel R.K. Roseblade. Inscribed on it were the words: to our friends the Royal Australian Corps of Signals, from the Signal Corps of the United States Army, 1960'. Both items came to occupy positions of honour in the Signal Corps Mess, while it remained at Balcombe.[119]

At one stage, the Mess silver was loaned to Signal units in other areas for use at special dinners, but the silver became damaged because of poor packing and handling when in transit. This was not surprising as it was crated as 'Signal Project Stores' to avoid the commercial freight and handling charges, which no party wanted to, or was in a position to pay. The School's Sergeants' Mess also followed this precedent gaining recognition as the 'Royal Australian Signals Corps Sergeants' Mess'.[120]

By the mid 1960s, the Corps also witnessed events that were significant in that they were symptomatic of a rising feeling of nationalism within the Corps. While desiring to retain past

118 Woollard, notes, September 1988; and Weir, letter.

119 'Plaque From U.S.' in Army, 6 October 1960, p. 11; Corps Memorandum No 5; and interview with Brigadier K.R. Colwill, October 1989.

120 Woollard, notes, September 1988.

affiliations with Royal Signals, there was a feeling that 'we were unique, we were differenet, we were Australian!'[121]

The Corps to the Rescue: Aid to the Civil Community

The Army provided a large reservoir of fit, disciplined and trained personnel with its own command and support infrastructure, the ability to operate in all environments an own independent communication system.[122] As a part of this team, the Signal Corps made a significant contribution to the community at large including assistance in search and rescue operations, flood relief and bushfire fighting assistance. Josiah Francis, the Minister for the Army, from 1949 to 1955, observed in Parliament that 'the 'signalling units...render every possible assistance to the community in flood, fire and tempest.'[123]

One of the first such tasks was the provision of communications for the coal strike of 1949, when the Government called out troops to maintain coal supplies. This task was assigned to the 2nd Infantry Divisional Signal Regiment. Some tense scenes were witnessed between striking coal miners, police and troops. Communications were established by WS 122 between Sydney and Newcastle with the Sydney station located inside the Moore Park drill hall, using only a 14-feet whip antenna, manned by Sergeant Tom Parker. The PMG and civil communications in general were badly affected by the strike as power was severely limited. In those days, no comprehensive civil emergency communi-

121 Woollard, letter, August 1990.

122 P. Mench (ed) *Armed Forces and Australian Society: The Next Decade*, Canberra 1974, p. 52, cited in McKernan & Browne, *Australia: Two Centuries of War and Peace*, p. 369.

123 J. Francis, in Commonwealth Parliamentary Debates, 1 Eliz II, Vol 218, 20:1:5, 4 September 1952, p. 979.

cations existed, so communications links maintained by Signal units were critical.[124]

1 Independent Infantry Brigade Signal Squadron also assisted the civil authorities with the fire-fighting in the central region of Victoria. As soon as a fire was reported, the squadron would send a vehicle to the scene of the fire and a remote control unit was connected at the home station terminal in Puckapunyal. On occasions, the wireless link was a direct means of saving acres of property and many head of stock. During floods the squadron mounted wireless sets on amphibioues vehicles (DUKWs), and, on one occasion, the use of a radio in the DUKW was instrumental in saving a family from drowning.[125]

The Corps also provided communications for searches. For instance, in April 1952, Army Headquarters Signals, in Melbourne, was tasked to provide a wireless net (using a WS122) for a search for lost hikers. A headquarters and five mobile outstations were required, with facilities for manpack wireless communications for cross country tasks, using the WS SCR536. A control station was established at Marysville and sub-stations at the three camps. Similar assistance had been provided by the School of Signals, under Captain J.P. Lockwood, for a fire in the north-east of Victoria in January 1952. The value of the Army's firefighting and communications assistance was out of all proportion to the personnel supplied because their presence showed practical outside help which kept people there rather than fleeing and leaving the fire to cause more damage.[126]

---

124 Weir, letter, November 1988, p.2; RASCM, List of Senior RA Sigs Corps Appointments, p. 1; MP 729/8, 37/431/81, 'ORBAT CMF Field Force 29 Sep '47'; and RASCM, Notes from interview with Lieutenant Colonel A.W. & Captain J.C. Ballantine, pp. 28-29.

125 RASCM, D.C. Gillian, '1 Sig Regt History Post 1947', p. 1; MP 897, 106/7/3108, '1 Inf Bde Sig Tp - Reorg as 1 Inf Bde Sig Sqn', 11 February 1953; RASCM, 'Corps History R Aust Sigs: 1 Indep Inf Bde Sig Sqn', March 1954 (in 'Corps History File'), p.3; and *Signals Bulletin,* Vol 2, No 1, December 1952, p.7.

126 MP 897, 166/1/398, 'Aid to the Civil Authority.

Assistance was also given to aid in the fight against insect plagues. From November to December of 1953, one officer (Lieutenant L. Christensen) and 14 other ranks were tasked to provide communications support to the Department of Agriculture's attack on a plague of grasshoppers along the line of the Murrumbidgee River in New South Wales.[127] A similar plague occurred in November 1955 and the School of Signals was tasked to assists with two trade courses having to be closed for four weeks to despatch radio vehicles and operators and general support items to the plague area.[128]

On occasions, the Corps also represented Australia and the Army at special events. In March 1952, the Australian Government decided to send to the UK an Army contingent for the Coronation of Her Majesty, Queen Elizabeth II. The Australian Coronation Contingent, as it was known, consisted of 250 all ranks, under the command of Brigadier D.A. Whitehead. Lieutenant General Sir Edmund Herring was not included in the contingent but was appointed leader, in which capacity he attended the Coronation Ceremony at Westminster Abbey and rode among the mounted senior officers in the Coronation Procession.[129] The Signals Corps was represented in the Contingent by Lieutenant J.A. McGreevy (later Brigadier), Sergeants E.P. Collins, J.V. Connor and N.C. Kerr, and Corporal R.S. (Stan) Conlon (later warrant officer Class One and Regimental sergeant Major at the School of Signals).[130]

The Contingent provided the Royal Guard at Buckingham Palace and St James' Palace for a day in May, 1953. On 2 June, 1953, the day after the Coronation, the Contingent marched from Earls Court to Buckingham Palace for inspection by Her Majesty the Queen and the presentation of Coronation Medals. On 14 June, after seven days leave in London, the Contingent boarded

127 *Signals Bulletin*, Vol 3, No 2, May 1954, pp. 24-31.
128 *Signals Bulletin*, Vol 5, No 1, June 1956, pp. 44-45; and Woollard, notes, September 1988.
129 Murphy, unpublished manuscript, p. 97.
130 CRS A2653, Vol 3, 1952, Military Board Proceedings, 'Australian Coronation Contingent.'

HMAS *Sydney* at Spithead off Portsmouth and the following day, the Queen reviewed the fleet in HMAS *Surprise*. The Contingent endured the return voyage to Australia via the USA, the Panama Canal, Hawaii and New Zealand.[131]

In 1954, the Queen visited Australia and the Signal Corps played a significant role in the conduct of the tour, as the Queen visited major cities in every state. Virtually every Signals unit in Australia was involved in street lining, Royal Guards of Honour, mounting guard at the various Government Houses, providing communications links for the opening of Federal and State Parliaments, and social events such as surf carnivals and official garden parties![132] In Sydney, communications links were provided by 1 ASSU around the harbour for the Queen's landing and for the harbourside fireworks display in the evening.[133] In Canberra, where the Queen opened Parliament in March 1954, accommodation and administrative support were provided for 4,000 troops at 'Camp Royal' - a temporary camp set up on ground adjoining Anzac Park in front of the Australian War Memorial. Contingent headquarters was established at the nearby Ainslie Hotel.[134]

Communications support for the Camp was provided by the Puckapunyal-based 1 Infantry Brigade Signal Squadron, under Major E.B. Starrett with assistance from Eastern Command Signals units. They provided line, wireless and despatch rider services for the various contingents as well as a public address system and a communications link for the saluting artillery battery. The timing was carefully worked out so that, after the Queen arrived at the airport, the sound of the first gun reached the official party as the third movement of the Royal Guard's 'Present' was completed. *The Canberra Times* observed that 'the National Anthem was played, the Royal Standard was unfurled and the hills echoed to the long, booming 21 gun salute of an artillery

131 Murphy, unpublished manuscript, p. 98.
132 *Signals Bulletin*, Vol 3, No 2, May 1954, pp. 67-84.
133 RASCM, '130th Signal Squadron: Unit History Notes'; and interview with A.W. & J.C. Ballantine, p. 27.
134 *Signals Bulletin*, Vol 3, No 2, May 1954, pp. 67-84.

battery on the far side of the airfield.'[135] A similar measure of success was experienced when the Queen reviewed the contingents from the three Services two days later, on 15 February 1954, at Parliament House, Canberra.[136] For the PMG, the Queen's visit to Australia focussed a need to strengthen and upgrade the overseas telephone links, while for the Army, it provided an excellent oppportunity to practise its skills and work on its public relations image.[137]

Apart from the glamour events, the Corps continued to assist in trying circumstances. 1 Air Support Signal Unit (ASSU), for instance, provided telecommunications support for the control of firefighting and rescue teams and for the maintenance of rear links during civil emergencies in New South Wales, such as the 1951 Kuringai bush fires, the 1954 Lismore floods, the 1955 Maitland floods and the 1956 Windsor floods.[138]

The 1955 floods, for example, were the worst on record in New South Wales and, on 25 February, the State Government declared a state of emergency in the area affected in the Hunter Valley. This was the day on which 1 ASSU's annual camp at Gan Gan ended but, instead of returning to its base in Sydney, the Squadron Commander, Major Stan Blackburn, was ordered to take the unit to Shortland Barracks in Newcastle and await further orders. There he was tasked with providing communications in the Maitland area where the flooding was severe. At that stage, he found himself in a quandary as he legally could not task his national servicemen, as their annual full-time service had expired on the last day of the camp. As a result, only volunteer CMF and ARA soldiers were employed on the tasks.

---

135 'Colourful Canberra Welcome to Royalty', *The Canberra Times,* 15 February 1954, p. 1.

136 *Signals Bulletin,* Vol 3, No 2, May 1954, pp. 82-84.

137 Moyal, Clear *Across Australia,* p. 185.

138 RASCM, '130th Signal Squadron: Unit History Notes'; 'Radio Men Test Emergency Sites', in *Army,* 8 September 1960, p. 10; and RASCM, interview with A.W. & J.C. Ballantine, p. 27.

Emergency Headquarters were established at the Maitland Town Hall, East Maitland Police Station and the local Gaol, while a radio channel was opened, using a WS 122, between the camp at Gan Gan and the police station, VKG, in Sydney. A detachment of 2 Divisional Signal Regiment, at Singleton Army Camp, provided a WS 62 voice link with the Police radio in Sydney for 12 days while also maintaining radio communications with 1 ASSU and supplying radio crews to augment the numbers on 1 ASSU detachments.

At Georges Heights in Sydney, a Signal Centre was opened to handle the traffic generated. Before the emergency had passed, the centre had a range of civil and military radios in action, as well as an amateur radio network opened by Lieutenant A.W.(Tony) Ballantine as a link between the amateur radios and the Army services. The media soon found that news could be easily obtained from there so there were usually two or three cars with reporters parked outside.

On 26 February, a DUKW carrying an ASSU detachment including Sergeant W. McGrath (CMF), Signalman E. Chard (ARA) and Constable B.A. Orrock of the Sydney Water Police, was travelling near the Maitland railway yards which were then under a record depth of water. The antenna of their WS 62 accidentally came into contact with some high tension power lines while they were passing underneath, and all three were electrocuted and thrown into the water. At the moment of contact, Orrock was holding the antenna, while McGrath and Chard were both wearing earphones attached to the radio. Later, it was found that the antenna insulator showed a black trace across it where the current had jumped. It was a piece of terrible irony that the killing of the three men was the only harm done, because no one else in the DUKW was affected, and the radio worked perfectly as soon as its power supply was reconnected. Their loss was remembered in the naming of the 1 ASSU Other Ranks Mess as the 'Chard-McGrath' Memorial Club.[139]

139 Barker, unpublished ms, pp. 408-409; and RASCM, nterview with A.W. & J.C. Ballantine, pp. 18-22.

The CMF personnel continued to serve for about six weeks until the emergency had fully passed. Not surprisingly, all these men suffered disruption to their personal affairs, with some CMF members even losing their civilian jobs due to the long absence from work. The Signals Corps opened a fund for the families of the two men of the Corps who were killed and, in May 1956, their widows were presented with cheques by Brigadier H.H. Edwards, CBE, ED, Colonel Commandant of the Corps in Eastern Command (New South Wales). Two members of 1 ASSU were decorated for their services in this flood emergency. They were Major T.G. Jackson, a UK exchange Signals officer, and Lance Corporal N. Marshall, who received an MBE and BEM respectively. During the 1956 floods in Windsor, New South Wales, similar conscientiousness was displayed by CMF and ARA personnel alike.[140]

The Army agreed again, in 1956, to provide communications support for a major event in Australia. This time it was the Melbourne Olympic Games. The holding of the Games in Melbourne gave striking impetus to all forms of telecommunications in Australia, including the installation of a telephone network by the PMG for the 15 scattered Games venues.[141] International circuits were provided by the Army to Nairobi, London, Singapore, San Francisco, Wellington and Kure, Japan.[142] A number of Foremen of Signals and senior technicians from 1 Signal Project Squadron were attached to Army Headquarters Signal Regiment at the Diggers Rest Transmitting and the Rockbank Receiving Stations to assist in maintaining the HF communications to London, San Francisco, Nairobi, Kure, Singapore, Wellington, Port Moresby and throughout Australia.[143]

While Army Headquarters Signal Regiment provided the international links, the School of Signals provided many of the internal communications links for the Games by ceasing all

140 Ibid.

141 Moyal, *Clear Across Australia*, p. 185.

142 RASCM, 'Minutes of the D Sigs Conference Held at "Grosvenor" on 6 September 1955', p. 3.

143 RASCM, '127 Signal Squadron - Unit History', p. 3.

trade courses for about six weeks. This encompassed the pre-games training, the games, and the wind up period at the end. The School provided teletype operators at each of the game sites and the central results and press rooms at the Melbourne Cricket Ground; mobile communications for outdoors events including the marathon run, cross country equestrian events and cycling; and a limited despatch rider service between sites. The School also provided orderlies on the main track and field area at the Cricket Ground. Afterwards the School was complimented by the Olympic Committee for the standard of its services and the behaviour of its members.[144]

The Corps was also represented on scientific expeditions in the Antarctic. In 1947, the Australian Government had begun sponsoring annual Australian National Antarctic Research Expeditions (ANARE) and, in 1948, the first ANARE station was established at Macquarie Island. Stations were opened at Mawson, Davis and Casey and staffed by scientists, volunteer technicians and others from government departments and the armed Services. They were employed in the Antarctic for approximately one year until they could be replaced by relief parties sent by ship from Australia.[145]

Members of the Royal Australian Signals Corps served with several of these parties, notably during the International Geophysical Year, in 1957 to 1958. The Corps was represented that year at Macquarie Base by Major Fred V. Baines, Officer in Charge of the Station, Captain Jim D. Creed, a radio physicist specialising in ionospheric research and predictions and Warrant Officer Class One 'Uncle' George E. Heinricks, the radio supervisor.[146] For that expedition, the WS 173 was used for communication with Mawson and Davis Bases and for daily and weekly schedules with Australia, New Zealand and the United States. One interesting assignment for part of the time was the

144 Woollard, notes, September 1988.
145 Notes from J. D. Creed, November 1984.
146 Ibid.

exchange of weather reports and data on the ionosphere with Mirny - a USSR base which was operating in another sector of the Antarctic continent.[147]

Another major service to the comunity was also provided - this time by CMF Signals units in Western Australia in 1962, for the Perth Empire (Commonwealth) Games. The Signal plan provided: wireless and other communications for the opening and closing ceremonies in the main stadium, control and standby circuits during sporting events, an administration service at the Games Village and a Results Reporting Centre. The Results Reporting Centre was the site for the teleprinter and telephone channels which were used by national and international reporters.[148] The Army provided ushers, orderlies, supervisors, messengers, hospital staff and results communication staff.[149] 405 Signal Squadron, stationed at Leederville, Perth, assisted by 1st Battalion, Royal Western Australian Regiment Signal Platoon, had the task of providing much of the Games communication network. Members relayed results of events in the various venues back to a central control point where they were duplicated and delivered to officials and the media. The Squadron operated the results reporting communications facilities from 9 am to 11 pm daily. Equipment used included teletypes, phone networks, and mobile wireless networks for marathon road cycling and rowing events.[150]

The whole project was reported to be a complete success with 310,370 messages having been passed with no delays. When the Games were over the Chairman of the Press Committee, speaking on behalf of the Games Council, expressed deep appreciation for the Signals effort. According to the *Signals Bulletin* report, correspondents of the *London Times, Daily Mirror* and *Sydney Morning Herald* left no doubt about how appreciative

147 Ibid.

148 RASCM, 'Signal Communications Provided for Empire Games in Perth 19 Nov- 1 Dec 62', in *Signals Bulletin,* Vol XII, No 1, March 1963, pp. 4.02-4.18

149 'Games Run as Military Operation', in *Army*, 29 November 1962, p. 7.

150 'Signals Seem Assured', in *Army,* 15 November 1962, p. 5.

they were. More important was the personal feeling of achievement that came from completing a worthwhile task. The training before the Games and the operation of the actual tasks performed was an unprecedented opportunity for the Signals units in Western Australia to work together and practise their skills.[151]

Commemorations

On Sunday, 25 June 1972, Melbourne's weather turned on a fine day to the more than four hundred who gathered at the School of Signals for the dedication of the two memorial paintings by the famous Australian war artist, Ivor Hele. A Guard of Honour was provided by 2 Signal Regiment, under Captain J.H. Honey, accompanied by the Southern Command Band. The parade was reviewed by the special guest, Lieutenant General Sir Edmund Herring, the Commander of the 1st Australian Corps in New Guinea from September 1942 to August 1943. He addressed the gathering saying

> We have entered into a goodly heritage but we must remember always that we have done so only because those who have gone before, have toiled and sweated and sacrificed themselves and when necessary, given their lives to preserve this heritage for their children. I trust that this memorial will help them to understand the need for preparedness.

Lieutenant Colonel Stan Watson, a distinguished officer who had served with the Signals Company of Engineers at Gallipoli and in France, read the messages which had been received from the Signal Corps' of Britain, the United States, Canada and New Zealand. Brigadier A.D. Molloy then recited the ode and led the gathering in prayer. The 'Last Post' was sounded and the National Anthem was played. The ceremony was reported as being 'a fit-

151 RASCM, 'Signal Communications Provided for Empire Games in Perth 19 Nov- 1 Dec 62', in *Signals Bulletin*, Vol XII, No 1, March 1963, pp. 4.02-4.18

ting climax to the great endeavours of a number of people to whom the Corps and its associations mean so much'.[152]

Another presentation was made, in March 1972, at Blandford in Dorset, England. There, doors for the newly constructed All Saints Church were presented to the Royal Corps of Signals. A gunmetal plaque, mounted on the hardwood doors, bears the inscription:

> These doors of Australian timber were presented to the Royal Corps of Signals on the occasion of their fiftieth anniversary by the Royal Australian Corps of Signals.

The presentation ceremony was conducted in front of the doors in the forecourt to the Church. The Royal Australian Corps of Signals was well represented by twelve members of the Corps, then serving in Britain, and the Australian delegation was supported at the ceremony by the Australian Army Representative (London), Brigadier L.I. Hopton and his wife. The Staff Officer (Signals), Australian Army Staff London, Lieutenant Colonel N.R. Bergin, addressed the gathering and read a message received from the representative Colonel Commandant, Colonel W.G. Clementson. The United Kingdom's Master of Signals, Major General P.E.M. Bradley, accepted the doors on behalf of the Royal Corps of Signals and expressed the gratitude of all ranks for the gesture of the close association that existed between both Corps.[153]

This ceremony occurred only months before the final withdrawal of Australian military forces from South Vietnam. The Royal Australian Corps of Signals, as part of the Australian Army Force sent to assist the Republic of South Vietnam in 1962, had made a significant contribution over the following ten years.

152 'Dedication Ceremony of the RA Sigs Memorial Paintings at Watsonia Barracks on Sunday 25th June 1972', in *Signals Bulletin,* Vol XVII, No 5, September 1972, pp. 2.5-2.9

153 'Royal Signals 50th Anniversary Presentation', in *Signals Bulletin,* Vol XVII, No 5, September 1972, pp. 2.10-2.11

CHAPTER ELEVEN

AN OVERVIEW OF THE IMPACT OF COMMUNICATIONS TECHNOLOGY ON THE CORPS, THE ARMY AND THE CONDUCT OF AUSTRALIAN MILITARY OPERATIONS 1947 -1972

THE STORY OF the Royal Australian Corps of Signals from 1947 to 1972 is complex and involved, but an understanding of it is vital to comprehend why it is the way it is today. This volume has sought to provide that understanding. It has also sought to highlight some of the key lessons learnt by the Corps along the way and, in so doing, perhaps assist in shedding light on the way ahead. Beyond this, the intention has been to provide the men and women who served with the Corps a fitting tribute to their work while providing members of the Corps and other interested readers with an insight into life in the Corps in the postwar years to 1972.

By focusing on how the Corps operated and why, this Volume also sheds light on the nature of the Australian Army as a whole. Explaining events in the Corps has served to contextualise developments in the wider Australian Army and to present another angle from which to understand the workings of the Army and its place in Australia's foreign policy and defence strategy during this period. For this reason, developments in the Corps have been placed within their wider Army context. The fate of the CMF Divisional Signal Regiments, for instance, was closely linked with the wider Army policies concerning National Service (which arose in part at least from a fear of possible global war), the move towards more readily deployable regular forces in 1957 and the introduction of the pentropic organisation in 1960 which served, among other things, to cement the change in the Army's equipment procurement orientation from the UK to the US. Also, the very existence of the strategic communications network known progressively as the Army Wireless Chain and AUSTCAN owes its existence to the creation of the BCOF in Japan, which was driven by strategic and foreign policy imperatives. These, in turn, had wide ramifications for the Army as a whole for the immediate postwar years, the Korean War and beyond.

Developments in the Royal Australian Corps of Signals did not happen in isolation. The pace of technological innovation, for instance, was largely determined by developments in the wider community, which were then adapted for military use, such as the introduction of the transistor. On the same token, developments in military communications technology also tended to benefit the wider community, as illustrated by the Corps' contributions to disaster relief and events of national significance.

The link between communications technology, the Corps, the Army and the conduct of Australia's military operations up to the end of World War II was established in Chapter One. The increased range, capacity and durability of communications equipment gave commanders a greater range of options and enabled them to take risks that otherwise would have been untenable. Obviously, the improvements in communications technology were not, in themselves, the only contributing factor to these changes, but they played a crucial role. Improvements in armoured vehicles and aircraft, as well as increases in ranges of weapons and the introduction of new weapon types, were significant determinants of corps' and armies' structures and methods of conducting military operations. However, their capabilities could not have been fully exploited without parallel developments in the field of communications technology.

Perhaps the most significant area in which communications technology impacted the conduct of military operations was in Signals intelligence. Although the techniques of Signals intelligence had been practised with great effect in World War II, the conflicts in Borneo and Vietnam presented unprecedented opportunities to exploit the Signals intelligence capability for operations at the lowest levels. In Borneo, for instance, individual patrols benefited directly from information received by electronic intercept. Their effectiveness was such that, in Vietnam, for the first time in the history of Australian military operations, a brigade-sized formation (1ATF) had its own Signals intelligence troop (547 Signal Troop) assigned to it. The expertise and effectiveness of 547 Signal Troop in providing accurate and timely information down to unit and sub-unit level have been amply demonstrated in Chapters Three and Five. Even the enemy in Vietnam appreciated the value of their own signals intelligence and exploited it more effectively than the US and allied forces were prepared to acknowledge.

For field force Signal units, the introduction of new and improved technology dramatically affected the service they could provide to commanders on operations. In Korea, the introduction of communications equipment in use by the US and UK allies helped overcome the problems of equipment compatibility between allies and gave greater control to commanders through more reliable equipment, which had a greater frequency range and thus more capacity. In Malaya, the introduction of rugged, reliable and man-portable HF radios enabled patrols to confidently venture out knowing that good communications could be established despite the difficult terrain and the distances involved. This was in marked contrast to the experience of Australian signallers in World War II, where undue reliance was placed on line communications because of difficulties in establishing effective radio communication. In Vietnam, the move away from reliance on line communications was convincingly illustrated by the selection of Nui Dat as the Task Force base - a site selected in part because of its potential for good radio communications. Vietnam was where several other significant improvements were experienced with the introduction of transistorised VHF radios, speech encryption facilities for use in conjunction with VHF radios, tropospheric scatter radios, multi-channel voice and telegraphy radio relay equipment (and the concomitant reduced reliance on line), communications shelters and eventually even satellite communications (although only with the US Signal Brigade). These items combined enabled commanders to reduce the combat ratio of forces required to ensure survivability in an encounter with the enemy. Commanders could rapidly deploy forces to remote locations with the assurance that, through reliable communications, combined with effective artillery support and the ability to rapidly reinforce with armoured personnel carriers and helicopters, their forces could be provided with an unprecedented amount of combat support in a very short time frame. This was most forcefully demonstrated in the battle of Long Tan in 1966. This combination of new, rugged and flexible communications technology also facilitated the repeated deployment of the Australian Task Force away from the Nui Dat Base, as communications support could be inserted and the link established wherever the Task Force deployed.

The strategic communications network also played a significant role in shaping the Corps, the Army and the conduct of Australian military operations from 1947 to 1972. The establishment of an international and inter-state military communications network in peacetime had been an unprecedented move driven initially by the commitment of forces to the BCOF in Japan. Thereafter, the important role of the network in providing a command and administrative link to Australia's forces deployed on operations overseas was demonstrated repeatedly. This helped the government to gain the maximum political advantage from the smallest of deployed elements whilst retaining national control and influence over those elements without having to rely on another nation's communications facilities to do so. By the time of the Vietnam War, the network was sufficiently sophisticated to provide the Chiefs of Staff and their political masters with minute-by-minute access to the decision-making process of commanders deployed in Vietnam, thus adding to the complexity of the deployed commanders' responsibilities. This access was facilitated by the introduction of new technology along the way - particularly the system upgrades with carrier frequency shift and single sideband working, as well as the automation of the network with STRAD.

For the Americans responsible for the higher operational and strategic level planning and conduct of the war, the complexity of communications arrangements was more problematic than for their Australian counterparts because of the magnitude of the issues faced. For although the US Army's communications network in Vietnam was the most extensive, expensive and sophisticated to that point in history, it proved in the end incapable of dealing with the 'bottomless pit' of information which it tended to generate.[1] This was because the new communications technology allowed an unprecedented process of centralisation to occur, which made effective worldwide command and control from Washington a practical technological proposition. This resulted in the decision-making threshold being raised as higher and higher-level decision makers were included in the process. This, in turn, generated a greater and more continuous demand for more information.[2] This demand for more information resulted in the generation of massive quantities of reports, which ended up overloading the system at critical junctures. It could also be argued that the heavy reliance on statistical data blinded decision makers to the real state of affairs in Vietnam.

In the meantime, parallel developments were occurring with the Australian civil telecommunications network, which was expanding and upgrading its services around Australia and overseas. These improvements made it difficult, at times, for the Army

---

1 General C.W. Abrams, the Commander US Military Assistance Command Vietnam from July 1968, quoted in Van Creveld, *Command In War*, p. 258.

2 Van Creveld, *Command In War*, p. 236.

to continue to justify having its own dedicated network. In the end, the economies achieved with the introduction of less labour-intensive and more sophisticated technology would lead to the Navy, Army and Air Force integrating their strategic communications networks to form a Defence Communications Network.

The Corps had to respond to the introduction of new technology by sending students overseas to keep abreast of developments and repeatedly restructuring the School of Signals while also changing the structure of courses and trade streams. Communications compatibility had been another ongoing concern throughout the period. Consequently, efforts were also made to improve the degree of compatibility between the ABCA armies through quadripartite efforts, as illustrated in the ill-fated but beneficial Project Mallard.

In conclusion, from the time that troops had been sent to Japan in 1946 and the requirement for the maintenance of the Signals Corps was confirmed as necessary in the postwar Army, the Corps underwent dramatic changes resulting from rapid technological developments, and changes in Australia's outlook. In the field of communications technology, developments outside the Army occurred at a startling pace. These changes took some time to affect the Corps directly as the Army's equipment procurement programme was slow to catch up with the most advanced technology available due to financial constraints and the abundance of leftover equipment from World War II. However, when they were introduced, their effect on the Corps, the Army and the conduct of Australian military operations was profound.

In this work, an effort has also been made to convey some sense of the identity and the experience of Signalmen, the people who are the Royal Australian Corps of Signals, providing swift and sure means of communications in such varied places as Japan, Korea, Malaya, Singapore, South Vietnam, Borneo, Indonesia, Papua & New Guinea and Australia. The Corps is, in the end, people; talented and dedicated people who operate and continue to operate equipment to support commanders' requirements to communicate.

APPENDIX

EQUIPMENT USED BY THE ROYAL AUSTRALIAN CORPS OF SIGNALS 1947-1972

AN IMPORTANT PREREQUISITE for an understanding of the developments in the Royal Australian Corps of Signals since World War II and the impact this had on the wider Army is an understanding of the basic workings of some of the equipment in use in this technologically complex corps. This Appendix describes some key facets of the technology in use at the time, and should assist in understanding the impact of communications technology on the conduct of Australian military operations.

Wireless - Radio

Wireless or radio communications can be divided into numerous frequency bandwidth areas.[1] In the forward battle area, the Very High Frequency (VHF) band began to be used, in the final stages of World War II by US forces and, more particularly, during the Korean War, for local radio communications. This was due to experiences of crowded airwaves in the narrower High Frequency (HF) band. The flexible HF band was used for local transmissions using ground wave signals, as the HF signals can 'bend' over terrain to a far greater extent than higher frequency

1 These include High Frequency (HF), Very High Frequency (VHF), Ultra High Frequency (UHF), Super High Frequency (SHF) and Extra High Frequency (EHF).

bands. HF was also (and more commonly) used for long-distance links using skywave signals as transmissions travel further after being reflected off the earth's ionosphere. The more direct Ultra High Frequency (UHF) and Super High Frequency (SHF) bands were increasingly used for Line of Communications links behind the battle area by means of line-of-sight links (radio relay) or by means of tropospheric scatter (signals reflected off the earth's troposphere).

In June 1948, the British Army adopted a new nomenclature for its new wireless sets, and Australia followed suit. No longer did the numeral code represent a specific tactical role. Instead, the new code was 'alpha-numerical': the letters A to E were used for the power bands 0-10, 10-100, 100-1000, 1000-10,000 watts, and above 10 kilowatts respectively.

Each letter was followed by a figure which indicated both the bandwidth in which it operated and the version of the set.[2] Thus, the Wireless Set (WS) A10 was the first version of a series of dry battery, low power (up to 10 watts) radios which operated in the HF band (numbers 10-39 indicated the MF/HF band).[3] US equipment used yet another lettering system in which the first letters indicated the principal users and the letters following a slant-line indicated the role of the equipment. Thus, the AN/PRC 10 was of US origin for use by the Army and Navy (AN) and was the tenth type of their **P**ortable **R**adio **C**ommunications sets.[4] This lettering system came into common usage in Australia by 1966 as the Corps placed increasing priority on interoperability with the US.[5]

---

2 'Signals Information Bulletin', No 34, 15 December 1948, p. 7; No 19, 15 July 1947, pp. 3-4; *Signals Bulletin*, Vol 1, No 1, September 1951, p. 20; and Coker & Rios, *A Concise History of the US Army Signal Corps*, p. 23.

3 40-69 indicated VHF/UHF, and 70-99 indicated SHF/EHF.

4 'Fifth Annual Glossary of Military Electronic Systems', in *Armed Forces Management*, Vol 13, No 10, July 1967, p. S-2.

5 *Signals Bulletin*, Vol XV, No 1, July 196, p.3.01.

TABLE OF SET OR EQUIPMENT INDICATOR LETTERS[6]

1ST LETTER INSTALLATION	2ND LETTER EQUIPMENT TYPE	3RD LETTER PURPOSE
A - Airborne	A - Invisible Light - Heat Radiation	A - Auxiliary Assembly - Not complete sets
B - Underwater submarine	B - Pigeon	B - Bombing
C - Air Transportable	C - Carrier (Wire)	C - Communications Receiving/ Transmitting
D - Pilotless Carrier	D - Radial	D - Direction Finding
F - Fixed	F - Photographic	
	E - Nupac	E - Ejection and/or Release
G - Ground - General Ground Use	G - Telegraph or Teletype (Wire)	G - Fire Control (Gun) or Searchlight directing
		H - Recording - Photo-Meteorological or sound
	I - Interphone - Public Address	
K - Amphibious	K - Telemetering	
	L - Counter measures	L - Searchlight control
M - General Mobile	M - Meteorological	M - Maintenance and test equipment
	N - Sound in Air	N - Navigation Aids
P - Pack or Portable	P - Radar	P - Reproducing - Photo and Sound
	Q - Sonar and underwater	Q - Special or combination of purposes
	R - Radio	R - Receiving
S - Water - Surface Craft	S - Special Types	S - Detecting and/or Range and Bearing
T - Ground - Transportable	T - Telephone (Wire)	T - Transmitting
U- General Utility		
V - Ground Vehicle	V - Visual and Visible Light	
W - Water Surface and Underwater	W - Armament	W - Remote Control
	X - Facsimile or Television	X - Identification and Recognition

6 Department of the Army *The Royal Australian Corps of Signals Reference Manual*, 1968, pp. 3-8 - 3-9.

The Corps' radio sets in the early post-war years were Morse code-based, heavy and bulky, complex to operate, had limited traffic carrying capacity, and were so often subject to interference and static noise that they could seldom be relied upon. The development of frequency shift keying (FSK), single sideband equipment, and improved componentry helped to overcome these shortcomings. With FSK, the output from the transmitter was changed (or shifted) in frequency, the marking condition being 425Hz (cycles) below the nominal channel frequency and the spacing condition 425Hz above. In this system, therefore, radio frequency energy was transmitted at all times, and the frequency shift process served to convey telegraphic information as mark or space signals of the five-unit 'Baudot' code. At the receiving end, this signal was subsequently converted into direct-current pulses (creating mark or space signals) to operate the receiving telegraph machine. One limitation of FSK was that it provided only one telegraph channel, operating at 60 Baud (the unit of modulation rate).[7]

Single sideband (SSB) equipment overcame the channel limitations of FSK. SSB provided six or eight combined FSK signal telegraph channels which could then be transmitted on a higher frequency to a receiver where the channels could be separated, reverted to the original lower frequency, and converted to direct-current mark-and-space impulses for the receiving telegraph machines.[8]

Improved components allowed further miniaturisation and reliability. Consequently, fewer frequencies and fewer people were used to perform the same tasks.[9] This was a feature that would impact the manning of the Corps and the span of com-

7 'An Introduction to 402 Signal Regiment', Royal Australian Signals, Information Booklet, pp. 9-10.
8 Ibid., pp. 10-11.
9 Molloy, 'Technical Developments in Recent Years', pp. 3-4. Examples of equipment which resulted from these developments included the WS88, WS A40, and WS A510 in the forward areas; WS 62, the US AN/GRC 3-8, and WS 153 for field formations; and WS E10 (SSB) for point to point links.

mand and control that commanders would be able to exercise on military operations.

Despite these advances, the role of the signalman remained unchanged. The requirement to get the message through remained, and equipment developments did not remove the need for a sense of urgency, priority and responsibility. The tasks of the Signals officer

TABLE OF RADIOS (referred to in text)

NAME	INTRO INTRO SVC	SIZE	DEPLOYMENT LEVEL	COUNTRY OF ORIGIN	FREQUENCY BAND	REPLACE
WS133/135	WWII	Vehicle mtd	Unit	UK	HF	
WS128/31	1944	manpack	Unit	UK	HF	WS108
WS88	1950	manpack	Unit	UK	VHF	WS38
C/PRC-26	1950	manpack	Unit	Canada	VHF	SCR 536
C11/R210	LATE 50'S	vehicle mtd	Unit	UK	HF	
C42/45	1960	vehicle mtd	Unit	UK	VHF	WS19 'A'
WS62	1953	manpack	Unit/Brigade	UK	HF	WS19
WSA510	1956	manpack	Unit/Coy	Australia	HF	
AN/PRC-10	1956	manpack	Unit/Coy	US	VHF	WS128/31
AN/PRC-47	1965	vehicle mtd	Unit/Brigade	US	HF	C11/R210
AN/VRC46/9	1965	vehicle mtd	Brigade	US	HF	C42/45
AN/PRC-25	1965	manpack	Unit/Coy	US	VHF	AN/PRC-10
PRC-F1	1969	manpack	Unit/Coy	Australia	HF	WSA510
AN/PRC-77	1970	manpack	Unit/Coy	US	VHF	AN/PRC-25

also remained essentially the same - to plan and co-ordinate his resources of personnel and equipment to meet the communication requirement.[10] If anything, the signalman's task was now more challenging as one could be hard pressed trying to set the correct frequency with a signal generator, tuning the transmitter to the receiver and loading the aerial with several switches and dials. The larger field equipment, such as the WS 133 and WS 135 (both usually vehicle-mounted), had power outputs of around 100

10 R. Clark, 'A Short History of the R Aust Signals Corps to equipment', RA Sigs Museum, p. 15.

watts with dozens of switches and dials. A mistake in the final stages of tuning could create an overload and 'send you back to scratch'.[11]

The first post-war generation of radios that came from the UK were the vehicular 'C' series. The principal radios of the 'C' series procured for the Australian Army were the C11/R210 (HF), C42 and C45 (VHF). The C42 and C45 were short-range radios designed to replace the WS 19 'A' set (a vehicular-mounted HF radio set widely used during World War II).[12] Although they were in production in the UK in the mid-1950s, they did not become available in Australia until 1960 due to financial and supply constraints. They had excellent tuning facilities compared with the slow to connect and difficult to operate sets which preceded them. When these vehicular radios were dismounted, however, as was often the case during jungle exercises, they became a 'nightmare'.[13] These sets were later augmented, and eventually replaced in the mid to late 1960s by the US range of VHF vehicular radio equipment (AN/VRC 46 and 49), when the need for compatibility with US forces became pre-eminent.[14]

The C11/R210 HF radios were replaced in late 1965 by the AN/PRC-47 SSB HF radio. This was about one-third the weight of its predecessor, and it was more reliable and gave better performance. It was a modular construction that could be carried by two people if necessary. It could be used for voice, Continuous Wave (CW or Morse) or radio teletype operation.[15] Increases in the range, reliability and capacity of communications equipment were to be critical factors in the conduct of dispersed and highly mobile operations in Vietnam in the years that followed.

11 Lieutenant Colonel B.G. Swan, ED RFD, (RL), p. 7; and Major K.M. Weir, letter, December 1988, p.2.

12 'A Short History of the Royal Australian Corps of Signals' (written for publication in the RA Sigs Newsletter, Vol 10, No 11, November 1981.

13 Weir, letter, December 1988, p. 2; 'Post War Developments in R Aust Sigs as at Nov '55', p. 8; and *Signals Bulletin*, Vol 1 No 3, March 1953, p.27.

14 'A Short History of the Royal Australian Corps of Signals' (written for publication in the RA Sigs Newsletter, Vol 10, No 11, November 1981.)

15 *Signals Bulletin*, Vol XIV, No 2, November 1965, p. 3.01.

In the 'man-pack' range of radios, the principal items used by Australians in the Korean War at platoon and company level were the British WS 88 set and the Canadian C/PRC-26. Clearly, interoperability with the other British Commonwealth battalions and brigades was critical for operations. Before the introduction of these sets, the VHF band had not been used for forward area communications. The WS 88 underwent trials in the UK in 1946 as a replacement for the ageing WS SCR 536 and WS 38 HF sets used in World War II. The WS 88 was a 6-pound, 'basic-pouch' sized set which used a 5.5-pound battery that was slightly smaller than a basic pouch. The WS 88 was capable of communicating with the similarly designed WS 31 (used mainly by the UK) and WS 128 sets used at battalion and company level by Australia.[16]

The AN/PRC 9 &10 sets were introduced in 1956 as replacements for the WS 31 and WS 128 at battalion and company level[17] It was also about this time that the term 'radio' was introduced to replace 'wireless'.[18] They were designed and built in the US (and in the UK as the WS A41) as a VHF short-range infantry manpack and weighed approximately 26 pounds. They were the first American VHF radios adopted by the Australian Army.[19] As Australia's commitment to South Vietnam escalated, the AN/PRC 9 and 10 sets were eventually replaced with the AN/PRC 25 sets in the early to mid-1960s. These had double the output of the older AN/PRC- 10s and nearly five and a half times the

---

16 Signal Information Bulletin, No 9, 15 September 1946, p.3; and *Signals Bulletin*, Vol 1, No 3, March 1952, pp. 40-44.

17 The WS 128, had only been introduced in 1947 as a replacement for the WS 108 Mk 3 of World War II vintage. The WS 128 was a lightweight, immersion proofed man pack set for tropical use. 'Signal Information Bulletin', No 9, 15 Sep '46, p. 4; No 18, 30 June '47, p. 4; and T. Barker, *Signals: a History of the Royal Australian Corps of Signals 1788-1947*, RA Sigs Committee, Canberra, 1987, p.174.

18 Major General R.P. Woollard, letter, August 1990.

19 *Signals Bulletin*, Vol 1, No 2, December 1951, p.24; 'Minutes of D Sigs Staff Conference' 6 Sep '55, p.5; 'The Royal Australian Corps of Signals Reference Manual', 1969, p. 3.29; and 'Post War Developments in R Aust Sigs as at Nov '55', p. 8.

number of frequencies available for use. This gave commanders significant added flexibility to exercise their commands. These, in turn, were replaced by the AN/PRC-77 set in the early 1970s. The AN/PRC-77 set was fully transistorised, had the same operating parameters as the AN/PRC-25 set and could be connected in pairs for 'retransmission'.[20] When speech security equipment was added, they could also be used for automatic ('online') encoded speech transmissions[21] - a significant feature given the increased awareness of the value of eavesdropping demonstrated by the American and allied forces in Vietnam, and their Vietnamese opponents.

Parallel developments were occurring in the HF range of radios. Although VHF was suitable for use in rolling plains, it proved ineffective at times in difficult terrain such as steep hills and gullies, or where there was thick vegetation. These limitations were noted both in Korea and Malaya. This prompted the Australian Army to develop a small man-pack HF radio set in conjunction with Amalgamated Wireless Australasia (AWA).[22] The WS A510 (previously known as the WS A10) was the result of this collaboration. It was a lightweight man-pack HF radio developed in Australia by AWA (under contract to the Army Design Establishment - ADE)[23] for use by long-range infantry patrols.[24] Work on the A510 commenced in May 1950, and by January 1952, laboratory models were being tested at the

20 That is, to relay, on a different frequency, between two remote stations out of communications range.

21 'The Royal Australian Corps of Signals Reference Manual', 1969, pp. 3.28-9 & 3.33; and Hon D. Fairbairn, *Defence Report, 1972*, p. 16. See also Directorate of Signals 'Field Expedient Antennas for VHF Radio', in *Australian Army Journal*, No 214, March 1969, p. 30.

22 'Notes on the A510 - Personal Recollections by K.G. Dean', AWA, Sydney, 1978; and 'The Malayan Trials of the WS A510', AWA Research Laboratories Report No 231, June 1952.

23 ADE became the Engineering Development Establishment (EDE) in 1974. See Australian Army, *Engineering Development Establishment Information Brochure*, May 1988.

24 Australian Military Forces, *User Handbook, A510 Wireless Station, 1956*; and Woollard Letter, August 1990.

School of Signals under Lieutenant Colonel A.W. (Arthur) de Courcy Browne (later an employee of AWA). This set was subsequently put through trials in Malaya in 1952 and 1953 under Lieutenant Colonel Don Small from ADE, Major Roy Coutts from the Directorate of Infantry and Major R.P. (Bob) Woollard (later Director of Signals and Major General).[25] The success of these trials influenced Army Headquarters to progress to the final version of the set rather than delay production by undertaking further trials as desired by ADE.[26] It was in quantity production by late 1955 and introduced into service in 1956. It was also purchased by the British and New Zealand armies - a significant achievement for Australian industry at the time.[27]

In contrast to the 'C' series, the radio was carried in a pair of basic pouches, the transmitter in one pouch (two valves operated by a modulated crystal oscillator) and the receiver in the other (using a 6-valve super-hetrodyne circuit including an amplifier, converter, detectors, coils and audio output). It was hoped that by arranging it in this manner, it would avoid the wireless operator being identified by his equipment and so become less of a prime target for enemy snipers. It was suitable for use within the infantry battalions and as a long-range patrol set, as it had a range of over 50 miles.[28] K.M. (Kel) Weir recalls that, when posted to a Field (Artillery) Regiment Signal Troop, they had trouble carrying radios and ended up borrowing WS A510s from Ordnance depots to avoid having to manpack the C series radios![29] The WS A510 eventually served as the starting point

25 'Notes on the A510 - Personal Recollections by K.G. Dean', ; 'The Malayan Trials of the WS A510'; and *Signals Bulletin*, Vol 1 No 4, pp. 11-24; Vol 3 No 1, pp. 26-34.

26 Woollard, letter, August 1990.

27 Brigadier K.P. Outridge, Director General of Logistics, letter to The Director Australian War Memorial, 10 May 1978.

28 Australian Military Forces, *User Handbook, A510 Wireless Station, 1956*; and *Signals Bulletin*, Vol 1, No 2, pp. 29-33; Vol 1 No 3, p. 16.

29 Weir, letter, December 1988, p.2; and 'Post War Developments in R Aust Sigs as at Nov '55', p. 8. The A510 was obsolescent by 1969. See 'The Royal Australian Corps of Signals Reference Manual', p. 3-20.

for AWA's later development of the WS A512 or PRC-F1 (as it was eventually named) HF radio, which was introduced into service in the Australian Army in 1969.[30]

The WS 62 was a standard British Army manpack HF radio developed during the latter stages of World War II, primarily for use at brigade to battalion level in jungle and airborne operations. Production of this set commenced in Australia in 1952 and delivery commenced in 1953. It was introduced to replace the WS 19 of World War II vintage, which had proved to have insufficient power for rear area communications in the Korean War. The WS 62 also later replaced the WS 22 and 122 as the major HF field radio in the Army. Moving a manpack WS 62 (30 pounds), with all its equipment, antenna, battery charger and spare batteries (50 or more pounds), requires fit and strong operators. The constant supply of wet batteries (each weighing 25 pounds) caused headaches for many a signals officer. Signalmen, using chargers, frequently ran afoul of staff officers of supported units and headquarters in the field because of the noise level. The immediate solution was to dig a pit about two feet (60 cm) deep, to muffle the charger's noise. Alternatively, batteries could be charged back at 'B' echelon to the rear, and moved back and forward constantly.[31]

In the late 1950s, the US electronics firm Collins had developed the AN/TRC-75 single-sideband radio transceiver for the United States Marine Corps. This is a 'simplex' facility that provides for communications between two points but only in one direction at a time. The view held by the Signals Directorate was that two transceivers could be connected to provide full 'duplex' operation (communication between two points in both directions simultaneously), and both transceivers, a small switchboard (SB22/PT), teletypewriter and telephone/ telegraph terminal

30 Outridge, letter, May 1978.

31 *Signals Bulletin*, Vol 1, No 3, March 1952, p. 22; Weir, letter,, December 1988, p.2; 'WS 62', pp. 1-2; 'Report on Visit to Korea by Major A.A. Mason, Dec '50', Appendix B to CSOs Branch Historical notes; and 'Post War Developments in R Aust Sigs as at Nov '55', p. 8.

could be arranged in the back of a long wheel-base Landrover.[32] This equipment was used extensively during the Vietnam War, as discussed in Chapter Six.

Radio Relay

Radio relay was first used by US forces in Europe in 1944[33] and began to play an increasingly important role in post-war military communications. It revolutionised line carrier work. Radio relay equipment generated a wide-band microwave beam in a given direction, which, with the use of channelling equipment, could transmit and receive a number of discrete channels on the one circuit. This reduced the requirement for line circuits between headquarters, was quick to establish and easy to redeploy without the repetitive heavy labour and stores required by line systems. The problem of security was overcome by the use of off-line encryption (encoding) telegraphy devices such as 'Typex' and 'Rockex', but these were bulky and cumbersome. Short-term speech security was only obtained by the use of voice codes such as 'Portex'. Portex was a hand-cranked portable machine about the size of a field telephone, but it was slow and prone to error. In operations, many users would act as though it provided a private person-to-person link invulnerable to interception. This was a real problem in Vietnam as some users believed that when they picked up the telephone, they were talking over a secure system.[34]

Radio relay equipment also served to integrate wireless and line circuits anywhere in the forward or rear communications networks. The wartime multi-channel UHF No 10 radio relay set had become obsolete by the mid-1950s, but new radio relay equipment (such as the UK-designed and built WS B70 and the

32 'AN/TRC-75', in *Signals Bulletin*, Vol X, No 1, October 1961, pp. 4.01-4.

33 Coker & Rios, *op.cit.*, p. 23.

34 Lieutenant Colonel R.E. P. Cowley, letter, July 1991.

US-designed and built AN/TRC-24) did not become widely available until the early 1960s.[35]

The WS B70 was 'man-portable' (rather than 'man-pack', it became 4 x 25-pound loads) SHF microwave transceiver (receiver and transmitter in one), that was easily identified by its twin dishes. It had a range of 25 to 30 miles that could be extended to 100 miles by the use of three relay stations. It was operated by Divisional Signals Regiment personnel and was used between brigades and divisional headquarters, and even at the battalion headquarters level. It could provide an engineering channel and 4 duplex telegraph channels, or an engineering channel and one speech channel - a significant achievement at the time.[36]

Line

The purchase of new and more advanced line equipment became less urgent with the advent of radio relay, as long-distance trunk links could now be provided by wireless. Telephone, switchboards and line equipment in use in 1948 were all surplus war stocks. Ten-line magneto switchboards were used in the field for ten telephone subscribers connected using an electric generator and permanent magnets. Cable layers, in use until the early 1960s, were of Australian World War II origin and used World War II 'D3' and 'D8' cable. A new cable known as 'D10' (referred to as 'Don 10') was developed in 1951 but not procured until the early 1960s.[37]

The all-purpose 'L' phones were retained for field use with 'F' phones for base areas. It was expected that a new 'O' phone would replace the 'F' phone, and the 'K' phone would replace the

35 'AHQ Administrative Symposium 'Sympex', 23 Oct '58 - Administrative Communications', p.2; Weir, letter, December 1988; 'Post War Developments in R Aust Sigs as at Nov '55', p. 8; Woollard letter, August 1990; and Mapson letter ('QRM - Yesterday's Log'), November 1989.

36 *Signals Bulletin*, Vol III, No 2, July 1959, p. 21; Weir, letter, December 1988, p.2; Interview with Lieutenant Colonel A.W and Captain J.C. Ballantyne, 26 November 1980, p. 31; 'New Sig. Equipment for More Mobility', in *Army*, 17 December 1959, p. 13; and *Signals Bulletin*, Vol II, No 1, Jan '54, p. 24.

37 Weir, letter, December 1988, p.2.

'L' phones. The British designed and Australian-built 'K' phone did not emerge until 1958. Production of the 'K' phone was delayed by 12 years, not by the tortuous development and procurement process (raising specifications, testing and retesting equipment, modifications, productions, etc.), but because raw materials and sub-assemblies had to be procured from overseas and the money authorised by the Treasury.[38]

Telegraphy[39]

Despite the drawn-out procurement process, the first batch of AN/PGC-1 lightweight automatic telegraph machines was ordered from Kleinschmidt in the United States in 1951, for use in field and static units. These progressively replaced the British-designed and built Creed EE97 teleprinters (weighing 200 kilograms) that were carried by seven men and had been in use from division down to battalion level. The new terminals weighed only 75 kilograms. The Kleinschmidt was waterproof, capable of being floated and was carried by three men. Tone Keyers and AC/DC bridges (equipment that provided a teletype operation over an audio circuit) were also introduced with this equipment.[40] In 1960, an order of Teletypewriter sets worth £105,000 was received from Kleinschmidt, but on a larger scale than before, and incorporating an agreement for their assembly in Australia from components manufactured in the USA. These were hardy machines, suitable for use within vehicle-mounted stations or shelters.[41]

38 'AHQ Administrative Symposium 'Sympex' 23 Oct '58 - Administrative Communications', p. 1; Barker, unpublished manuscript, pp. 380-381; and Weir, letter, December 1988, p.2.

39 Telegraphy involves the transmission by wire or radio of alpha-numeric codes.

40 Barker, unpublished manuscript, p. 382; and 'Signals,Bulletin', Vol 1, No 2, December 1951, p. 34; Vol 2, No 1, December 1952, pp. 21-22; and Woollard, letter, August 1990.

41 'Latest Sig. equipment delivered', in *Army*, 28 January, 1960, p. 1; and *Signals Bulletin*, Vol 9, No 4, December 1960, pp. 4.01-4.03.

In the meantime, Australia used two types of telegraph machines in the Army's field communication system. These were the page printer (a lightweight field teletypewriter TT-4A/TG), and the tape machine (teletypewriter reperforator-transmitter TT-76/GGC). Depending on the estimated rate of flow, combinations of this equipment were used throughout the division to handle the volume of telegraph traffic. By grouping the page printing and tape machines a field-type tape relay system, similar to that used for the strategic communications network, was available within the division. Because of the standard voice frequency telegraph terminals used, it was possible to set up teletypewriter nets as and when required. This then gave a degree of operating flexibility to the telegraph system which had not been previously available in the Australian Army.[42]

The July 1960 *Signals Bulletin* illustrated the way an unclassified message could be passed through the newly arranged network by tracing its path from rear divisional headquarters to two forward battalions (or 'battle groups' as they were called in the pentropic system), with the divisional trunk communications connecting the rear headquarters to the Main and to the battalions.

> The Signal centre at Rear may transmit the message to Main using either a TT-4A/TG (page printer) or a TT-76/GGC (tape machine). It will depend entirely on the local machine availability. In the AN/MGC-17 [a 'minor signal centre' containing a page printer and two tape machines], a small 12-line switchboard allows the operator to select any of the machines and connect them to any outgoing circuit available. Probably this message would be sent on a tape machine, giving a local copy of a printed and perforated tape.

42 Ibid., pp. 27-28.

The incoming message at the Main Signal centre perforates and prints a tape in one of the TT-76/GGC in the AN/MSC-29 [a similar but larger configuration of the AN/MGC-17, used at main divisional headquarters]. This tape will be examined by an operator who will:

- register the message in;
- connect the two circuits to the battle groups on his VF telegraph switchboard and insert the tape in the transmitter of a TT-76/GGC; and
- transmit the message tape simultaneously to both battle groups, and complete the register entry showing the message transmitted.

At each battle group (battalion), the message will be received on TT-4A/TG in hard copy, ready for delivery.[43]

Shelters

There had been discussions in the Directorate of Signals and the Army Design Establishment, in the 1950s, concerning the possibility of acquiring telephone switchboards and radio relay facilities that were mounted in containers or 'shelters' and which could be transported by road or by air. These were being developed by the US Army Signal Corps to provide fully fitted, purpose-built, transportable, weather-proof communications facilities.[44] By 1963, 1 Division Signal Regiment and 3 Line of Communication Signal Regiment had received a limited number of such shelters,

43 Ibid., pp. 28-29.

44 *Signals Bulletin*, Vol 10, No 2, December 1961, p. 4.03.

including the communications patching panel SB-611/MRC[45], the manual telephone central office AN/MTC-3, the radio relay shelter AN/MRC-69, the teletype central office AN/MSC-29, and others described below.[46]

The first exercise deployment for some of these new shelters was during 'Exercise Sky High'. This exercise was conducted in the Colo-Putty Training Area in October-November 1963 under the command of the Deputy Commander of the then Pentropic Division, Brigadier A.L. MacDonald.

On Exercise Sky High, the SB-611/MRC communications patching panel was the start point for fault checking or circuit testing, as all possible circuits in the headquarters area passed through this shelter.[47] The linemen and clerks who ended up using the AN/MTC-3 shelters found that working inside them was a vast improvement in terms of the light, display and comfort space for diagrams provided.[48]

The radio relay shelter, AN/MRC-69, was used throughout the combat zone to the point of entry and back to Australia to provide trunk circuits, in theory, down to divisional headquarters level. It could be employed as a repeater station, a double terminal detachment or a 'carrier termination centre' for one or two

45 'The Royal Australian Corps of Signals Reference Manual: Part 3 - Equipment', p. 3-102. The communications patch panel (SB-611/MRC) was a switching or 'patching' shelter which provided facilities for circuit patching, routing and re-routing, circuit testing and monitoring of speech circuits, telegraph circuits and the internal engineering of the shelter. The shelter consisted of a shelter housing, a manual switchboard (SB-22); a terminal telegraph (TH-5/TG); a teletypewriter (TT-4A/TG); a telephone set (TA-312/PT) and patching cords.

46 Captain B.D. Phillips, 'Communication Shelters', in *Signals Bulletin*, Vol XIII, No 2, December 1964, pp. 4.01-7.

47 Ibid., p. 4.02.

48 The manual telephone central office (AN/MTC-3) shelter was used both as the main switchboard for the headquarters area and as the clerking office of the signal centre (sigcen). As a switchboard shelter, the AN/MTC-3 proved itself to be a vast improvement on the previous field switchboard facilities. Phillips, 'Communication Shelters', pp. 4.02-3.

14-channel carrier systems.[49] The AN/MRC-69 also included two AN/TRC-24 line-of-sight, VHF FM radio sets suitable for 'multi-channel' systems, although only two sets could be prepared for use at a time.[50]

The AN/MSC-29 telegraph terminal was capable of terminating telegraph circuits on tele-typewriters, auto-transmitters/reperforators or switchboards. It was vehicle-mounted and air portable. The AN/MSC-29 was supposed to be deployed down to the divisional headquarters. It also acted as a tactical 'tape relay' station (a station which received and transmitted messages in tape form). On exercises, it was found that, although the terminal was supposed to work as a tape relay centre, it was also required to become its own signal centre for messages passing through it.[51]

The AN/TRC-90A, transportable, tropospheric scatter terminal, was also used for radio trunk bearers in the communications zone. This equipment bounced radio waves off the troposphere to communicate over approximately 240 miles. It was also housed in a shelter, but operated in a higher frequency band (SHF) than other shelters, such as the AN/MRC-69. The terminal contained 'multiplexing' equipment, for the simultaneous use of several channels on a single circuit, and radio equipment to provide 24

49 'The Royal Australian Corps of Signals Reference Manual: Part 3 - Equipment', p. 3-63. The major components of the AN/MRC-69 were three radio sets (AN/TRC-24), four electrical filter assemblies (F-98/U), eight telephone terminals (TA-5006A/U), three group modems (F36700), a watt-meter (ME-28/U), a telephone set (TA-312/PR), and an intercom set (LS-147B/F1).

50 'The Royal Australian Corps of Signals Reference Manual: Part 3 - Equipment', p. 3-34; Captain G.W. Arbogast, 'Radio Communications in Vietnam', in *Signal*, April 1967, p. 24; and Phillips, 'Communication Shelters', pp. 4.04-5.

51 'The Royal Australian Corps of Signals Reference Manual: Part 3 - Equipment', p. 3-109; and Phillips, 'Communication Shelters', pp. 4.04-5. The terminal consisted of a shelter, telephone switchboard (SB-22/PT),12 telegraph terminals (TH-5/TG), eight signal converters (TA-182/U), four filter assemblies (F-98/U), four teletypewriters (TT-4A/TG), eight re-perforator-transmitter teleprinters (TT-76/GGC), an intercom station and cables.

'duplex' channels. It provided voice, teleprinter, data, facsimile, telemetry and remote control functions.[52]

Vehicles

The range of vehicles in use by the Corps throughout this period also changed. Ford and Chevrolet 'Blitz Buggies' were in use for a variety of communications purposes, including radio, cipher and technical workshops. Despatch Riders rode twin-cylinder 750cc Harley-Davidsons or single-cylinder 500cc sidevalve BSA bikes. World War II vintage left-hand drive Jeeps were still in use until 1959, when, following changes made in the UK, they were replaced on a trial basis by the less reliable and less popular Austin Champ. Many people were sorry to lose the old Jeeps, as they gave tremendous service to the Royal Australian Corps of Signals and were adapted by the Corps for roles including radio, line laying and Signals Despatch Services (SDS). The unpopular Austin Champs were soon replaced in 1960 by the Corps' first Land Rovers, once again following the UK changeover.[53]

Each long wheel base Land Rover variant was designed to carry equipment for a specific task, power-packs for the equipment and operators (including their rations and personal gear). In practice, however, when loaded up, the bodies would sit hard on the wheels, and extra three-ton vehicles accompanied units going on exercises.[54]

52 'The Royal Australian Corps of Signals Reference Manual: Part 3 - Equipment', p. 3-65. The main components of the terminal were one VF telegraph terminal, Northern radio model 325, carrier terminal, an amplifier modulator, one SHF power amplifier, two FM receivers, 3 synthesizers, one de/multiplexer, one deviation and level monitor, one voice monitor, one teletypewriter, 2 antenna assemblies, one telegraph terminal and an HF radio (AN/PRC-47). The AN/MRC-69 was eventually replaced, in 1972, by the more advanced AN/MRC-127, which performed a similar role. See Fairbairn, *Defence Report*, 1972, p. 16.

53 Weir, letter, December 1988, p. 3, and November 1988, p. 2; and Colonel G.J. Mapson, letter, November 1989.

54 Major B.G. Saunders, letter, August 1990.

BIBLIOGRAPHY

Archival Records

Australian War Memorial

AWM 95 War Diaries - Australian Observer Unit, Malaya and 709 Signal Troop South Vietnam.
AWM 98 Headquarters Australian Force, Vietnam
AWM 102 Vietnam - Miscellaneous
AWM 103 Headquarters 1 Australian Task Force (Nui Dat)
AWM 114 Written Records Japan (BCOF) and Korea
AWM 116 Headquarters 1st Australian Logistic Support Force Group (Vung Tau)
AWM 121 Directorate of Military Operations and Plans
AWM 123 Special Collection, Defence Committee Records, 1939-1957
AWM 200 Headquarters Far Eastern Land Forces (FARELF)
AWM 206 ANZUK Force Headquarters
AWM 207 ANZUK Force Headquarters
OW 84/5 Australian Observer Unit (Malaya) War Diary

Australian Archives

CRSA462/1 Prime Minister's Department Correspondence Files
CRS A816 Department of Defence Correspondence Files
CRS A2107 Department of Defence 'K' Series Correspondence Files (Korean War)
CRS A2031 Minutes of the Meetings of the Defence Committee (1929-).
CRS A2653 Military Board Proceedings (1905-1976)

CRS A5954/1 Defence Records collected by Sir Frederick Shedden (1937-1971)
CRS A6059 Correspondence Files, Multiple Number Series, Jan 56 - Dec 64
MP 742/1 Department of the Army Correspondence Files (1927-62)
MP 729/8 Department of the Army Classified Correspondence Files
MS 987 Papers of the Hon J.J. Dedman, National Library of Australia. (785)

Royal Australian Corps of Signals Museum

Papers of the Royal Australian Signals Corps Museum (RASCM): ex-Directorate of Signals files).

Published Works

Adams, R.M. *Through to 1970,* Royal Signals Institution, London, 1970.
Alexander, F. *From Curtin to Menzies and After: Continuity or Confrontation,* Nelson, 1973.
Atkinson, J.J. *The Kapyong Battalion,* New South Wales Military Historical Society, 1977.
Ball, D. *Australia's Secret Space Programs*, Canberra Papers on Strategy and Defence, No 40, Strategic and Defence Studies Centre, ANU, Canberra, 1988.
Ballard, Geoffrey, *On ULTRA Secret Service: The Story of Australia's Signals Intelligence Operations During World War II*, Spectrum Publications, Richmond, Victoria, 1991.
Bannister, Colin, *An Inch of Bravery: 3RAR in the Malayan Emergency 1957-59*, Directorate of Army Public Affairs, Canberra, 1994.
Barclay, G. St J. *Friends in High Places: Australian American Diplomatic Relations since 1945*, Melbourne, OUP, 1985.
Barclay, G. St J., *A Very Small Insurance Policy: The Politics of Australian Involvement in Vietnam, 1954-1967,* University of Queensland Press, St Lucia, 1988.
Barker, Theo, *Signals: A History of the Royal Australian Corps of Signals 1788-1947*, Royal Australian Signals Committee, Canberra, 1987.

Barker, Theo, *Craftsmen of the Australian Army: The Story of RAEME*, Crawford House Press, Bathurst, 1992.

Barnett, C. *Britain and Her Army 1509-1970: A Military, Political & Social Survey*, Allen Lane The Penguin Press, 1970.

Blaxland, J.C. *Organising an Army: The Australian Experience 1957-1965*, Canberra Papers on Strategy and Defence, No 50, Strategic and Defence Studies Centre, ANU, Canberra, 1989.

Breen, R.J. *First to Fight: Australian Diggers, N.Z. Kiwis and U.S.. Paratroopers in Vietnam, 1965-66*, Allen & Unwin, Sydney, 1988.

Buckley, R. *Occupation Diplomacy: Britain, the United States and Japan 1945-1952*, Cambridge University Press, 1982.

Burstall, T. *The Soldier's Story*, University of Queensland Press, St Lucia, 1986.

Carver, M. *War Since 1945*, Putnam's Sons, New York, 1981.

Clutterbuck, R. *The Long Long War: The Emergency in Malaya 1948-1960*, Cassell, London, 1967.

Coates, John, *Suppressing Insurgency: An Analysis of the Malayan Emergency 1948-1954*, Westview Press, Boulder, 1992.

Cubis, R. *A History of 'A' Battery*, Elizabethan Press, Sydney, 1978.

Firkins, P. *The Australians in Nine Wars: Waikato to Long Tan*, Rigby, 1971

Forward R. & Reece B. (eds), *Conscription in Australia*, University of Queensland Press, 1968.

Frost, F. *Australia's War in Vietnam*, Allen & Unwin, Sydney, 1987.

Gander, T. *Encyclopaedia of the Modern British Army*, 3rd edition, Stephens, Wellingsborough, 1986.

Gelber H.G. (ed), *Problems of Australian Defence*, Melbourne, OUP, 1970.

Gelber, H.G. *The Australian American Alliance: Costs and Benefits*, Penguin, 1968.

Greenwood G. & Harper N. (eds), *Australia in World Affairs 1961-1965*, Cheshire, Melbourne, 1968.

Grey, Jeffrey, *A Military History of Australia*, Cambridge University Press, Cambridge, 1990.

Grey, Jeffrey, *Australian Brass: The Career of Lieutenant General Sir Horace Robertson*, Cambridge University Press, Cambridge, 1992.

Grey, Jeffrey, *British Commonwealth Armies & the Korean War: An Alliance Study*, Manchester University Press, Manchester, 1988.

Harper N. (ed), *Pacific Orbit: Australian-American Relations Since 1942*, Cheshire, Melbourne, 1968.

Hawkins, D., *The Defence of Malaysia and Singapore: From AMDA to ANZUK*, Royal United Services Institute, London, 1972.

Hilsman, R. *The Politics of Policy Making in Defense and Foreign Affairs: Conceptual Models & Bureaucratic Politics*, Prentice Hall, New Jersey, 1987.

Hopkins, R.N.L. *Australian Armour: A History of the Royal Australian Armoured Corps 1927-1972*, AWM & AGPS, Canberra, 1978.

Horner, David, *The Gunners: A History of Australian Artillery*, Allen & Unwin, Sydney, 1995.

Horner, David (ed), *Duty First: The Royal Australian Regiment in war and Peace*, Allen & Unwin, Sydney, 1990.

Horner, D.M. *SAS: Phantoms of the Jungle*, Allen & Unwin, Sydney, 1989.

Horner, D.M. (ed), *The Commanders: Australian military leadership in the twentieth century*, Allen & Unwin, Sydney, 1984.

Horner, D.M. *Australian Higher Command in the Vietnam War*, Canberra Papers on Strategy and Defence, No 40, Strategic and Defence Studies Centre, Canberra, 1986.

Horner, D.M. *High Command, Australia and Allied Strategy 1939-1945*, Allen & Unwin & AWM, Sydney, 1982.

Hudson, W.J. (ed) *Australia in World Affairs 1971-75*, Allen & Unwin, Sydney, 1980.

John, A.W. *Uneasy Lies The Head That Wears A Crown: A History of the British Commonwealth Occupation Force Japan*, Gen Publishers, Cheltenham, Victoria, 1987.

Johnson, F.A. *Defence by Committee: The British Committee of Imperial Defence 1885-1959*, London, OUP, 1960.

King, P. (ed) *Australia's Vietnam*, Allen & Unwin, Sydney, 1983.

Lee, David, *Search For Security: The Political Economy of Australia's Postwar Foreign and Defence Policy*, Allen & Unwin and RSPAS, Canberra, 1995.

Lee, Air Chief Marshall Sir David, *Eastward: A History of the Royal Air Force in the Far East 1945-1972*, Her Majesty's Stationery Office, London, 1984.

Lider, J. *British Military Thought After World War II*, Gower Press, Aldershott, 1985.

Lider, J. *Military Theory, Concept, Structure, Problems*, Gower Press, Aldershott, 1983.

Madden A.F. & Morris-Jones, W.H. *Australia and Britain: Studies in a Changing Relationship*, Sydney University Press, 1980.

Mahon J.K. & Danysh R., *Army Lineage Series - Infantry Part I: Regular Army*, OCMH, Washington, D.C., 1972.

McAulay, L *The Battle of Long Tan*, Arrow Books, London, 1986.

McAulay, L. *The Battle of Coral*, Hutchinson, Melbourne, 1988.

McCormack, G. *Cold War Hot War: An Australian Perspective on the Korean War*, Hale & Iremonger, Sydney, 1983.

McKay, G. *In Good Company: One Man's War in Vietnam*, Allen & Unwin, Sydney, 1987.

McKernan, M. & Browne, M. (eds) *Australia Two Centuries of War and Peace*, Australian War Memorial & Allen & Unwin, Sydney, 1988.

Maclear, M. *Vietnam: The Ten Thousand Day War*, Thames Mandarin, London, 1989 (reprint).

McNeill, I.G. *The Team: Australian Army Advisers in Vietnam 1962-1972*, Australian War Memorial and University of Queensland Press, Canberra, 1984.

Millar, T.B. *Australia in Peace and War: External Relations 1788-1977*, ANU Press, Canberra, 1978.

Millar, T.B. *Australia's Defence*, Melbourne University Press, 1969 (2nd edition).

Modelski G. (ed), *SEATO: Six Studies*, Cheshire, Melbourne, 1962.

Moir, J.S. (ed) *History of the Royal Canadian Corps of Signals 1903-1961*, Corps Committee, Royal Canadian Corps of Signals, Ottawa, 1962.

Morton, Peter, *Fire Across the Desert: Woomera and the Anglo-Australian Joint Project 1946-1980*, AGPS, Canberra, 1989.

Moyal, A. *Clear Across Australia: A history of telecommunications*, Nelson, Melbourne, 1984.

Nalder, R.F.H *The History of British Army Signals in the Second World War*, Royal Signals Institution, London, 1953.

Nalder, R.F.H *The Royal Corps of Signals: A History of its Antecedents and Development (circa 1800-1955)*, Royal Signals Institution, London, 1958.

Oliff, Lorna, *Colonel Best and Her Soldiers: The Story of the 33 Years of the Women's Royal Australian Army Corps*, Ollif Publishing, Sydney, 1985.

Pemberton, G. *All The Way, Australia's Road to Vietnam*, Allen & Unwin, Sydney, 1987.

Pizer, V. *The United States Army*, Praeger, New York, 1967.

Reese, T.R. *Australia, New Zealand, and the United States: A Survey of International Relations 1941-1968*, OUP, London, 1969.

Rintoul, S. *Ashes of Vietnam: Australian Voices*, Mandarin, Melbourne, 1987.

Richelson, Jeffrey T, and Ball, Desmond, *The Ties That Bind*, Unwin Hyman, Boston, 1990.

Sexton, M. *War For The Asking*, Penguin, Ringwood, 1981.

Short, A. *The Communist Insurrection in Malaya 1948-1960*, London, 1975.

Singh, R. *Official History of the Indian Armed Forces in the Second World War, 1939-1945: Post War occupation Forces- Japan and South East Asia*, Combined Inter Services Historical Section, Kanpur, 1958.

Skennerton, Ian D., *Australian Service Longarms*, Margate, Queensland, 1976.

Smith, E.D. *Counter Insurgency Operations: 1 - Malaya and Borneo*, Ian Allan, London, 1985.

Starke, J.G. *The ANZUS Treaty Alliance*, Melbourne University Press, 1965.

Stone, G.L. *War Without Honour*, Jacaranda Press, Sydney, 1966.

Thayer, T.C. *War Without Fronts: The American Experience in Vietnam,* Westview Press, Boulder, 1985.

Thompson, R (ed) *War In Peace: An Analysis of Warfare Since 1945*, Orbis, London, 1983.

Van Creveld, M. *Technology and War: From 2000 BC to the Present*, MacMillan, New York, 1989.

Van Creveld, Martin, *Command in War*, Harvard University Press, Cambridge, 1985.

Wah, C.K. *The Defence of Malaysia and Singapore: The transformation of a security system 1957-1971*, Cambridge University Press, 1983.

Waller, P.B.G. *A Study of the Emergency Regulations of Malaya, 1948-1960*, Stanford Research Institute, Menlo Park, California, 1967.

Warner, P. *The Vital Link: The Story of the Royal Signals, 1945-1985*, Leo Cooper, London, 1989.

Watt, A. *The Evolution of Australian Foreign Policy 1938-1965*, Cambridge University Press, 1967.

Welburn, M.C.J., *The Development of Australian Army Doctrine 1945-1964*, Strategic and Defence Studies Centre, Canberra, 1994.

Weller, Stuart, *The Cream of Her Youth*, private publication, Sydney, 1988.

Winterbotham, F.W. *The Ultra Secret*, Weidenfeld & Nicholson, London, 1974.

Facts on File: US - Communist Confrontation in Southeast Asia, Volume 5 1970, Facts on File, New York, 1973.

Journal Articles & Newsletters

Arbogast, G.W. 'Radio Communications in Vietnam', in *Signal, Journal of the Armed Forces Communications and Electronics Association*, April 1967, Vol XXI, No 8, pp. 14-21.

Babbage, R. 'Australian Defence Planning, Force Structure and Equipment: The American Effect', in *Australian Outlook*, Vol 38, No 3, 1984, pp. 163-168.

Barnes, Ken, 'The Defence Signals Directorate - Its Role and Functions' in *Defence Force Journal*, No 108, September/ October 1994, pp. 3-7.

Botwright, G. 'Forty Years of Army Design', in *Signalman*, Vol 5, 1980, pp. 9-11.

Cain, F. 'An Aspect of Post-War Australian Relations with the United Kingdom and the United States: Missiles, Spies and Disharmony', in *Australian Historical Studies*, Vol 23, No 92, April 1989, pp. 186-202.

Carey, K.P. 'Electronic Counter Measures', in *Australia Army Journal*, No 180, May 1964, pp. 5-9.

Dennis, Don, 'Of Shush Missions, Voice Sorties and Pucker Factors', in *Australian Aviation*, May 1991, pp. 39-41.

Eayrs, J.W. '133 Signal Squadron: The City of Penrith Signal Squadron', in *Signalman*, Vol 11, 1983, pp. 20-21.

Edwards, P. 'Malaya and Vietnam: Some preliminary thoughts by the official historian', in *Journal of the Australian War Memorial*, No 3, October 1983, pp. 10-12.

Edwards, P. 'The Australian Commitment to the Malayan Emergency, 1948-1950', in *Historical Studies*, Vol 22, No 89, October 1987, pp. 604-618.

Eggensperger, J.D. 'Packing to Go Home', in *The Jagged Sword*, Vol 2, No 2, Publication of the 1st Signal Brigade, 1971, pp. 11-13.

Garland, R.S. 'Operations in Malaya', in *Australian Army Journal*, No 119, April 1959, pp. 25-31.

Grant, A.R. 'Improved Communications: Some Ideas for the Design of the Signal System', in *Australian Army Journal*, No 140, January 1961, pp. 5-10.

Gration, P.C. 'Reflections on 1ATF in Vietnam', in *Journal of the Australian War Memorial*, No 12, April 1988, pp. 45-46.

Gregorian, Raffi, 'CLARET Operations and Confrontation, 1964-1966' in *Conflict Quarterly*, Vol XI, No 1, Winter 1991.

Grey, J. 'A Military Alliance at Work? Commonwealth Forces in the Korean War', in *Journal of the Australian War Memorial*, No 9, October 1986. pp. 39-46.

Groom, K.G. 'ABCA: The American-British-Canadian-Australian Standardisation Programme', in *Australian Army Journal*, No 177, February 1964, pp. 30-35.

Hopkins, R.N.L. 'History of the Australian Occupation in Japan, 1946-50', in *Journal of the Royal Australian Historical Society*, Vol XL, 1954, pp. 93-116.

Horner, D.M. 'A complex command: the role of the Commander, Australian Force, Vietnam', in *Journal of the Australian War Memorial*, No 13, October 1988, pp. 19-30.

Horner, D.M. 'The Australia Army and Indonesia's Confrontation with Malaysia', in *Australian Outlook*, April 1989.

Jones P.J. & Wood, R. E. 'Brigade Aviation: Anywhere, Anytime', in *The Jagged Sword*, Vol 2, No 2, Publication of the 1st Signal Brigade, 1971, pp. 2-4.

Lawson-Baker, C. '126 Signal Squadron, 1 Commando Regiment, A Different Signal Unit: Unit in Focus', in *Signalman*, Vol 9, 1982, pp. 13-16.

Mason, J.F. 'Vietnam- Electronics in the War', in *Electronics*, Vol 39, No 10, May 16, 1966.
Mitchelmore, M.J. 'The Closure of 146 Signal Squadron (1906-1987)', in *Signalman*, Vol 19, 1987, pp. 11-15.
McNarn, 'M.R. 146 Signal Squadron', in *Signalman*, Vol 10, 1982, pp. 8-9.
McNeill, I.G. 'An Outline of the Australian Military Involvement in Vietnam July 1962-December 1972', *Defence Force Journal*, No 24, September-October 1980, pp. 42-53.
McNeill, I.G. 'Petersen and the Montagnards: An episode in the Vietnam war', in *Journal of the Australian War Memorial*, No 1, October 1982, pp. 33-46.
Raspa, Dario, 'Vale - 134 Sig Sqn Communication Tower' in *Signalman*, Vol 25, 1990, pp. 17-18.
Rienzi, T.M. 'Management Tools for dynamic communications-electronics', in *Signal*, Official Journal of the Armed Forces Communications and Electronics Association, October 1989, pp. 24-27.
Roberts, C.M.A.R. 'Operations in Borneo', in *Australian Army Journal*, No 194, July 1965, pp. 28-33.
Robson, L.L. 'Behold a Pale Horse: Australian War Studies' in *Australian Historical Studies*, Vol 23, No 90, April 1988, pp. 115-126.
Serong, F.P. 'Defence of New Guinea's Land Frontier', in *Australian Army Journal*, No 2, January 1951, pp. 5-15.
Stretton, A.B. 'On Active Service in Malaya, 1962', in *Australian Army Journal*, No 165, February 1963, pp. 10-20.
Swifte, L.B. 'SEATO Today', in *Australian Army Journal*, No 204, May 1966, pp. 40-49.
Trotter, J.E. 'The Integrated Communications System in Vietnam', in *Signal, Journal of the Armed Forces Communications and Electronics Association*, April 1967, Vol XXI, No 8, pp. 22-25.
Van Ryzin, W.J. 'Communications Equipment Requirements in Vietnam,' in *Signal, Journal of the Armed Forces Communications and Electronics Association*, April 1967, Vol XXI, No 8, pp. 28-36.
Walters, W.H. 'Nuclear Warfare and Signal Communications', in *Australian Army Journal*, No 108, May 1958, pp. 25-31.

Willis, J.W. 'The British Commonwealth Occupation Force: An Experiment in Inter-National and Inter-Service Integration', in *Australian Army Journal*, No 81, February 1956, pp. 12-15.

Pacific Islands Regiment, 'The Value of CW Transmission in Long Range Patrolling', in *Australian Army Journal*, No 162, November 1962, pp. 38-39.

'Modern Radio for the Modern Army', in *Australian Army Journal*, No 111, August 1958, pp. 43-48.

'1st Signal Regiment on the Move', in *Signalman*, Vol 7, 1981, pp. 39-41.

'2 Signal Regiment (City of Heidelberg Regiment): Unit in Focus', in *Signalman*, Vol 10, 1982, pp. 37-39.

'127 Signal Squadron: Unit in Focus', in *Signalman*, Vol 14, 1985, pp. 28-29.

'Unit in Focus: 152 Signal Squadron, Special Air Service Regiment', in *Signalman*, Vol 13, 1984, pp. 9-11.

'The Pentropic Division: Signals', in *Australian Army Journal*, No 129, February 1960, pp. 26-28.

Directorate of Signals, 'Field Expedient Antennas for VHF Radio', in *Australian Army Journal*, No 214, March 1967, pp. 26-30.

'Equipment for the Pentropic Division: Signals', in *Australian Army Journal*, No 134, July 1960, pp. 46-51.

'British Commonwealth Occupation Force-Japan(1946-1951)', in *Sabretache, Journal of the Military Historical Society of Australia*, Vol XVII, No 1, July 1975, pp. 34-37.

'Army Reorganization', in *Australian Army Journal*, No 274, March 1972.

'Fifth Annual Glossary of Military Electronics Systems', in *Armed Forces Management*, July 1967, Vol 13, No 10.

103 Sig Sqn 66-67 Newsletter.

Royal Australian Corps of Signals Newsletter, Vols 7-16.

'The Thurunka(The Message Stick)', Royal Australian Signals Association (New South Wales), Vol 2 No 2.

Official Publications

Australia

Commonwealth of Australia, *Commonwealth Parliamentary Debates* House of Representatives, Canberra.

Commonwealth of Australia, *Parliamentary Papers*, House of Representatives, Canberra.

Department of External Affairs, *Current Notes on International Affairs*, Canberra.

Department of Defence, *Royal Australian Corps of Signals : Corps Memoranda,* Canberra, 1989.

Department of Defence, *Engineering Development Establishment Brochure*, 1988.

Department of Defence, *Historical Listings, DRB 2*, June 1984 edition.

Department of Defence, 'Key Elements in the Triennial Reviews of Strategic Guidance Since 1945', Canberra, April 1986.

Department of the Army, *Communications: Royal Australian Signals*.

Department of the Army, *The Army List of Officers of the Australian Military Forces*, Vol 1, 1970.

Department of the Army, *GOC's Exercise Hindsight 1971*, Exercise Papers.

Department of the Army, *Training Information Bulletin, No 19*.

Department of the Army, *Signal Training (All Arms) Pamphlet No 6, Communications Security, 1971*

Department of the Army, Australian Military Forces, *User Handbook, A510 Wireless Station*, 1956.

Department of the Army, *The Royal Australian Corps of Signals Reference Manual*, 1968.

Department of the Army, Royal Australian Signals, *An Introduction to 402 Signal Regiment*.

Department of the Army, Royal Australia Corps of Signals, *Signals Bulletin*, 1946-1973

Barnard, L.H. *Defence Report 1973*, Canberra.

Edwards, P. with Pemberton, G., *Crises and Commitments: The Politics and Diplomacy of Australia's Involvement in Southeast Asian Conflicts 1948-1965*, Allen & Unwin and Australian War Memorial, Canberra, 1992.

Fairhall, A. *Defence Report 1967*, Canberra.

Fairhall, A. *Defence Report 1968*, Canberra.
Fairhall, A. *Defence Report 1969*, Canberra.
Fairbarn, D.E. *Defence Report 1972*, Canberra.
Fairfax, D. *Navy in Vietnam: A record of the Royal Australian Navy in the Vietnam War 1965-1972*, Department of Defence, AGPS, Canberra, 1980.
Fraser, J.M. *Defence Report 1970*, Canberra.
Gorton, J.G. *Defence Report 1971*, Canberra.
Hasluck, P. *Vietnam: Efforts For Peace*, Department of External Affairs, Canberra, 1967.
McNeill, Ian, *To Long Tan: The Australian Army and the Vietnam War 1950-1966*, Allen & Unwin and Australian War Memorial, Canberra, 1993.
Millar, T.B. *Committee of Inquiry into the Citizen Military Forces*, Canberra, 1974.
O'Neill, R.J. *Australia in the Korean War 1950-53, Vol II: Combat Operations*, Australian War Memorial & AGPS, Canberra, 1985.
O'Neill, R.J. *Australia in the Korean War 1950-53, Vol I: Strategy and Diplomacy*, Australian War Memorial & AGPS, Canberra 1981.

United States

Aggression From the North: The Record of North Viet-Nam's Campaign To Conquer South Viet-Nam, Department of State, Washington, D.C., 1965.
Coker, K.R. & Rios, C.E, *A Concise History of the U.S. Army Signal Corps*, Office of the Command Historian, U.S. Army Signal Centre and Fort Gordon, 1988.
Command Communications, July 1970, Headquarters United States Army Vietnam, USARV Pam 105-12.
Command Communications, October 1968, Headquarters United States Army Vietnam, USARV Pam 105-4.
Rienzi, T.M. *Vietnam Studies: Communications-Electronics 1962-1970*, Department of the Army, Washington, D.C., 1972.
VC/NVA Signal Equipment Identification Handbook, Combined Materiel Exploitation Center, MACV, Saigon, May 1969.

United Kingdom

Director of Operations, Malaya, *The Conduct of Anti-Terrorist Operations in Malaya*, (3rd edition), 1958.

The War Office, *Signal Training, Volume I: Signal Organization and Tactics, Pamphlet No 6, Infantry Divisional Signal Regiment*, 1950.

The War Office, *Signal Training (All Arms), Pamphlet No 10: Signal Tactics, Part II, The Infantry Battalion*, 1952.

The War Office, *Signal Training, Volume I: Signal Organization and Tactics, Pamphlet No 9, Air Support Signal Unit*, 1951.

The War Office, *Signal Training Vol 1, Signal Organization and Tactics, Pamphlet No 11, Signals for the Royal Artillery*, 1955.

The War Office, *Signal Training Vol 1, Signal Organization and Tactics, Pamphlet No 1, The Higher Organization of Signals*, 1950.

Gravely, T.B. *The Second World War, 1939-1945 - Army: Signal Communications*, The War Office, 1950.

Unpublished Sources

Barker, T. 'Signals, The Royal Australian Corps of Signals, 1947-1980', draft manuscript of the Royal Australian Signals Corps Committee.

Dean, K.G. 'Notes on the A510 - Personal Recollections', AWA Research Laboratories, Sydney, 1978.

Dean K.G. 'The Malayan Trials of the WS A510', AWA Research Laboratories, Report No 231, June 1952.

Fraser, M. 'Address by the Minister for Defence to the United Services Institute of New South Wales on Tuesday 27th October 1970.'

Lawrence, G.J. 'Royal Australian Corps of Signals - Corps History: The South Vietnam Campaign 1962-1972.'

Murphy, J.E. 'History of the Post War Army', unpublished manuscript of the Military Board, 1955.

Shelton, H.R. 'The United States Infantry Division and the Australian Pentropic Division - Similarities and Differences', Master of Military Art and Science Thesis, Fort Leavenworth, Kansas, 1964.

Unit Histories

'A Short History of the Royal Australian Corps of Signals.' (unpublished)
'Post War Developments in R Aust Sigs as at November 1955.' (unpublished)
'7th Signal Regiment Unit History.'(unpublished)
'6 Signal Regiment Unit History.'(unpublished)
'1 Signal Regiment Unit History.'(unpublished)
'The History of 1st Signal Regiment.'(unpublished)
'4th Signal Regiment Outline Unit History.'(unpublished)
'History of the RA Signals in Western Command.'(unpublished)
'126 Signal Squadron Unit History.'(unpublished), Sergeant K.E. Larner (ed).
'127 Signal Squadron Unit History.'(unpublished)
'104 Signal Squadron Unit History.'(unpublished)
'130th Signal Squadron Unit History.'(unpublished)
'History of 135 Signal Squadron.' (unpublished)
'130th Signal Squadron, Tactical Air Support Squadron Brief.'(unpublished)
'134 Signal Squadron Unit History.'(unpublished)
'123rd Signal Troop Brief History.'(unpublished)
'Unit History 700 Signal Troop.'(unpublished)
'A History of 547 Signal Troop in Vietnam', Davies, M.J. (ed). (unpublished)
'547 Sig Tp : June 1971 to December 1971.'(unpublished)
'133 Signal Squadron - History.' W. Salisbury (ed). (unpublished)
' History - 139 Signal Squadron.'(unpublished)
'143 Signal Squadron.'(unpublished)
'The History of 1 RAR: From Morotai to Townsville 1945-1980.'(unpublished)
'The Royal Australian Regiment: Standing Orders.'

Newspapers

Age
Australian
Australian Army
Brisbane Telegraph
National Times

Newcastle Morning Herald
(28) Sunday Tiger Standard (Malaya)
Sydney Morning Herald
Utusan Melaya (Malay language paper)

Correspondence

Correspondence with Author, Lieutenant Colonel G.J. Lawrence, Lieutenant Colonel S.P. Dossetor, Colonel T.D. Davies and Mr T. Barker: Captain E. Belcher; Brigadier K.R. Colwill; Lieutenant Colonel R.E.P. Cowley; Major General P.A. Cullen; Staff Sergeant J.E. Danskin; Major G.N Donley; Lieutenant General D.B. Dunstan; Lieutenant Colonel P.S. Emery; Lieutenant Colonel B.D. Evans; Brigadier P.J.A. Evans; Captain A. Findlay; Major General S.C. Graham; Major A.R. Grant; Major A.F. Hanley; Lieutenant Colonel S. Hart; Captain P.N. Hill; Major S.R. Hinton; Major General B.H. Hockney; Lieutenant Colonel D.M. Horner; Major General R.L. Hughes; Major D.E Huxley; Brigadier O.D. Jackson; Warrant Officer Class One C. King; Warrant Officer Class Two A.W. Lane; Major General K. Mackay; Colonel G.J. Mapson; Colonel J.H. Marsh; Colonel R.J.S. McDonald.; Lieutenant Colonel K.M. McDonald; Mr J.K. McKenzie; Lieutenant Colonel J.S.D. Mellick; Major R.A.H. Murray; Lieutenant Colonel N.C. Munro; Warrant Officer Class Two G. Munro; Captain S. Perijmibida; Major G.F. Powell; Lieutenant Colonel D.J. Randall; Major P.D. Rothwell; Staff Sergeant Salisbury; Major B.G. Saunders; Lieutenant Colonel E.B. Starrett; Lieutenant Colonel B.G. Swan; Major General K.J. Taylor; Mr N.E. Tonkin; Major K.M. Weir; Brigadier S.P. Weir; Lieutenant Colonel I. Willoughby; General Sir John Wilton; Major General R.P. Woollard.

Interviews

Staff Sergeant J.W. Bell; Brigadier K.R. Colwill; Lieutenant Colonel R.E.P. Cowley; Staff Sergeant J.E. Danskin; Major General B.H. Hockney; Captain A.J. Johnson; Brigadier K.P. Morel; Lieutenant Colonel P.T. Murray; Warrant Officer Class One R. Ratchford;

Lieutenant Colonel R.L.C. Twiss; Major General D. Vincent; Lieutenant Colonel K. Whyte; Major General R.P. Woollard.

NOTES ON SOURCES

A NUMBER OF published works covering the period of this study have served to place this work in its proper context. *The Royal Corps of Signals*, by Major-General R.F.H. Nalder, was published in Britain in 1953 and this proved helpful in adding depth to my research, as did the *History of the Royal Canadian Corps of Signals*, (edited by J.S. Moir and published in 1962) and *A Concise History of the US Army Signal Corps* (K.R. Coker and C.E. Rios, published in 1988). Another more recent work by P. Warner, entitled *The Vital Link*, was published in Britain in 1989, and this anecdotal history also helped to give me a more balanced perspective.

Barker also prepared a sketchy, unpublished manuscript which dealt with the post-war Australian Corps of Signals, covering the period up to the late 1970s. It has provided helpful guideposts for areas that required research. His work did not contain acknowledgments of his sources, although I managed to locate most of his source material and acknowledged it where I used it. Lieutenant Colonel G.J. Lawrence wrote a short history of the Signals Corps' involvement in the Vietnam War (also unpublished), and this work, like Barker's, is helpful and well written but limited in scope and depth of research. I am, nevertheless, indebted to him for helping me formulate a framework for the Vietnam chapters. I also managed to uncover most of his source material in preparing this manuscript. Colonel R.A. Clark wrote 'A Short History of the Royal Australian Corps of Signals' (unpublished), as did others, including Brigadier A.D. Molloy, and these provided further similarly helpful but brief overviews. J.E. Murphy's unpublished 'History of the Post War Army', written in 1955 for the Australian Army's Military Board, provides a helpful, albeit brief, picture of the Signals Corps in the years immediately following World War II.

The Royal Australian Corps of Signals Museum at Simpson (previously Watsonia) Barracks, near Melbourne, contains further valuable information. This has been collected over the years by the Honorary Curator, Lieutenant Colonel Keith D. Munro (RL). In this collection was found a series of volumes of historical notes collected since the end of World War II, dealing with the major facets of the Corps. This collection included monthly reports written to the Signals Directorate at Army Headquarters by unit commanders stationed around Australia and overseas, both ARA and CMF. Also in the collection were the monthly *Signals Bulletins* covering the entire period from 1948 to 1972. These Bulletins were prepared by the Signals Directorate and provided a wealth of information on unit, personnel and training matters, as well as equipment developments and reports from various exercises and operations.

These sources were supplemented by Department of Defence and Department of the Army files held by the Australian War Memorial and the Australian Archives in Canberra and Melbourne. I was also given special access to closed period files held in these repositories and within the Department of Defence's offices in Canberra. Items of particular import were the unit war diaries and files, as well as proceedings of the Military Board and the Defence Committee, as well as its sub-committees, such as the Joint Communications Committee. Unfortunately, no written records of the 28th Commonwealth Brigade Signal Squadron were located. The official historians working with Dr P.G. Edwards have found a similar dearth of material on this predominantly British unit in Malaya. The Prime Minister's Department Correspondence files (CRS A462/1), covering open period material, and the National

Library were also searched and confirmed my expectations of not finding significant detail concerning the Signals Corps.

It appears that matters concerning military telecommunications rarely reached beyond the levels of the Defence Committee. Exceptions were the policies of the Occupation Force in Japan concerning women (discussed in Chapter Two), the repeated studies covering the proposed amalgamation of the strategic communications networks of the three armed services (discussed in Chapter Nine), as well as the deliberations concerning the ill-fated Project 'Mallard' of the late 1960s (discussed in Chapter Ten). Numerous letters were received from ex-service personnel, the names of whom are listed in the bibliography and to whom I am greatly indebted. These provided many helpful anecdotes and explanations for events that would have otherwise remained colourless or unexplained. In doing this, I have been aware that oral history is problematic because people tend to create or idealise events. My role, therefore, has been to understand their experience, critically appraise it and put it into a broader historical context. Moreover, the text has been scrutinised by numerous retired members of the Corps, in particular members of the Royal Australian Corps of Signals Committee. Their reviews have helped to reduce inaccuracies. A number of interviews with various staff officers and commanders were also conducted to deal with certain remaining aspects of the book, and these also proved to be very helpful.

In the arena of published material, several works, listed in the bibliography, have served to add depth to the account.[55] Works that helped to provide the strategic and political context in which events occurred include Edwards' *Crises and Commitments*, Barclay's *Friends in High Places*, Buckley's *Occupation Diplomacy* and Pemberton's *All The Way*. The works by Millar, Gelber, Harper, Hawkins, Wah, Watt and McCormack,

---

55 For a broad review of literature on Post 1945 conflicts see P.G. Edwards, 'The post-1945 conflicts: Korea, Malaya, Borneo and Vietnam', in McKernan & Browne (eds), *Australia: Two Centuries of War and Peace*, pp. 296-309.

among others, have also proved valuable. The military context was enhanced by works such as McNeill's *To Long Tan*, Frost's *Australia's War in Vietnam*, Carver's *War Since 1945*, Rienzi's *Vietnam Studies: Communications Electronics* and O'Neill's *Australia in the Korean War*. The works by Horner, Breen, Adams, and Waller, among others, were also of value. Moyal's *Clear Across Australia* also proved useful in contrasting the civil telecommunications facilities within Australia. A number of government publications listed in the bibliography were also useful. Journal articles found in *Signalman, Australian Outlook, Australian Army Journal, Journal of the Australian War Memorial,* and *(Australian) Historical Studies*, as well as various newspapers, were also significant.[56] Commonwealth Record Series files listed with a CRS prefix. Nowadays, these are categorised as being records of the National Archives of Australia, or NAA.

56 All these works are listed in the bibliography.

www.ingramcontent.com/pod-product-compliance
Lightning Source LLC
LaVergne TN
LVHW052335100826
845147LV00020B/1067

* 9 7 8 1 9 2 3 6 8 9 9 0 9 *